Unless

VOLUME 1

Foundations of
FUTURES
STUDIES

Human Science for a New Era

VOLUME 1
History, Purposes, and Knowledge

Foundations of
FUTURES
STUDIES

Human Science for a New Era

WENDELL BELL

TRANSACTION PUBLISHERS
New Brunswick (U.S.A.) and London (U.K.)

Second printing 1997
Copyright © 1997 by Transaction Publishers, New Brunswick, New Jersey
08903. This is volume 1 of *Foundations of Futures Studies*. Volume 2, *Values,
Objectivity, and the Good Society*, published by Transaction Publishers © 1997.

This book is printed on acid-free paper that meets the American National Standard for Permanence of Paper for Printed Library Materials.

Library of Congress Catalog Number: 96-22496
ISBN: 1-56000-271-9
Printed in the United States of America

Library of Congress Cataloging-in-Publication Data

Bell, Wendell.
    Foundations of futures studies : human science for a new era / Wendell
Bell.
        p. cm.
    Includes bibliographical references and index.
    ISBN 1-56000-271-9 (v. 1 : acid-free paper)
    1. Forecasting. I. Title.
CB158.B45 1996
303.49'09'04—dc20   96-22496
                                                                      CIP

For my wife Lora-Lee,
whose art enriches me,
whose friendship I cherish, and
whose love I will return, always.

Contents

List of Figures

List of Tables

Acknowledgements

I wrote the first words of this book in Canberra, Australia in January, 1985, more than ten years ago. Although I take full responsibility for what I have written, this book is primarily a work of synthesis of other people's work. It could not be otherwise in an attempt to describe an entire field of inquiry. To organize and structure modern futures studies I rely on the many futurists who have provided its foundations and continue to contribute to it.

I believe that to provide a synthetic overview of a specified field is an important task that is too seldom carried out. Such an overview is of particular importance to subject matters that are both relatively new and fragmented, such as futures studies is. Moreover, it is an antidote to the "infoglut" that now threatens to engulf us, and, indeed, it can give the structure and meanings without which information cannot be called knowledge. But very little in this book is based on any original research of mine. I herewith acknowledge my enormous debt to the hundreds of authors whose works are the basis of what I have written here.

Most such authors are acknowledged by being listed in the references given at the end of this book. Some may not be so acknowledged simply because I read something of theirs, heard them speak, or heard someone speak about them and tucked away some bit of information in my mind without remembering the source. If I use anyone's work and fail to cite it, I apologize. It is not a deliberate oversight, because I wish to acknowledge in full my intellectual debts.

Many people played a role in helping me write this book. Among the most important is Frank L. Jones who arranged for a Visiting Fellowship for me at the Institute of Advanced Studies, Australian National University in 1985. It was there that I was able to devote my full attention to this book for the first time and to write its beginning chapters. He and his colleagues—and their spouses—not only were gracious hosts, introducing my wife and me to the many wonders of Australia, but also gave me encouragement and constructive criticism as I struggled with some ideas that were then less than half-formed.

Over the years many of my (sometime) colleagues at Yale gave me the benefit of their critical judgments, suggestions for improvement, and expertise. They include Neil Bennett, Steven Brint, Michele Dillon, Paul J. DiMaggio, Leonard Doob, Kai T. Erikson, Robert E. Lane and his wife Helen (the novelist Helen Hudson), Wayne A. Meeks, Jonathan Reider, Albert J. Reiss, Jr., Lloyd H. Rogler, Ian Shapiro, Steven B. Smith, John J. Stanfield II, and Peter C. Yeager. I thank each for his or her generosity and friendly collegiality.

At Yale, I taught a course on social change and the future, under one title or another, from the late 1960s until the mid-1990s. Some of the ideas in this book come from discussions with both undergraduate and graduate students. In recent years, my students read some of the drafts of this book, and their comments were useful as I revised the manuscript. Although it is impossible to recognize them all by name here, I am happy to acknowledge their assistance. Former Yale graduate students who made significant contributions to my thinking about futures studies include Steven F. Alger, Menno Boldt, Bettina J. Huber, and Jeffrey K. Olick.

I am also indebted to the UCLA graduate students who worked with me in the Caribbean in the early 1960s where my interest in futures studies began. I especially wish to acknowledge the contributions of Anthony P. Maingot, James A. Mau, Charles C. Moskos, and Ivar Oxaal.

Other scholars and scientists have given kindly of their time and thought as they responded to various parts of this book. I particularly thank biologist Peter Bellinger, philosopher Keekok Lee, political scientist Richard C. Snyder, and my college professor of English literature, the late Earl Lyon, who did what he could throughout most of his life to improve my writing. I thank, too, another former professor and mentor, sociologist Leonard Broom, whose standards both of professional excellence and social justice have been a life-long inspiration to me.

In the last few years particularly, more and more of my intellectual peer group has been composed of futurists. They have been most generous in giving me their support and help on this book, even though in some cases it competes with their own projects and in nearly all cases intrudes onto their intellectual turf. I wish to thank especially futurists Joseph F. Coates, Howard F. Didsbury, Jr., Theodore J. Gordon, Bernd Hamm, Richard L. Henshel, Michael Marien, Eleonora B. Masini, Max

Singer, Richard A. Slaughter, Bart van Steenbergen, Lorne Tepperman, and Robert B. Textor.

Four other futurists served as official reviewers of my manuscript and took an extraordinary amount of time from their own busy schedules to read and comment on it at length. Moreover, each has made additional personal contributions to my education as a futurist over many years for which I am grateful. They are James A. Dator, Gary Gappert, Reed D. Riner, and Allen Tough. Although I was not able to follow all of them, their editorial suggestions have made this a better book than it would have been otherwise.

No manuscript ever gets off to a publisher without logistic support. This manuscript is no exception. The staff of the Yale University Department of Sociology most ably provided much of the managerial and secretarial assistance needed to complete this book. I thank Pamela B. Colesworthy, business manager, Ann Fitzpatrick, senior administrative assistant, and, particularly, Nancy Hopkins, registrar, who served as my administrative and secretarial assistant through many drafts of this book from the first to the last page.

* * *

I wish to thank the following publishers for permission to reprint in this Volume (usually in revised form) some short pieces that I had published earlier:

The World Future Society for:

"Is the futures field an art form or can it become a science?" *Futures Research Quarterly* 3 (1) (Spring 1987): 27–44; and

"The rise of futures thinking in the new states: the decisions of nationhood." Pp. 169–182 in Howard F. Didsbury (ed.), *The Future: Opportunity Not Destiny*. Bethesda, MD: World Future Society, 1989.

Butterworth-Heinemann journals, Elsevier Science Ltd. for:

"An epistemology for the futures field: Problems and possibilities of prediction," *Futures* 21, No. 2 (April 1989): 115–135 (with additional thanks to my co-author, Jeffrey K. Olick); and

"H. D. Lasswell and the futures field: Facts, predictions, values, and the policy sciences," *Futures* 25, No. 7 (September 1993): 806–813.

The American Sociological Association for:

"What do we mean by 'paradigm'?" *ASA Footnotes* 18, No. 9 (November 1990): 17.

Also, I have used several figures and tables from other authors and I wish to thank the publishers for their permission to reprint them here (figure and table numbers are from my chapters):

Oxford University Press for:
Figure 6.1 from p. 138 in Elisabeth Noelle-Neumann. 1989. "The public as prophet: Findings from continuous survey research and their importance for early diagnosis of economic growth." *International Journal of Public Opinion Research* 1, No. 2 (Summer) (with additional thanks to Elisabeth Noelle-Neumann for updated data).

Cambridge University Press for:
Figure 6.3 from p. 121 in K. Holden, D. A. Peel, and J. L. Thompson. 1990. *Economic Forecasting: An Introduction.* Cambridge: Cambridge University Press.

The RAND Corporation for:
Table 6.1 from p. 23 in Gordon, Theodore and Olaf Helmer. 1964. *Report on a Long-Range Forecasting Study.* Santa Monica, CA: RAND, P-2982.

Jossey-Bass, Inc., Publishers for:
Figure 6.6 from p. 48 in Wagschall, Peter H. 1983. "Judgmental forecasting techniques and institutional planning: an example," in J. L. Morrison, W. L. Renfro, and W. I. Boucher (eds.), *Applying Methods and Techniques of Futures Research.* San Francisco, CA: Jossey-Bass.

Chelsea Green Publishing Company for:
Figures 6.7, 6.8, and 6.9 from p. 133, p. 199, and p. 205 in Meadows, Donella H., Dennis L. Meadows, and Jorgen Randers. 1992. *Beyond the Limits.* White River Junction, VT: Chelsea Green Publishing Company.

* * *

Working with the people at Transaction Publishers has been a delight. They understand ideas, are sensitive to the sometimes unreasonable hopes and fears of authors, and treat an author's "baby" with as much, if not more, care than the author himself. My special thanks to Irving Louis Horowitz, Mary E. Curtis, and to my editor, Laurence Mintz, whose competent and caring craftsmanship rescued many of my sentences from incoherence.

Finally, I wish to thank my wife, Lora-Lee, to whom I have dedicated these two volumes. With her tolerance, good will and love of life, she has made the last decade—as she has made all of our life together—a period of joy, while also allowing me the time to do the work that had to be done.

W.B.

Preface

On Becoming a Futurist

I became a futurist sometime between the late 1950s and mid-1960s. It's hard to say exactly when, because it was not a deliberate decision but rather a process that took place over several years. Trained as a sociologist, I went to Jamaica in 1956 to do urban research, specifically intending to do the social areas of the city of Kingston to compare with earlier work I had done in Los Angeles and San Francisco. At the time, Jamaica was in transition from being a British Crown Colony to becoming a politically independent nation-state, a change that was, of course, part of the post-World War II breakdown of the European and American empires and the formation of the new states (that I briefly describe in chapter 1).

It was a heady time in Jamaica. Everyone was looking forward, toward the future. People of all walks of life talked of little else than the coming political independence. What had to be done to create a new Jamaica? What would Jamaica be like—and what ought it to be like—after independence? As they asked themselves such questions, some Jamaicans had high hopes for the future, while others had terrible fears. They wrote a new constitution. They designed a new government. They envisioned a new national future. Some of them questioned and thought of redesigning the entire society. No longer dependent on the British proconsuls, Jamaicans became citizens and leaders of their own country, and they shouldered the tasks of making the decisions of nationhood.

This was history-in-the-making and I found it irresistible. I changed my research plans to focus on it and, over the next twenty-five years and with the help of both American and West Indian graduate students and colleagues, I studied political development, social change and the future in most of the thirteen new states in the Caribbean, from Jamaica in the north to Trinidad and Tobago and Guyana in the south.

Studying how the decisions of nationhood were made in the new states of the Caribbean, I came to see the importance of images of the future in shaping present behavior and in shaping the future itself. Although I continued my studies of nationalism and leadership in the Caribbean, more and more I began to try to understand the general principles of futures thinking and the role they play in all human behavior everywhere, both in the individual decisions of everyday life and in the momentous collective decisions of national and world history. It is quite clear that people generally, like the West Indians I studied, quite naturally engage in futures thinking, but they do so only more or less well. Thus, there was—and is—a need for intellectual tools that will allow people to imagine alternative futures more deliberately and act on them more effectively.

Early on, I turned to the emerging futures field and began reading some key works. In my first attempt to bring together futurist ideas in some systematic way, James A. Mau and other colleagues joined me in writing *The Sociology of the Future* which was published in 1971.

Since then, futures studies has continued to develop, not without its ups and downs, of course, but trending toward increases in the number of futures-related publications and organizations, and toward methodological sophistication, practical experience, conceptual and theoretical complexity, clarity of purpose, ethical awareness, substantive depth, quality of research and scholarship, and professionalization.

Also, since then, much of the human community has continued to lose faith in the dominant world views and values that guided human beings in the past. Today, many people are aware that the life-support system of our planet may be endangered by human actions and they are conscious of their mutual dependence on other human beings, of the possibility of a common human fate. Increasingly, people are coming to realize that they must take responsibility for the future, both for their own individual futures and, collectively, for the shared future of all humankind. Such people seek the knowledge and wisdom to speak for the well-being of future generations, including those of the far future hundreds or thousands of years hence.

This book is my second attempt to summarize and expand the contributions of futurists to the envisioning power and well-being of humanity. I bring together futurist intellectual tools, describing and explaining not only the methods, but also the nature, concepts, theories, and exemplars of futures studies.

Purposes

In this book, I aim to achieve five different, though interrelated, purposes:

1. *To show that futures studies exists as an identifiable sphere of intellectual activity that has already made—and continues to make—important contributions to the knowledge base of modern society; to show that futures studies—just like other fields from anthropology to zoology—has a body of sound and coherent thought and empirical results that can fit within the covers of a book and that can be the basis of a serious course of study.*

Emphasizing its unity, I bring together some of the important work of futurists and of futures-oriented scholars and scientists, going beyond some of the pioneering efforts to describe futures studies such as those of Edward S. Cornish (1977), Bertrand de Jouvenal (1967), Eleonora B. Masini (1993) and Allen Tough (1991). I describe the recent origins of futures studies, its major purposes and assumptions, a theory of knowledge appropriate for knowing the past and present and for making warranted assertions about possible and probable futures, some detailed methods of futures research, and some exemplars of futures studies.

As part of this synthesis, I also show how modern futures studies is linked to a long history of social thought by describing its connections to a sample of key utopian writings. From this perspective, modern futures studies is a continuation of the age-old human quest to understand the nature of the good life and to find the correct values and norms of conduct that will lead to the flourishing of human society. I discuss human values, how to put them to use in judging the desirability of alternative futures, how to evaluate objectively the values themselves, and what major values appear to be nearly universally held.

Moreover, I suggest that there are some traditional human values that people ought to consider changing so as to meet the challenges of the twenty-first century. For the new millennium, surely, will bring a future startlingly different in many ways from the past and present and will require reassessments not only of age-old beliefs but also of value priorities.

In sum, I describe some of the work futurists have done as they study possible, probable, and preferable futures and as they contribute to the guidance mechanisms of both individuals and society. Doing so, I show how futures studies has a long and distinguished intellectual history. I show how, today, futures studies is not a wild "crystal-ball-gazing" enterprise as it is sometimes misportrayed by the popular press, but is a worthy academic field deserving of respect and support. Finally, I show how it is also innovative and anticipatory as it transcends the limitations of the past and forges new practical guidelines to prepare people for the coming future.

2. *To create a teaching instrument that can be used, especially in colleges and universities, both as a basic text for core courses in futures studies and as supplemental reading in other courses.*

For both undergraduate and graduate students in futures studies courses, I give an overview of the field, identify some major futurists, explain what they do as futurists, and describe contributions they have made to the methods and theories of futures research as well as to social policy and change. Also, I describe past trends and current issues in the field, identify some major futurist institutions and publications, and explain the major foundational principles of futures studies. This book, then, can serve as a basic introductory text for the futures field and is recommended to instructors of core courses in futures studies.

The book is also appropriate for use as supplemental reading in a variety of courses outside of futures studies itself, especially in the human sciences. For example, students taking any of the traditional social sciences courses that deal with political, economic, social or cultural change, decision-making and deliberate human action, human values, or social policy will find the description and analysis of futures thinking given in this book useful for their purposes. For futures thinking is necessarily involved in all of these subjects—if they are to be properly understood.

Also, other students can benefit from this book, including those taking courses that deal with applying knowledge for the design of social action whatever their specific subject matter—from forestry and environmental studies, international relations, management studies, and communication and transportation studies to education, public administration, epidemiology and public health, urban studies, and even to architecture and applied engineering. In all such cases, images of possible, probable, and preferable futures are relevant, if not absolutely essential. Although it is often done badly, futures thinking is always part of a conscious decision to act. The consequences of decision-making and human action always occur in the future. Indeed, futures studies itself can be accurately described as an "action science."

Please note, however, that this book is not intended to be encyclopedic. There is simply too much futures work now available to discuss everything that is worthy of mention. Thus, I have tried to be selective, focusing on pioneering works on some occasions, on particularly excellent or influential examples of futures thinking on others, occasionally even on bad examples (because we can learn what not to do), and at least once on neglected or forgotten work (e.g., the studies of Nathan Israeli mentioned in chapter 1). I focus selectively, too, on basic futurist presuppositions and concepts, and on major themes that tend to unify the field. I have tried to provide a useable map of futures studies giving its major features in an intelligible and meaningful way, a map that highlights the forests yet also describes some of the trees.

If readers want a comprehensive literature summary, I recommend that they look at the monthly abstract of futures-related publications, *Future Survey*, and the *Future Survey Annual* for various years. Both are publications of the World Future Society.

3. *To futurize the thinking of specialists in other disciplines, both because their disciplines would benefit from expanding their time perspectives to include the future and because futures studies would benefit from the related futures-oriented work of scholars and scientists from other disciplines.*

Specialists in disciplines outside of futures studies will find in this book a convenient summary of futures studies as it is today and as it may be becoming. My hope is that they will see how futures thinking can be of help in their own disciplinary research. Even history—one of the most past-oriented of disciplines, could benefit from a more developed and sophisticated orientation to the meaning and functions of time in general and to an expanded conception of the future in particular. Historian Paul M. Kennedy in his bestselling *Preparing for the Twenty-First Century*, for example, has demonstrated how identifying past trends may help to understand present problems and to direct the search for possible future solutions.

Thus, I hope to encourage disciplinary specialists to spend more time looking forward than they have in the past, discovering the alternative possibilities for the future in the present, and forecasting the probable consequences of present actions and events in the subjects that they study. There are few disciplines that could not benefit from becoming more future-oriented and building bridges to futures studies.

Additionally, futurizing the standard disciplines not only would benefit them, but it would benefit futures studies too. Futures studies, which is distinctively transdisciplinary, is still being shaped and its own future would be enhanced in methods, concepts, theory, and substance by contributions from other disciplines.

In addition to the social sciences which already overlap a good deal with futures studies, philosophy is of special concern. The philosophy of knowledge could help advance futurists' efforts to warrant their assertions about the future, while moral philosophy could help futurists deal more adequately with justifying their value judgments about preferable futures. I deal at some length with these topics in this book, trying to give a satisfactory way of grounding both futures-oriented, truth-like propositions on the one hand and value-judgmental, "ought" propositions on the other. Moreover, I give examples of what direction future collaboration between philosophers and futurists might profitably take, especially in volume 2, chapter 2 where I rely on the work of philosopher Keekok Lee.

4. *To contribute to the further development and improvement of futures studies.*

In this book, I try to advance futures studies in three different ways. The first is by reinforcing the emerging consensus among futurists about the nature of futures studies and about its importance to both the organization of knowledge and the policymaking process. Toward this end I draw together diverse strands of the field into a meaningful whole, highlight key futurist concepts, and identify a community of futurists. Where possible, I codify those substantive parts of futures studies now amenable to systematic organization.

Of course, some parts of futures studies can be more confidently reported than others (e.g., the recent origins of futures studies and the links to the history of utopian thought) and some are more fully codified and agreed upon than others (e.g., the methods of futures research, the purposes, and many of the assumptions of futures studies), while some parts contain only emergent glimmerings of organization and consensus (e.g., theories of change).

The second is by filling some of the lacunae in futures studies as presently conceived. Some parts of futures studies remain somewhat hazy. Sometimes this is because they are not yet fully developed. For example, futurists make assertions about preferable futures, but have had no adequate methods for justifying or testing them. I fill this gap in volume 2, chapter 2 by using Lee's method of epistemic implication to test value assertions objectively.

Then, too, some parts of futures studies are difficult to generalize about because they remain controversial. That is, different futurists say contradictory things about them. In such cases, I have tried to describe conflicting views fairly and accurately. Yet I have not been coy about letting the reader know where I stand. One example is whether or not futures studies is an art or a science. Another, related example is the question of what theory of knowledge is appropriate for futures studies. Still another is whether or not prediction plays much of a role in futures work. In each case, there is, I believe, a preferred answer, so I give it. More important, I give reasons that will help readers judge for themselves whether or not the answers that I give are adequate.

The third is by suggesting some future directions for futures studies itself. Thus, I speak to professional futurists, both to mature scholars and to those students who plan careers in futures studies. Although I focus primarily on describing how futures studies was and is, in this book I also occasionally describe, prospectively, how it may be becoming and, prescriptively, how it ought to be—as any good futurist would and should. One example is in volume 2, chapter 3 where I propose some professional ethics for futurists.

5. To provide tools to empower both ordinary people and leaders to act so as to create better futures for themselves and their societies than they otherwise would be able to do.

A major feature of futures studies is its problem-solving approach. It aims to be practical, to be of use in the real world. It can be used by people and organizations to help them navigate through time into the future. Thus, this book is intended, too, for the general reader, for people who want to learn how futures thinking can be used to enhance their individual lives.

For example, futures studies can be useful as people consider the real possibilities for their own personal futures. It does not have to be used in any formal way, as in the future workshops described in chapter 6. Quite informally, people in their everyday lives can use the futures perspective to discover and create choices for themselves, to forecast roughly the consequences of their contemplated actions, to judge the desirability of the various consequences and to make decisions among them, and to design and plan specific action.

Perhaps, the most typical activity of futurists is to work as a researcher/consultant for some client to help the client achieve his or her specific goals. Such a futurist may be a regular staff member of some organization or hired as an independent consultant. The clients range from large corporations and small businesses to universities, hospitals, and governmental agencies. Sometimes, futurists work for "think tanks," like the RAND Corporation described in chapter 1, that have many different clients simultaneously. Some of this work is proprietary and confidential and, thus, the available futurist literature is only a sample of the actual futures work being carried on. Some of it, however, is focused on the narrow concerns and special interests of particular clients.

But there is more to the practical side of futures studies than this. Other futurists, taking a long-range and holistic view of the world, have much broader and more public-oriented goals. They aim to raise the level of human understanding and consciousness about the interrelatedness of all people to each other. Despite the apparent cultural diversity in the world, many futurists believe that humans everywhere, both as biological and cultural beings, have much in common. They see similarities not only in basic needs, but also in human goals and values. They see, too, the global growth of mutuality, and the need to define the collective aspirations of all humanity.

Such futurists explore the meanings of human betterment and invite the human community to engage in public discourse about a global ethic. Can we humans, they ask, learn from the errors of the past and, through effective actions, create a future world of compassion, peace, justice, and happiness for all? Ought we to want to do so? Moreover, can we—and should we—work to make such a world sustainable indefinitely into the future? Can we—and should we—create a world in which both present and future people have a good chance of living long and satisfying lives? Can we—and should we—try to create a world moral community?

As we end one millennium and begin another, such questions are especially timely. So far, the twenty-first century is virginal. It is, as yet, untouched and, hence, unspoiled by human acts. No doubt, it is unrealistic to expect perfection, to anticipate a future world of endless individual happiness and social harmony to be created without—or even with—pain and suffering. But it is certainly not unrealistic to pledge our efforts to work toward perfection, to act in order to make the world a better place than it otherwise would become. Such lofty aspirations, too, are part of the futurist enterprise.

Although I became a futurist largely as a result of the circumstances in which I found myself in Jamaica in the late 1950s and 1960s, I remain a futurist because the emerging futurist intellectual program offers a solid standpoint from which to engage in constructive social criticism and action. Moreover, despite both its sometimes doomsday rhetorical style and widespread use by special interests, futures studies offers hope for the future of humanity and some concrete ways of realizing that hope here in the real world of our daily lives. As a futurist, I want to share that hope with others and to encourage the use of futures studies to serve the public interest. I want to share, too, the sense of excitment and adventure that comes with transforming such hope into effective acts that create the future.

Thus, I have written this book in which I explain what futures studies is and describe some of the ways that it has been used—and could be used—to change human lives for the better, both in the immediate and the distant future.

Wendell Bell
Bethany, Connecticut
31 August 1995

1

Futures Studies: A New Field of Inquiry

Thinking About the Future

Recently, at a Chinese restaurant, my fortune cookie revealed these words of wisdom: "The smart thing to do is to prepare for the unpredictable." We might all agree that that's smart. But exactly how do we go about preparing for the unpredictable? If something is totally unexpected, then it seems impossible to prepare for it because we don't know what "it" is. "Be prepared" may be good advice, but be prepared for what? Shall we be prepared to abandon ship, escape fire, move to the country, shovel snow, take a surprise examination in chemistry, spend the million dollars we'll win in a lottery, be run over by a truck, write our acceptance speech for the Nobel Peace Prize, or what? The possibilities— likely and unlikely—are too numerous to contemplate if we try to prepare for what is truly unpredictable. Such an effort could become an endless and fruitless search.

The future—fortunately or unfortunately, depending on whether or not we like surprises—is like that. The best laid plans can go awry. Unexpected things can happen to us. The future contains an element of uncertainty. Nonetheless, we do try to prepare for the future and to deal with its uncertainties. Moreover, in our everyday lives we do so surprisingly well, although some people seem to do it better than others.

We also try to control the future, not only to "prepare" for it by adapting to what we think is coming, but to make things happen that we want to happen. We map out the next day or the next few years; we plan our future marriage or occupational career, knowing, of course, that things may not turn out exactly as we have planned.

Most of us do not give much conscious thought to how we do this, so we don't have ready answers to such questions as: What procedures do we use

1

when we think about the future? How do we prepare to carry out our plans and projects? What makes us successful in shaping or adapting to the coming future? At any given time, what alternative courses of action are open to us? What will be the future consequences of choosing to do one thing rather than others? What ought we to want the future to be?

These are central questions of futures studies, a new field of inquiry that involves systematic and explicit thinking about alternative futures. It is a growing body of work that is based on distinctive perspectives and assumptions and that utilizes specific theories, methods, and values. It aims to demystify the future, to make possibilities for the future more known to us, and to increase human control over the future. In the broadest sense, futurists hope to inform people's expectations of the future and to help make their efforts to shape the future to their worthy values and purposes more effective. In some sense, thus, futures studies helps us to "prepare for the unpredictable."

The Universality of Time Perspectives

Thinking about the future, of course, is not new. It is a universal phenomenon that can be traced back to the dawn of human prehistory. In every known society, people have conceptions of time and the future, even though some of their conceptions appear diverse, with different emphases on past and future and different degrees of elaboration and detail.

Divination, for example, is found in some form in every society. Sometimes it involves, as its root meaning suggests, discovering the will of the gods, but, more generally, it refers to finding obscure or secret things, including discovering the future by eliciting a divine response. In some cultures, a belief in fatalism predominates. In others, the hope of taking action to avoid a calamitous predicted event or to bring about a desirable one is common. Although it often concerns the everyday interests of particular individuals, divination also may involve grand prophecies dealing with the destinies of whole tribes and nations.

The range of techniques of divination is truly staggering. Fortune-telling, as done in one culture or another, has included the examination of the configurations of such animal organs as lungs, gall-bladders, intestines, hearts, stomachs, and livers. Particular configurations of such organs were correlated with presently observed events and, then, taken to portend similar consequences for the future. Ancient diviners made observations, continually corrected them, and produced "a huge corpus of cuneiform handbooks devoted to every possible configuration" (Hallo 1993: 6).

Other methods of divination include observing the patterns a fire makes as it burns (pyromancy), consulting the patterns of fire-cracked shoulder blades (scapulimancy), and watching cheese coagulate (tyromancy). They include, also, interpreting the patterns made by pouring oil on water, observing human facial features and bodily functions, and finding meanings in spittle, flowers, knotted threads, hemp seed, the whites of eggs, apples in water, ashes, salt, or cabbages. In fact, any phenomena in the world have been pressed into service to foretell the future (Hallo 1993). Oneiromancy (dream interpretation), for example, existed long before Sigmund Freud; William W. Hallo (1993: 5) believes that it may have been the oldest means of forecasting and gives an example of its use in Sumer as early as the twenty-fifth century B.C.E. In one society or another, all sorts of sorcerers, witches, shamans, sacrificial victims, and other oracles and readers of omens have been consulted in order to discover and to control the future (EB 1974: 916-20; Turner 1968; Yalman 1968).

The underlying belief in the ancient Near East, for example, was that "both omen and consequence were alike the expression of the same divine intent—the one as its announcement, the other as its fulfillment" (Hallo 1993: 6). More generally, however, anthropological studies of divination are not neatly separated from the study of magic, because it is not always a simple matter to distinguish an act which is meant to foretell the future from one which is meant to control it through intervention with supernatural powers.

Distinctions between past and future can also be seen in rites of passage. People recognize through ceremonial activities the major changes in individuals' lives, such as birth, coming of age, marriage, and death. Such changes represent transitions from past to future social roles and alterations in rights and duties as individuals move through the life course (Goody 1968). Peoples of nonliterate societies typically appeal to gods and goddesses, or perhaps to less elevated spirits of some kind including those of their ancestors, to increase the availability of game animals, the yield of the harvests, and the fertility of the tribe—all future-oriented thoughts.

In modern industrial societies many of us take for granted, and largely ignore, the fact that divination of various kinds exists in our midst and is sincerely believed in by some otherwise sophisticated people and followed by many others who may be skeptical but do not completely reject it. Thus, ouija boards are consulted, forked sticks used to decide where to dig a well for water, bumps on the head examined by phrenologists, *I Ching*—the Chinese *Book of Changes*—studied, tea leaves and Tarot cards examined, black cats seen as omens, and, most widespread of all, astrological horoscopes are read—in daily newspapers and even university bookstores—to discover what the stars, planets, sun, and moon and their various interconnections have to tell us about what the future has in store for us.

The ability to anticipate the future begins early in life, as soon as a newborn child learns that his crying results in reactions from other people. At first, such anticipation reaches only into the immediate future, the briefest of moments beyond the present, perhaps without any conception that the immediate future is a different time than the present. As a child gets older, his or her time horizon begins to develop and then expand. Consciousness of the past is pushed farther back in time, eventually well beyond an individual's personal experiences into those of his parents and grandparents, then still farther back into the oral traditions and charter myths of the tribe or the written histories and archaeological remains of earlier cultures and civilizations. So, too, time consciousness is pushed forward, beyond the immediate future farther into the time that is yet to come.

As children learn a language, of course, they also learn the time perspectives of their culture (Fraisse 1968). But such learning does not occur all at once. Children begin by talking solely about the present, about objects and events in the here and now. Later, they acquire the notion of nonpresent, grasping first the idea of the past, things that they have already experienced. After that, they begin to understand the notion of the future, things that they have not yet experienced. In learning about the nonpresent, children at first tend to confuse words like "yesterday" and "tomorrow." Harner (1975), for example, in his study of two- to four-year-old children found that they began by interpreting "yesterday" as both past and future, that is, as "not present." Next they got the meaning of "yesterday" right, then, finally, they learned what "tomorrow" meant (Clark and Clark 1978: 251).

Although the vast majority of languages have tenses, there are a few tenseless languages. There are, however, no timeless languages. For even in a tenseless language, the ideas of past, present, and future can be understood and communicated. For example, in a tenseless language, conveying ideas about time is achieved by other means such as adverbials, the aspectual contrast of perfect or imperfect, and context (Gell 1992).

The scale of time perspectives—that is, how far into the past and how far into the future an individual's thinking goes—may vary from one society to another. It varies also among individuals within the same society, especially within large and complex societies characterized by considerable social inequality. But some time perspectives, some conception of past and future, and some effort to predict or control the future appear to be part of the language and culture of all—or nearly all—peoples.

Time, like space, is an inevitable aspect of individual experience and social interaction. Everything occurs at some time and in some place. Although at quite different rates, nearly everything changes from one time to another. Thus, it is not surprising to find that time is in some degree a part of human consciousness everywhere. After critically reviewing the anthropological literature on time, Gell (1992: 315) concludes that, despite some claims to the contrary, nowhere do "people experience time in a way that is markedly unlike the way in which we do ourselves." Everywhere, there is past, present and future. Everywhere, time moves on and on. Everywhere, time "is always one and the same" (see chapter 3).

The Utility of Divination

The modern approach to the study of the future, like primitive divination, includes an effort to discover and often, if possible, to control the future, to bend it to human will. Unlike divination, though, it is not based on beliefs in the supernatural, magic, mystification of methods, superstition, or the secret powers of particular individuals or groups. Rather, the opposite is the case. Futures studies is part of modern humanism, both philosophical and scientific. It is secular. Futurists, the practitioners of the futures field, aim to demystify the future, to make their methods explicit, to be systematic and rational, to base their results on the empirical observation of reality where relevant, and to test rigorously the plausibility of their logic in open discussion and intellectual debate. They also use creativity and intuition. Although some futurists occasionally abuse these values, they remain the ideals that most futurists strive to fulfill.

Despite the enormous qualitative differences between divination and modern futures research, we should not underrate divination nor dismiss it too quickly as utter nonsense. As James Frazer understood in his classic anthropological work, *The Golden Bough*, primitive practice is often based on "excellent observation of natural phenomena and involves a theory of causality" (Yalman 1968: 521). Today, many of our theories of causality are different and, therefore, so are many of our observations since they are guided by them. Especially, the meanings that we assign to our observations are different than those used in divination.

At least, divination could eliminate paralyzing indecision when an individual was faced with the uncertainties of the future. It could help a group in conflict decide what future course of action to take, thereby resolving the dispute and legitimating a given decision. Flipping a coin, of course, can—and for some people does—serve the same function of permitting decisive action in perplexing situations with no greater knowledge of the future than a fifty-fifty chance of being right. At most, divination did more. For example, the diviner does better than mere chance in guessing the future—or whatever secret was at issue—where the diviner could rely on his past experience of similar situations or ferret out the implicit knowledge revealed by the anxieties of his clients that lay below the level of their own consciousness. Additionally, in deciding such questions as where to hunt, divination sometimes provided a more or less randomized exploitation of the habitat and, thus, was adaptive.

In fact, divination may have at least one major advantage over modern futures research, since for either to be of much use people have to believe in it. There is no question but that in many nonliterate societies people had great faith in their diviners, more, no doubt, than many people in modern, industrial societies generally have in their futurists.

Recent Origins of Futures Studies

Faith in modern futurists may be increasing, though, if it is accurately measured by the rapid, even astonishing, growth of futures studies during the last few decades. Of course, as I show in subsequent chapters, futures studies can be viewed as a recent culmination of a long, sometimes interrupted, sometimes waning as well as waxing development of ideas stretching as far back as the Greeks and, beyond that, even to the primordial diviners we have just discussed. Yet the specific origins of the modern futures movement are much closer to us.

It is risky, of course, to pick a particular date and say that is when something began, because it is easy to question the choice with examples of precursors of various kinds. For example, a good case can be made for dating the recent origins of futures studies from the publication of H. G. Wells's *Anticipations of the Reaction of Mechanical and Scientific Progress upon Human Life and Thought* that first appeared in 1901 in the British magazine, *The Fortnightly Review* (Wager 1991). Among other things, Wells proposed "a science of the future" and, in a radio talk over the B.B.C. in 1932, he claimed that not only Professors of Foresight were needed, but entire Faculties and Departments of Foresight to anticipate and prepare for the coming future (Wells 1987). Another good case might be made for tracing the modern futures movement to the futuristic novels of the French writer, Jules Verne, which were published in the 1860s and 1870s.

Although in volume 2, chapter 1, I discuss some early writers—from Thomas More to Karl Marx—whose works have provided many of the themes, purposes, and exemplars that have helped shape modern futures studies, my aim in this chapter is more modest. Here, I limit myself to a few recent illustrative strands of intellectual history that appear to have led to—or in one case *might* have led to—the creation of futures studies as it is today.

Ogburn and Technology

One such strand is that of sociologist William F. Ogburn and his co-workers, especially S. Colum GilFillan. U. S. President Herbert Hoover appointed a President's Research Committee on Social Trends in 1929. The committee, headed by Ogburn, produced the most comprehensive description of social change in American society up to that time. The report, *Recent Social Trends in the United States*, was published in 1933. After the election of Franklin D. Roosevelt as president of the United States in 1932, Ogburn was once again called to public service. He served as an active member of the U. S. National Resources Committee and helped shape its report, *Technological Trends and National Policy, Including the Social Implications of New Inventions* (1937).

Ogburn's method included forecasting the future by quantitatively determining long-term trends concerning the past, and, then, projecting them into the future for a number of decades. He considered such trends, for example, as the "increasing role of government and the growth of large businesses" (Jaffe 1968: 279).

Ogburn's theory of social change emphasized the role of invention. For him, change in the modern world typically followed a causal sequence beginning with some technological invention or innovation. The technological change, in turn, produced change in economic organization, which produced change in social institutions—such as the family or government. Finally, according to Ogburn, changing social institutions produced change in people's social philosophy, that is, in their beliefs, attitudes, and values. He would also agree that the sequence was sometimes circular, with social philosophies altering the demand for certain types of inventions and, thus, leading to technological change and starting the causal sequence over again (Jaffe 1968: 279).

Ogburn made many contributions to the study of the social effects of technology and he cofounded and was the first president of the Society for the Study of Technology. Thus, there are ample justifications for the claim that some people make for him that he was the father of technology assessment (Cornish 1977: 74). Technology assessment has become one standard approach to futures research, has been institutionalized in a variety of organizations, such as the U.S. Congress's Office of Technology Assessment, and has been a career path for prominent futurists, such as Joseph F. Coates and Vary T. Coates.

Also, Ogburn's idea that a society should produce a quantitative picture of itself as a way of knowing where it had been, where it was going, and how to make sound decisions about social policy grew into the "social indicators movement" in the 1960s. The idea had never died, but during World War II it was largely suspended as the war mobilized nearly all activities, including statistical work, directly for the war effort itself. After the war, the idea of the need to monitor the state of the society by using a variety of social indicators—from population, labor force participation, and technological change to crime, education, and health—took hold again (Innes 1990).

The Twentieth Century Fund published a report on "America's Needs and Resources" in 1947 and again in 1955. President Lyndon B. Johnson in 1966 set up a group in the then-Department of Health, Education, and Welfare to develop social indicators and to use them to monitor societal welfare. That same year Raymond A. Bauer published *Social Indica-*

tors. The Russell Sage Foundation advanced the social-indicator movement further with the 1968 publication of *Indicators of Social Change: Concepts and Measurements*, authored by leading scholars (Sheldon and Moore 1968). By 1972, Wilcox et al. found over 1,000 published articles and books on the subject, and, by 1973, the U.S. Executive Office of the President had published *Social Indicators 1973* (Innes 1990).

The social-indicator movement, of course, was not limited to the United States, but occurred in many other countries also. In the United Nations, starting soon after the end of World War II, efforts were made to develop a variety of measurements, partly as a correction to the almost exclusive focus on economic growth in developing countries, that have come to constitute what we now call "the quality of life" (Innes 1990).

GilFillan should receive some mention as well, because he was not simply a disciple of Ogburn, although he worked for him as a junior partner on some of his projects. GilFillan wrote an essay for his master's degree at Columbia University in 1920 in which he evaluated the predictions of four writers from the eighteenth century, showing that they were reasonably accurate. Condorcet, for example, about whom we'll learn more in volume 2, chapter 1, got the highest score for accuracy, 76.4 percent. GilFillan (1920: 3) aimed to show that prophecy of social realities "has already been *proved* possible by its successes in the past." Also, at this early date, he proposed the term, "mellontologist," for a student of the future. GilFillan continued his interest in the future and social change throughout his career, and, among other things, wrote *The Sociology of Invention* (1935).

Israeli and a Forgotten Start

Ogburn is well known among futurists today and he seems destined to occupy a secure place in any proper history of futures studies, because there is a demonstrable continuity between his work and some aspects of current futures research, including, as we have seen, the study of social trends, technology assessment, and social indicators. Other prototypical futurists may be unrecognized and their identities now lost to us, perhaps awaiting the day when some modern futurist will discover their pioneering work. When a thorough history of the origins of futures studies is written, such scholars may be identified and honored for their pioneering roles.

In my own efforts to reconstruct a small bit of the past, I discovered one such lonely worker in the futurist vineyards who deserves to be rescued from obscurity. He is Nathan Israeli who appears to have carried out research on the social psychology of time and the future singlehandedly. Fortunately, although the fruits of his labors lie forgotten, many still exist buried in the library stacks.

In the early 1930s, Israeli carried out a number of experiments, reasonably sophisticated for his day, using students at the University of Maine as subjects. Israeli, among other things, anticipated some of the features of the Delphi technique to be discussed in chapter 6, for example, in questioning his subjects about their estimates of the probability of the occurrence of future events and choosing among alternatives. In addition to some general works trying to establish a social psychology of the future, Israeli studied students' perceptions of the factors producing social change; their predictions of future divorce rates and other phenomena; the factors that influenced their predictions; and the students' emotions and attitudes toward the past, present, and future (see items for Nathan Israeli listed in the references).

Of course, other individuals were working in what we would now call futures studies between the two world wars as well. Some, like Buckminster Fuller, Arthur C. Clarke, and Clifford C. Furnas, have been treated better by history than was Israeli, although they were more "freelance prophets," as John McHale (1978: 8) calls them, than empirical social researchers.

National Planning: Military, Economic, and Social

When Nathan Israeli offered a social psychology of the future to the world, the world was not ready for it. Yet the seeds of futures studies had been sown in the national mobilizations of the combatant nations of World War I. They were being sown, as Israeli was publishing, in efforts to cope with the economic crises of the Great Depresssion, to create a communist, industrial state in the Soviet Union, and to develop a corporate state in Fascist Italy and a mobilized society in Nazi Germany. Moreover, the seeds were to sink deep roots during World War II which was still to come. All of these events required making plans, formulating policies, and designing blueprints for the future.

Of course, planning itself—just as futures thinking—is as old as human society. Even tribal groups engage in some community projects,

both ceremonial and utilitarian; some hunting and fishing are cooperative and future human effort must be coordinated; and both agriculture
and herding require that some things be done in the present with the
intention of reaping future benefits. As the complexity of society increases,
or decreases as it sometimes does, so does the potential for an increase in
the scope of planning flow and ebb. Collecting taxes, managing estates,
irrigating the land, and waging wars require planning.

Alexander of Macedonia didn't conquer the Greek city-states and
the Persian empire without planning, nor did Hannibal, the Carthaginian,
cross the Alps to invade Italy, nor Genghis Khan conquer much of Asia
and Eastern Europe without it. It took planning to build the Great Wall
of China, the library at Alexandria, the Colossus of Rhodes, the Olym-
pian Zeus of Phidias, and the Taj Mahal. Yet it was not until the twentieth century that economic and social planning grew into the
comprehensive activites that reach into the everyday lives of nearly
every individual. It was not until the creation of the modern state that
everyday life became so fully under conscious regulation, encouragement or direct control.

World War I. Before 1914, organizations or groups drawing up large-
scale, comprehensive collective goals and ways of achieving them on the
level of the whole society were most likely to be military general staffs.
They planned the character and the movement of entire armies and navies,
and, because warfare increasingly depended on an industrial base for
material, supplies, transportation, and communication, they began planning also for key aspects of the entire economy and society.

The national mobilizations of World War I also brought nonmilitary
national leaders into the picture. Among other things, the war demonstrated the inadequacies of the nation to meet the national emergency
produced by the war. National levels of health and education, for example, were below what many leaders thought were needed. Economic
and social reforms, it was thought, should be introduced (Madge 1968:
125). Furthermore, the mobilization efforts themselves required complex planning, by civilian as well as military leaders, from the allocation
of material and human resources in industry to the distribution of food
and clothing to the civilian population.

Thus, the mobilizations brought on by World War I enlarged the organizational capabilities for establishing futures thinking in the institutional
structures of modern societies and created a favorable social psychological disposition for doing so.

The Great Depression. During the interwar years, the Great Depression contributed to the beliefs that something had gone wrong with the economy and that something should be done about it, especially through governmental intervention. Then current socialist thinking, along with economic ideas involved in national accounting, in Keynesian theories, and in trying to eliminate the troughs of business cycles, encouraged the idea that the economy should be regulated or controlled, at least to alleviate such negative effects as unemployment, high inflation, and lack of capital investment (Tinbergen 1968: 102). In the United States under the presidency of Franklin D. Roosevelt, the New Deal ushered in a period of economic and social engineering by the federal government, including the massive planned development of the Tennessee Valley that began in 1933.

Responses to the Great Depression, thus, called forth many of the elements in a rudimentary way that are now commonly understood to be part of modern futures studies: an analysis and interpretation of the recent past and the present, projections of future developments if no intervention occurs, a description of possible alternative actions and the different futures each will lead to, an evaluation of alternative futures as to which are most desirable, a selection of specific policies to implement in order to achieve a desirable future.

Communist Russia. Speeding across Germany in a sealed train, crossing Sweden and Finland, Vladimir Ilyich Lenin arrived in Petrograd on the evening of April 16, 1917, just a few months before he and his Bolshevik adherents were to seize power in the "October Revolution." Neither Lenin nor any of the other leading revolutionaries were prepared to run the country, to establish and manage the day-to-day mechanics of decisions and activities that would replace the systems of production and distribution they were determined to destroy.

The literature of communist revolution from Karl Marx's *Das Kapital* to Lenin's own *Imperialism: The Highest Stage of Capitalism* was largely a critical examination of the capitalist system, describing and explaining not only its rise but its inevitable fall (Barnard 1966: 46–47). Although it gives the image of the coming communist future in general terms, it mostly defines what it wants to create negatively: the opposite of some major features of capitalism. It gives few details of the future administration and management of the communist system. "The shape of things to come" includes a dictatorship of the proletariat, the abolition of private property, and control over the means of production (Whitney 1962: x).

What positive thinking the revolutionaries had done was mostly focused not on responsible governing after attaining power, but rather on the strategy and tactics of achieving power. There were no precedents. No one before had actually tried to create, maintain, and manage a "dictatorship of the proletariat." These Russian revolutionaries were least prepared of all. Most of them had believed, for what they considered to be good Marxist theoretical reasons, that the revolution would come not in economically backward Russia, but rather in an advanced capitalist country such as Germany.

Their visions of the future, however, began taking shape in concrete plans and actions almost from the beginning of their coming to power. Amid war and civil war, an invasion by Poland, the chaos of rapid change and bloody turmoil, and boundless revolutionary hope, the Bolsheviks nationalized nearly all industry and land and seized the state bank. By March 1920, they had created one of the first planning organizations, GOELRO, the State Commission for the Electrification of Russia. In February 1921, they formed the state planning commission, Gosplan (Munting 1982: 45–46). By 1926, the new ruling group had made a political commitment to planning with the goal of creating a self-sufficient industrial economy, planning projects with long gestation periods. Before the end of 1928, Gosplan had presented two versions of the first five-year plan, which was to run for the period 1928–1933. In April 1929, the Sixteenth Party Congress unrealistically accepted the most optimistic variant of the plan, "the maximum possible achievement" (Munting 1982: 70, 73).

Drafting the first five-year plan took four years, almost as long as the plan was intended to cover, and, of course, there were changes in midstream. The revolutionaries-turned-ruling group worked out the "principles of planning...by trial and error. There was no theory; practice came first." For them, planning "was a complex and pioneering experiment" (Munting 1982: 85, 87). During the early years of rule, the revolutionaries' view of planning changed from the idea of planning as action necessary to cope with immediate situations to the idea of planning as setting future goals beyond the horizons of the present and specifying the means for achieving them. Eventually, they developed a system of long-range planning (ten or more years), medium-range planning (five years), and short-range, operational planning (one-year or quarterly) (Nove 1977: 31). Also in the early years, they had to give up the idea that revolutions would soon occur in the more economically advanced countries and that the working classes there would help them solve Russia's economic problems.

Both the voluntarist elements of Marxism stressed by Lenin and the idea of the divorce of the future from the past through action permeated Bolshevik thinking. Lenin developed the notion that a conscious socialist minority could seize power and the revolutionaries did not have to await the time when the economic and social situation was, according to Marxist theory, ripe for revolution (Nove 1969: 37). Furthermore, in drawing up the first five-year plan, the planners shifted away from the "genetical" school that based planning on past data projected into the future, that is, using the method of extrapolation (see chapter 6). They shifted toward a "teleological" school which argued that "the Russian proletariat had, with the Social Revolution," moved from necessity to freedom. No attention had to be paid to the past. The point was to define a great purpose and then go about accomplishing it.

As the political economist, S. G. Strumilin, who helped formulate the first five-year plan, said, planning actively "created a new world for itself" (Brutzkus 1935: 124). In other words, the past is not a guide to the future. The past can be transcended. Conscious decisions and efforts to achieve great purposes can shape the future. "Political intention could be fulfilled...by the strength of political will" (Munting 1982: 104). Many of these ideas, although with some important modifications, are part of the thinking of modern futurists.

Considering the first and second five-year plans together, that is, the first decade of planning, 1928–1937, we see the creation of a "command economy," centralized and enforced with the power of the state. The Soviet Union transformed itself into a great power and an industrial nation, nearly catching up with the advanced capitalist economies. Huge increases in the input of labor and capital produced nearly unparalleled rates of growth (Munting 1982: 105).

The astonishing achievements captivated the imaginations of some foreign observers. For others, the generally low living levels of the people, the squalor of the mass of the peasantry, the large and inefficient agricultural sector, and, most of all, the use of terror as an economic weapon tarnished any possibility of enchantment (Munting 1982: 104-6).

Some of the terror may have been—surely was—an essential part of the early consolidation of Bolshevik power and transformation of the economy and society, for example, the civil war against the Mensheviks on the one hand and the collectivization of agricultural production on the other. Yet, just as surely, much of the killing, imprisonment, and intimidation was unnecessary. In fact, such brutal repression actually interfered with the proper functioning of the economy and the achievement of the goals of the economic plans. During the great purge of 1937–38, for example, the political police removed thousands of people, some of whom were effective economic managers, in Stalin's efforts to establish his personal control over the growing totalitarian system. Such terror, of course, was hardly necessary and probably was counterproductive (Munting 1982: 85).

Even if repression, murder, torture, and terror had been effective instruments of development, they were—and are—morally wrong. The tragic failures of the Soviet system can be more fully understood in the context of Marxist theory that I discuss in volume 2, chapter 1.

Fascist Italy. The Fascists came to power in Italy in 1922 and ruled until 1943. During this time, Italy became, at least on paper, a state-controlled society. The Institute for Industrial Reconstruction, begun in 1931 as an antidepression effort, became a central agency controlling major financial institutions and most heavy industry. A central governmental bureaucracy exercised control "over both labor unions and employers' associations." The Fascists pursued the images of future national power, military and economic; territorial expansion; and brave, new centralized governmental control over national resources and energies. One basic Fascist tenet, the supremacy of the technical expert over the politician, foreshadowed a main feature of the societies many futurists were later to envision developing near the end of the twentieth century: the preeminence of the professional and technical class and a technology focused on information and knowledge (Einaudi 1968: 337).

The experience of Fascist Italy presaged the future in two other ways as well. One was the power shift in the modern mass-based polity from the legislative to the executive branch of government. Another was the innovation of the mass-based one-party state, the prohibition of any political opposition within a system of legitimacy nonetheless based on the popular will. The one-party state, of course, has appealed to some of the leaders of the new states that have been created from the former colonial territories since the end of World War II, although few such leaders—if any—would acknowledge its Fascist origins as their inspiration (Einaudi 1968).

Nazi Germany. In Germany, the National Socialist German Workers' (Nazi) party came to power in March, 1933, under the leadership of Adolf Hitler, who in the next year, as we all know, became dictator. The lightning moves in territorial expansion are well known: the Rhineland occupied in 1936, the Sudetenland taken and Austria annexed in 1938, Czechoslovakia occupied and Poland invaded in 1939, Norway and Denmark invaded in April, 1940, and Holland, Belgium, Luxembourg, and France invaded in May, 1940. Germany held most of the European continent before the end of 1940.

At the same time, the economic crisis that helped bring the National Socialists to power was overcome. The entire nation was being mobilized into a unified, dynamic, and spiritual community. Well, not quite the *entire* nation, because dissenters, recalcitrants, Jews, and others who did not fit the image of the Aryan master race or who were not enthusiastic believers of the Nazi ideology were intimidated into silent conformity, forced to flee, imprisoned, or murdered. Yet for the great bulk of the population, the early years of National Socialism brought full employment, increased productivity, new roads and buildings, consumer goods, an end to the inflated, worthless currency which, in November of 1923, had made one U.S. dollar equal to 4.2 billion marks, and, of course, the promise of national power and international respect (Passant 1966). The early years also brought the process of *Gleichschaltung* (political coordination), by which all German political, economic, and social life was to be brought in step with Nazi thought (Broszat 1966: 142).

There was little that was "socialist" about this except the early party program and speeches. The "Twenty-five Points" of 24 February 1920 had included the nationalization of trusts, profit-sharing by workers in industry, communal ownership of department stores, land reform, and other unquestionably socialistic proposals. But these were no longer part of the Nazi program after 1931 and none of them was ever implemented under Nazi rule (Broszat 1966). There was, however, a similarity to Marxist-Leninist movements in organizational principles and strategy: regimentation throughout the whole party structure down to the system of street cells (Bracher 1971: 149), cadres of well-trained professionals (Merkl 1966: 7), and strict discipline of party members.

Although the economy of the Third Reich was hardly a smoothly functioning system of planning and controls, it would be a mistake not to recognize the important lessons of the potential of planning and futures thinking that the rest of the world learned—and often feared—as they watched the rise of Nazi Germany (Bracher 1971: 334). The Nazis created an involved and dynamic system of superagencies, planning councils, controls, and guidance mechanisms for the economy and the society. Contrary to the view of some of its admirers at the time, it did not prevent "waste, jurisdictional conflict, corruption, or faulty planning" (Bracher 1971: 333–34). But it did create a largely unified, mobilized society in which, for a brief period, the total energies of the people were multiplied and focused on collective goals that sometimes seemed nearly impossible given the economic and social chaos of post-World War I Germany out of which the Third Reich had grown.

Many of the developments in Germany could be called, as Marxist scholars tend to do, "state-monopolistic capitalism," but the system went beyond that. The first Four-Year Plan was announced in 1933. Goring's Four-Year Plan, the second, began in 1936 and involved drawing up comprehensive plans for everything from wage and production controls to the nature of working conditions. Although the existing capitalist structure was often used, the economy was nonetheless largely under the control of the Nazi regime (Bracher 1971: 335). Private firms were influenced or forced to take certain actions, for example, Krupp was forced to finance a synthetic rubber project (Grunberger 1971: 178). At the same time the government got directly involved in far-flung enterprises, such as the Hermann Göring (steel) Works. It is no accident, perhaps, that Albert Speer, an architect skilled in creating plans and building small-scale models in the present of what often became life-sized realities in the future, ended up running most of the German economy.

The educational system was reorganized and manipulated, with the party and the government controlling education from kindergarten through the university. The Hitler Youth replaced a plurality of youth groups and furthered the indoctrination of the young. The government-sponsored Labor Front replaced independent labor unions and strikes were prohibited. The "German faith" was promoted to replace Christianity. Concentration and extermination camps were built. Children were "encouraged to inform on their parents and unmarried women to breed a new *Herrenrasse* (master race) out of wedlock" (Ebenstein 1968: 46). The Nazis created a new social order, *ein Volk, ein Reich, ein Fuhrer*, an image of a future "great united society of all patriotic Germans working together in harmony for the glory and prosperity of the Fatherland" (Passant 1966: 179).

Yet it wasn't until after the German defeat at Stalingrad in 1943 that the German economy was totally nationalized (Grunberger 1971: 180). By then, of course, the beginning of the end had been reached. In 1945, the adventure was over: Germany surrendered, unconditionally.

Futures thinking and acting with foresight—like any other type of thinking and acting—can serve many purposes and values, both good and evil. Regrettably, in Nazi Germany, they ended up serving, as they did in Fascist Italy, the values of national power and social exclusion, while violating other values, such as justice, freedom, and human dignity. Perhaps, if the Nazi leaders had foreseen the future consequences of their acts more accurately, they would have behaved differently.

World War II and its aftermath. In the Soviet Union, the goals of the second five-year plan shifted away from agricultural production and consumer goods and from higher levels of living and toward heavy industry where the priority had been in the first plan, in part because of the rising threat of Germany. The Nazis made no secret of "their determination to make war on the Soviet Union" (Nove 1969: 227). Mobilization for war became increasingly pervasive by the late 1930s, not only in Italy, Germany, the Soviet Union, and, of course, Japan, but also in the Western democracies, headed by Britain and the United States.

During World War II, both on the military and home fronts, the requirements of massive planned change toward the greater organization of economic and social life forced leaders and their functionaries to make plans for the future, both for the short and the long run. Imagine the magnitude of the managerial tasks involved in inducting, training, transporting, clothing, housing, and feeding millions of men and women as they poured into the various military services. Imagine reorganizing entire industries for the war effort, predicting how much of what material and equipment, manpower, food, and so on should be produced and when and where such things should be transported. Imagine even the relatively minor task of planning a gasoline rationing system to get enough fuel when and where it was needed for industrial and military purposes, while keeping enough fuel available to allow the national workforce to move daily from home to office and factory and back. Such tasks characterized the conduct of the war, the economy, and even much recreation and entertainment. In the United States, for example, planning for economic recovery from the Depression gave way to planning for war.

Beyond that, there were larger questions that the war forced leaders to face even while the war was raging, questions that probed farther into the future and into socioeconomic and political issues that were to shape the coming postwar world. The Marshall Plan for the recovery of Europe, for example, was one important result, as was the political and social transformation of Japan under the command of General Douglas MacArthur. Rebuilding countries, rebuilding cities, rebuilding political, economic, and social institutions after the holocaust in Europe and after nuclear annihilation in Japan raised the questions of what kind of country, city, or institutions exactly ought to be built. Both the war effort and demobilizing for peace—for example, switching from planes and tanks to automobiles and refrigerators—required futures thinking on a scale and an intensity seldom, if ever, demanded before in human history.

After World War II, national planning blossomed nearly everywhere. As Lewis (1968: 199) says, wartime economic controls over such things as consumer goods, raw materials, and foreign exchange gave a respectability to the idea of planning in largely private enterprise economies that it did not have before the war. Now, public expenditures were reviewed several years ahead and policies formulated, usually to raise the rate of economic growth. The plan not only set goals concerning the desirable direction and amount of economic development for a country, but also specified policies for achieving them and a standard of appraising actual performance (by comparing later results with earlier goals) sometime in the future (Tinbergen 1968: 103). Britain had begun national planning during the war. Norway and the Netherlands soon followed and, by 1946, so did France (Tinbergen 1968: 102).

France is of particular interest, both because long-term planning has been widely used there (Tinbergen 1968: 106) and because futures research and planning appear to have developed more conjointly there than anywhere else. Additionally, joint roots can be readily found in the positivism of Auguste Comte, if we are willing to look that far back in history, and before that in the *École Polytechnique* that was established in Paris after the French Revolution (Madge 1968: 126). As I argue later, however, the French *philosophe* Condorcet may be an even better candidate than Comte for the honor of the father of the futures field, and the utopian socialist ideas of Henri Saint-Simon, for whom the young Auguste Comte at one time worked as a secretary, remain influential in the *grandes ecoles* where many of France's top civil servants are educated.

By the 1950s, France was clearly an incubator of the modern futurist movement. In 1957, for example, Gaston Berger, acting on his growing interest in the study of the future, founded a Centre International de Prospective in Paris and in the following year published a journal dealing with the future, *Prospective*. Sadly, Berger was killed in an automobile accident in 1960.

Berger's work and the Prospectives group, however, were continued, with Pierre Masse, then general commissioner of the French plan, playing a large role. Masse, among others, promoted interaction and integration of the growing ideas of the futurists with the practical concerns of planning faced by the technocrats. For example, Masse in 1963 appointed a committee to look forward to the year 1985 and to consider the future of French economy and society (Cornish 1977: 78–83; Masini 1978: 18).

Also in France, Bertrand de Jouvenel had long been concerned with the future consequences of present action and the possibility of guidance and had published a book as early as 1928 in which he proposed a "directed" economy. In 1960, he and his wife, Helene, founded the Association Internationale de Futuribles, which still functions today, in Paris with branches in Lille and Toulouse, as an international clearing house on futures studies. It also publishes a journal, *Futuribles: Analyse, Prevision, Prospective*, under the editorship of the Jouvenels' son, Hugues.

With financial support from the Ford Foundation, Jouvenel and his colleagues went to work in the early 1960s obtaining and analyzing speculations about the future from experts in a variety of fields, creating a communications network of futures-oriented scholars, and organizing a series of conferences on the methodology of futures thinking: in Geneva in 1962, in Paris in 1963, and in New Haven, Connecticut in the United States at Yale University in 1964. A fourth meeting held in Paris in April 1965 was devoted to more general topics in futures research and brought together 120 participants—"perhaps the largest assemblage of future-oriented scholars up to that time" (Cornish 1977: 139).

The purpose of the Association is to "instigate or stimulate efforts of social and especially political forecasting" in the conviction "that the social sciences should orient themselves toward the future" (Jouvenel 1967: viii). But, as we shall see, the purposes of futures studies soon included much more than forecasting, as Jouvenal, in fact, went on to say in his book, *The Art of Conjecture*.

This book, first published in French in 1964, was—and is—a key work in the development of modern futures studies. In it, Jouvenal describes the perspectives of this new field in a systematic way, explains the nature and purposes of what was to become futures studies, and enunciates some of its philosophical foundations, all in a context of linking the emerging field to the practical tasks of both short-term and long-term planning.

In Eastern Europe, nationalization of large-scale industry, banking, and foreign trade occurred rapidly after World War II in Czechoslovakia, Poland, and Yugoslavia. By 1946, it was largely complete there. In East Germany, Hungary, Rumania, and Bulgaria, it occurred somewhat later, in 1947 and 1948. Local industry and retail trade moved under state management between 1949 and 1951 and farming did so last, mostly in the late 1950s.

Although there were variations in the various Eastern European countries, for example Yugoslavia had considerable decentralization in its system through such institutions as the Workers Councils, the principal problem faced by the central planners concerned "preparing an internally coherent set of balanced estimates for materials and equipment" (Montias 1968: 112). In a planned economy, the tasks of predicting future needs where everything must be coordinated with everything else and is to some extent dependent on everything else are enormous. Although the laboriousness has been reduced with the use of electronic computers, dealing with the complex uncertainties and interrelationships of various parts of an emergent future remains a formidable intellectual task, yet an indispensable one if informed decisions are to be made.

For example, in the former Soviet Union, if a change was made in the plan, the result was that hundreds, perhaps thousands, of additional changes had to be made. Thus, a decision to make more tanks called for more steel; more steel required more iron ore and coking coal; more iron ore and coking coal required more mining operations and transport; these required more machinery, power, construction, wagons and rails; in turn, these required still more steel and many other things too (Nove 1969: 264).

Thus, national planners both in the Western capitalist countries and the Eastern communist countries became involved in the nuts and bolts of forecasting and, more generally, in futures thinking as a necessary part of the planning and decision-making process. Setting goals, making projections into the future, selecting policies, monitoring results, making new projections, altering policies, reassessing goals became part of the normal activities of making and revising a national plan. Even though the Western democracies had less direct control over the economy and society than the Eastern dictatorships and often had to regulate, prevent, or persuade—for example, private investors, owners, and managers—rather than positively control them, the intellectual tasks of national planning in both cases encouraged the rise of futures studies.

Nation-Founding and Nation-Building

Near the end of World War II, another great transformation began that helped pave the way for the development of systematic approaches to futures thinking. This was the formation of the new states from the former colonial territories of European countries and the United States. About 120 new states were created, the great flood having occurred in the 1950s and early 1960s, with the trend continuing more recently as the former Yugoslavia and the Soviet Union broke up into separate independent states. The international system is still rocking from the impact of this growth, which tripled the number of nation-states within a few decades and increased the interrelationships and potential subgroups of coalitions among states many times over that. Today, there are about 200 independent states in existence, the greatest number in world history.

After World War II, in the African, Asian, Caribbean, and Pacific colonies, emergent local leaders grasped the reins of national power, sometimes by bloody armed struggle as in Algeria, Indonesia, Mozambique, and Zaire, sometimes by peaceful means as in Jamaica, Barbados, and Trinidad and Tobago. This tremendous global reorganization fostered futures thinking because it opened to question the existing political, economic, and social structures and revealed possibilities for alternative futures. It did this in a number of ways.

First, during years of colonial rule, the imperial masters had imposed many forms of political, economic, and social organization that the bulk of the local people considered alien, that they sometimes saw as symbols of despised foreign domination. As French Antillean, Frantz Fanon (1967: 170, 230,) bluntly puts it, "I am Negro, and tons of chains, storms of blows, rivers of expectoration flow down my shoulders." Colonialism, he says further, not only holds people in its grip and "empties the native's brain of all form and content," but also it takes "the past of the oppressed people, and distorts, disfigures and destroys it." When formed, the new states had cultures in which the cup of custom had been broken by foreign domination. From it, their peoples could no longer drink the meaning of existence (DeVos and Romanucci-Ross 1975).

At best, there was an ambivalence, an appreciation of some things British, French, Dutch, Belgian, or American and a smoldering anger at other things, especially the exclusions and injustices based on class, race, and nationality that invariably were enforced by colonial political rule. In the Caribbean, where colonialism lasted nearly 500 years and included more than 300 years of African slavery, this ambivalence between attraction and rejection is captured in the words of a former British colonial subject, "the twin orbs of empire, the cricket ball and the black ball."

As new nationalist leaders came to power and had the opportunity to decide their new states' futures for themselves, they asked: What ought to be preserved of the period of colonialism? What parts of the colonial social structures would be useful to the politically independent future? What new social institutions ought to be created to achieve the new state's aspirations for its own independent future?

Second, for many of the new nationalist leaders who had received Western educations, either abroad in the capitals of Europe or America or at home in schools dominated by Western thought, the traditional past of their own peoples generally could not serve exclusively as a model of the new nationalist future any more than could the former colonial social system of oppression. The traditional past was suspect, partly because it was often diverse, encompassing the pasts—sometimes conflicting—of different and historically antagonistic ethnic or racial groups who were becoming members of the same new national citizenries and who were trying to learn to live together.

The traditional past was also suspect because the values of modernity and development, including those of secular humanism and Western science and technology that many—if not most—of the new nationalist leaders had acquired, often clashed with it.

Although the traditional past was frequently used as rationalization to justify proposed courses of action in the present, it was usually selectively drawn upon so as to reinforce the future-oriented perspectives of Western-educated elites. There were exceptions, of course, and we certainly have not yet seen the end of conflict between tradition, exclusion, and fundamentalism on the one hand and modernizing, inclusive, and universal ideologies on the other. In the 1980s, to take only one example, we witnessed the consolidation of a past-oriented traditional elite in Iran, although even there, we find factional conflict and modernizing dissenters.

For most new nationalist leaders, colonialism made both the colonial and the traditional ways of life subject to debate and conflict. As they came to power, they questioned received truths of all kinds and searched for alternative future possibilities for their emerging nation's own distinctive authenticity, identity and meaning.

Third, most of the new national leaders accepted the goals of economic and social development and believed that they could be facilitated by governmental action. Thus, they set about manipulating the present in order to achieve future development. This was encouraged, of course, by leaders, advisers, and teachers from the economically advanced countries and various international organizations. Not only money and equipment were sent to the developing areas and occasionally onerous requirements as conditions of foreign loans and aid, but also experts bearing theories of development, the techniques of planning, and images of a future good society. By the mid-1960s, nearly all the new states then formed had national development plans and central planning units (Madge 1968: 125).

Planning, as we have seen, directly involves futures thinking. Additionally, to have a goal, such as economic growth, dominating national thinking provides a framework to judge and question everything else in the society that may bear on it. For example, is economic growth being enhanced by the way children are educated in schools (to make them willing workers or risk-taking entrepreneurs), by the nature of kinship obligations (to keep managers from giving their unskilled or lazy relatives jobs), by how many women are using contraceptive devices (to keep the birth rate down so income per capita will go up), or by how newspapers report and comment on the news (to spread faith in the regime and encourage people to work hard)? The list of questions could be extended into every aspect of social life, and many new national leaders, vigorously pursuing the goal of economic growth (and, of course, the consolidation of their own political power), could—and often did—make nearly every aspect of society problematic by asking how it could be changed so as to maximize a positive contribution to their country's goals as they had defined them.

Fourth, equally pervasive was the way the transition to nationhood itself raised questions that in the old states tend to be taken for granted. Imagine yourself to be a leader in a colonial territory about to become an independent nation-state. You and your fellow leaders face some decisions that must be made, or so it is thought, if you are going to create what people consider to be a proper nation-state. The foreign, colonial masters have gone. The future of the new state is now in your hands.

What ought the new national flag look like? What ought you to choose as the national motto, national anthem, or national song? Will you have a contest to pick the national bird, flower, tree—or whatever? What will you name your country? Will you use the old colonial name, such as Northern or Southern Rhodesia, or will you give it another name of your own choosing, such as Zambia or Zimbabwe?

These are obvious, and perhaps trivial, examples compared with other things that must be decided and that I have called "the decisions of nationhood" (Bell 1964; Bell and Oxaal 1964). They include: How much national sovereignty ought the new state to have, full self-government or partial as in the case of Puerto Rico? What should the geographical boundaries of the new state be, that is, should it be one large state or several states as India, Pakistan, and Bangladesh? What form of government ought the new state have, for example, a one-party or multiple-party system, a democracy or not? How much of a role should the government play in the economy and in the society, for example, ought the new state to have a command economy of a socialist variety or a regulated capitalist economy or only minimal government control? Should the state in the narrow political sense be coterminous with the nation in the broader cultural sense or ought the government of the new state encourage cultural plurality within the state's boundaries?

Psychologists, sociologists, and anthropologists generally take the national character or basic psychology of a people, the nature of society and its social institutions, or a culture as givens. That is, they ask what exists, what has existed, or what has changed in the past, and they use the tools of their disciplines to measure, describe, and explain the psychological, sociological, or cultural phenomena as they did or do exist. Of course, in your role of new national leader you, too, will need such information. But—and here is the crucial difference—you will also need to know more and to do more, because you are faced with deciding and acting to create the future. You must formulate and implement policy—or do nothing which, of course, is also a policy choice.

Thus, in your role as a new national leader, for example, you may also have to face such enormous questions as what kind of people the new state ought to have (e.g., loyal, tolerant, hard-working, trustworthy, highly educated, etc.?), what the social structure of the new state ought to be (e.g., egalitarian, inegalitarian?), and what the cultural traditions of the new state ought to be (e.g. in Jamaica key questions in rewriting the new state's social and cultural history dealt with how much emphasis ought to be placed on the African origins of the majority of the new citizenry and the period of slavery).

Finally, since no state exists in isolation, you, still in your role as new nationalist leader, must also decide for your new state what its external affairs with other states should be. That is, the new state must have a foreign policy. With what other states should it align itself? What international organizations should it join? How should it vote in the United Nations? In what other countries should it establish consulates and embassies? How should its foreign service officers be trained? There are, obviously, many additional questions of foreign policy as well, as details get piled onto details.

After such role-playing, you may believe that the tasks involved in making and implementing the decisions of nationhood are staggering. If so, then you are right. They were—and they are—staggering. You have correctly glimpsed some of the social realities that were faced by new nationalist leaders. Furthermore, there are limits that existing psychological, political, economic, social, and cultural realities place on whether or not efforts toward deliberate change can be successful and how much change can occur within a given time period. Yet the decisions of nationhood faced all new national leaders as they came to power. In the 120 or so new states since World War II, new national leaders wrote new constitutions, constructed economic plans, and designed the futures of their societies and cultures. Such leaders, like it or not or ready or not, were necessarily involved in planning and futures thinking.

Some aspects of Aldous Huxley's *Brave New World* had become reality. In both democratic and authoritarian states, in capitalist and socialist and mixed economies, in Asia, Africa, the Caribbean, and the Pacific— and then in the former USSR and Yugoslavia—the future had become a realm for which self-conscious designs were made and deliberate historical actions were taken. Images of the future increasingly came to cause present action.

Of course, the future did not often turn out as planned. Natural disasters such as hurricanes, floods, or volcanoes created setbacks. Fierce debate about what the future ought to be sometimes erupted into fundamentalist reaction and violent conflict that produced social paralysis. Strategies of conflict often won out over strategies of cooperation with the result that all participants ended up worse off than they might have been.

Even under the best of circumstances, planning was sometimes badly done. Unintended and unanticipated consequences frequently made a mockery of hopes for a better life. Forecasts on which actions were based often proved to be wrong. Alternative possibilities for the future were not usually fully explored. Values and goals were not always considered in their relationship to each other. Planning was too often short-sighted. Implications of changing technology were sometimes ignored. In sum, there was a clear and urgent need, as H. G. Wells had claimed, for a science of the future.

Operations Research and the Think Tanks

World War II produced still another strand in the development of the futures field. One story, probably apocryphal since it has been told in so many different versions, concerns a team of civilian researchers assigned to observe military operations and make suggestions for improvements. Watching a six-man gun crew in the British Army, they saw that one man had nothing whatever to do. Making inquiries, they discovered that in World War I, when cannons were pulled by horses, the sixth man's job was to handle the horses. Even though there were no horses anymore, the sixth man had been retained in the crew out of the sheer inertia of past practice. As a result of the inquiry, so the story goes, the size of the gun crew was reduced to five and a new discipline was born: operations research (Dickson 1972: 22).

A more critically important and comprehensive problem of incorporating a technological innovation into military practice than that illustrated above, one that literally was to save the life of a country, confronted the British war managers in 1939: How to incorporate the then new technology of radar (*RA*dio *D*etection *A*nd *R*anging) into routine procedures for the air defense of Britain? This task was among the first assignments that military managers gave to a group of scientists (Ackoff 1968: 291). When the Battle of Britain began (by official post hoc decree) on July 10, 1940, the Germans threw the *Luftwaffe* at Britain with daily attacks that began in early August of up to as many as 1,500 aircraft per day. At first, the German targets were airfields of the Royal Air Force's Fighter Command. Then, in mid-September, the targets shifted to include London and other major cities and ports, and the Blitz was on.

But the RAF had a secret weapon. A chain of ground radar stations spotted approaching bombers before they got to Britain, sometimes as they were joining up over the continent for the flight across the English Channel. British fighter planes scrambled into the air and met the German bombers as they arrived. Following instructions from ground controllers with radar information, the British pilots flew to a point behind the bombers from where they intercepted and attacked them. Later, when the Germans switched to nighttime bombing to avoid visual detection and interception, the British countered with airborne radar in the fighter planes that could take over from the ground stations as the fighters approached within a mile or so of the bombers.

By then, the British fighter pilots, who had been vastly outnumbered at the start, had won the air war over Britain. They won it with the crucial assistance of an operational system based on radar that included a series of predictions and revised predictions of the future course of the German bombers that were used to direct the fighter pilots, not to where the bombers had been or then were, but to where the bombers *were going to be* when the fighters reached them. More relevant to the discussion here, the operational system itself was designed to maximize the effectiveness of the fighters in a range of future contingencies. Although the German bombing continued until April 1941, the Battle of Britain was effectively won by October 31, 1940 (its official end).

The success of the scientific teams in this and other instances led to the adoption of a similar approach to war management problems by the United States, Canada, and France. Because these teams "were usually assigned to the executive in charge of operations," their work "came to be known in the United Kingdom as 'operational research' and in the United States by a variety of names," including "operations research" and "systems analysis" (Ackoff 1968: 291).

By the end of the war, such teams had demonstrated their worth to both the civilian and military war managers by increasing the effectiveness not only of air defense, but also of bombing and naval operations. Such a team, of course, had created the atomic bombs that were dropped on Hiroshima and Nagasaki before the end of the war with Japan. In order to keep some of the "operational researchers" together and working on future military technology, General H. H. Arnold, Commanding General of the Army Air Corps in late 1945 made an arrangement between the Douglas Aircraft Company and the Army Air Corps (later the U.S. Air Force) that created Project RAND, the acronym standing for *Research ANd Development.*

As it turned out, there was to be research, but less development, because RAND became a "think tank." RAND was not the first and by no means the only such organization. Others whose stories we might tell include those of Systems Development Corporation and Stanford Research Insitute—now SRI International. SRI, for example, had a futures research group whose work resulted in an influential book, *Seven Tomorrows* (Hawken et al. 1982). But RAND, certainly, is the most famous and influential of the think tanks.

Like other think tanks, RAND primarily produced paper. That is, even though much—if not most—of its work was data-based, it did thought research. Mostly, the results were written reports. Included were "policy alternatives, evaluations, designs, theories, suggestions, warnings, long-range plans, statistics, predictions, descriptions of techniques, tests, analyses or, simply, new ideas" (Dickson 1972: 23–26).

RAND personnel went to work on the future of military technology, strategy, and operations as well as on the containment of communism. They explored the possibilities of scientific satellites and outer space (which in the 1940s was more a topic of science fiction than serious scientific thought), the use of rocket engines for missiles, nuclear propulsion, inflight refueling, various locations for Air Force bases, the metallic element titanium that later became widely used in aviation and space flight, intercontinental ballistic missiles, deterrence policy that determined American defensive strategy, the prevention of accidental detonation of a nuclear weapon by the U.S. military, the economic and military potential of the Soviet Union, the prediction of the behavior of Communist leaders, and hundreds of other military-related phenomena (Dickson 1972). Additionally, RAND workers made important contributions to developing the tools of realistic cost analysis for the Pentagon.

By 1970, RAND had added nonmilitary projects to its agenda as well, that, by then, amounted to about 35 percent of its activities. RAND researchers, among other things, studied possibilities for creating a stroke detection center; making intensive-care units for heart attack victims more effective; improving the organizational effectiveness of urban hospitals;

solving urban problems; affecting population growth and birth control in underdeveloped areas; curing cancer; using the computer for medical research (including simulations of lungs, kidneys, the brain, the blood system, and the eye so that experiments could be done without using humans); and creating a computer model to simulate global climatic changes (Dickson 1972).

The essential characteristics of operations research include a systems approach to problems (a holistic view and a view of the different parts in relation not only to each other but also to the whole itself), an interdisciplinary team of researchers coordinating their efforts while viewing the problem from different perspectives (the theoretical, the practical, the academic, and that of the ordinary operator or observer), and a methodology that often includes building an abstract, that is, symbolic, model of the system being studied, so that the model can be experimented with in a variety of ways and with a speed that would be impossible in the real organizational situation itself. The basic aim is to improve the functioning of the system by locating variables that affect the performance and that can be manipulated by management. Operations research incorporates the investigation of organizational structure, communications, and control with the aim or increasing the effectiveness of the executive's decision-making (Ackoff 1968).

One thing that RAND researchers learned was that expected future events should be a major factor in policy research. Long-range forecasting and futures thinking became a hallmark of RAND's style; and they also became part of the activities of most of the hundreds of other think tanks that now exist (Dickson 1972: 47, 57–58).

Another thing that the RAND researchers did was to innovate, especially in creating new methods or in modifying and routinizing methods already available. Some of their contributions to futures research include both noncomputer and computer-assisted games involving role-playing; computer simulations; mathematical techniques such as linear and nonlinear programming and the Monte Carlo method; technological forecasting, including the Delphi technique; program budgeting, cost-effectiveness analysis, and systems analysis (Dickson 1972). Some of these methods are discussed in chapter 6.

RAND also served as a school for futurists and as a model, both by emulation and contrary reaction, for futures research organizations. For example, one of its most famous alumni was the late Herman Kahn, who in the 1970s was perhaps the futurist best known to the public after the bestselling author Alvin Toffler. Kahn went to UCLA and then to the California Institute of Technology where he received a master's degree in physics in 1948. Then he went to work at RAND as a research analyst. Kahn became famous—or some people might say infamous—in 1960 with the publication of his book, *On Thermonuclear War*. Part of the book was drawn from a RAND Report on which Kahn and others had worked that had been published in 1958. Kahn followed this with the publication in 1962 of *Thinking About the Unthinkable* which was in part a rehash of the first book, though written in a more lively fashion, and in part an answer to the objections that had been raised to his earlier work.

Among other things, Kahn asked the American people to consider the aftermath of World Wars III through VIII which might take place at various times in the future. On the one hand, he forced upon them the gory details of what came to be called "megadeaths," and, on the other hand, he made them think about the nearly-as-scary life after a thermonuclear war involving the United States. Also, he forced them to consider the limited, chancy, and, perhaps, unlikely alternatives to it. Such a future war, he said was possible. It won't go away by not thinking about it. Failure to think about it, he said, may make it more probable.

He may not have alienated some of his bosses at RAND, as some writers have claimed, because he was so full of fun that it was hard not to like him even if you disagreed with him (as one of his then co-workers recently told me). But he no doubt annoyed the president of RAND at the time who never much liked the civil defense work and some top leaders at the Pentagon whose official strategy—later to be termed Mutually Assured Destruction (appropriately labeled "MAD")—Kahn said would not work. He did alienate much of the public and some members of the intellectual community, because they wrongly read him as encouraging the chances of a nuclear war—the very thing he said he was against—by suggesting that the United States could survive it and live to prosper. Some leading peace activists who properly understood his argument, however, supported his views.

For example, some people concluded that Kahn's claim that 40 million Americans dead are better than 80 million and that 20 million dead are better than 40 million would encourage the war managers to begin thinking of actually carrying out such a war. Such calculations, to their minds, rationalize the irrationality of war and lead to the acceptance of 20 million dead (the lower, "reasonable," "rationally better" number). Such thinking was—and is—anathema to many people who aren't part of the military-intellectual complex. It expresses the idea that in a nuclear war "there are significant differences between victory, stalemate, and defeat" (Kahn 1960: 564), when they wish to believe that it is—and ought to be—unthinkable.

Yet Kahn did a service for the American people. He made public the kind of strategy discussions and alternative thinking that had been going on behind the security clearances at the Pentagon and RAND for some time. He himself criticized, analyzed, and proposed alternative strategies. And he became a leader in changing the strategy discussions. The public, in an informed way, could now enter into the argument, an argument, afterall, that affected their future greatly—if, indeed, they were to have a future at all. In the U. S. Air Force, the impact of Kahn's work beneficially resulted in a greater awareness of the horrors of nuclear war.

Even Kahn's opponents at the Pentagon must have agreed with the bottom line in *On Thermonuclear War*: "It is impossible for me to believe that we can meet our current military problems satisfactorily without increasing the national budget by 10 to 20 per cent" (Kahn 1960: 564). Some of that money, moreover, was going to be Herman Kahn's, because the Department of Defense was to become a major client of the Hudson Institute which he, Max Singer, and Oscar M. Ruebhausen co-founded in 1961 in Croton-on-Hudson, New York (Pickett 1992).

By now, Kahn had left RAND and was a full-fledged futurist, and by 1970, with the aid of about thirty permanent staff and research workers and about a hundred outside consultants, the Hudson Institute had become a leading futures-research center. It studied everything, according to Dickson (1972: 91), not only thermonuclear strategies of attack and defense, but the development of whole continents and the future of the the entire Western world.

One of the Hudson Institute's principles, following Kahn's strategy in thinking about the unthinkable, was that all the alternatives in a given situation should be examined no matter how inconsistent with conventional wisdom or, we should add, no matter how unpalatable (Dickson 1972: 116). A major contribution of Kahn and the Hudson Institute was to legitimate asking such questions and studying them seriously.

But Kahn occasionally may have liked to challenge conventional thinking too much; the farfetchedness of some of his Hudson Institute proposals invited parody and disdain from some quarters, as when a Hudson
Institute proposal for large canals in South Vietnam designed to breakup
the area and make military control easier was characterized as the near
irresponsible proposal of "digging a moat around Saigon" (Dickson 1972).

Kahn died in 1983, but the Hudson Institute continued its policy-ori-
ented futures work, relocating to Indianapolis, in 1984. Max Singer,
Kahn's junior partner at the Hudson Institute, produced an independent
work on the future in 1987 with the publication of *Passage to a Human
World*, most of which he wrote after leaving Hudson, first at the Russell
Sage Foundation and then at his own firm, the Potomac Organization,
Inc. where he remains president.

Theodore J. Gordon, another pioneer of the futures field, had been a
young aeronautical engineer at Douglas Aircraft, a test conductor of the
Air Force Thor missiles at Cape Canaveral, and, then, project engineer
for the Delta rocket. He was chief engineer for Douglas's Saturn stages
and director of Advanced Space Stations and Planetary Systems. In the
early 1960s, he was on the leading edge of the futures field, writing a
book entitled *The Future*. At least part of the future, he claimed, would
be formed by the results of scientific research underway in the present.
To gain material for the book, he interviewed scientists about their work
and its consequences. This put him in touch with researchers at RAND,
in particular Olaf Helmer, a mathematician who, with Norman C. Dalkey,
Bernie Brown, and others, was developing the Delphi method, a systematic way of gathering and analyzing judgments about alternative futures
(see chapter 6).

Gordon was invited to consult for the RAND mathematics department
on their planned long-range forecasting study, the first external application of the Delphi method. The topic: future scientific and technological
breakthroughs. The resulting report, co-written by Gordon and Helmer,
was a RAND all-time bestseller. As much as any single report or publication can be a milestone in the development of a field, this report helped
focus and galvanize the futures movement.

Gordon, Helmer, and others at RAND visualized the time when Delphi
and operations research techniques, in general, could be applied to serious social issues of the time. The 1960s were tumultuous and concerned
scientists such as Gordon and Helmer reacted by seeing the need for

clear, calm, and systematic thinking. Kahn was a speculative "genius" forecaster, a brilliant futurist but little interested in methodology. The Gordon/Helmer group had a strong interest in methodology and hoped to provide more valid evidence, objective analysis, and systematic thinking. Helmer coined the term "social technology" and published a book by that name that included the long-range forecasting study.

Gordon, Helmer, and others established the Institute for the Future in 1968 in Middletown, Connecticut, moving it to Menlo Park, California some years later (Cornish 1977: 87). Looking five to fifty years into the future, the Institute's staff aimed "to forecast socio-economic trends and their future implications, to warn of significant departures from existing trends, and to develop and refine forecasting techniques." It set itself apart from both RAND and the Hudson Institute in that it stayed away from military research, although its fellows were definitely interested in doing research designed to contribute to peace through nonmilitary means. Also, unlike RAND, the Institute's founders resolved to make the results of all their studies available to the public. Furthermore, they agreed not to engage directly in policy formulation, but instead to focus their efforts on constructing alternative futures useful to policymakers (Dickson 1972: 323).

RAND became a grandparent of futures research when Gordon, then a vice president of the Institute for the Future, left the Institute in 1971 to set up The Futures Group in Glastonbury, Connecticut (Cornish 1977: 87). Serving as president, Gordon proceeded to build a research and management consulting firm serving both government and corporate clients that by 1990 had eight offices in five countries around the world, including Indonesia and Togo, and grossed about $20 million per year. The work of The Futures Group has focused on population growth and family planning in third world countries and many other topics such as studies of market prospects for microcomputer software services, trends of travel to and from the United States, socioeconomic possibilities for the future of the United States, the future eating environment, the future of work at home, and the future demand for fiber optics to the future social and economic environment for business in Brazil, a regional planning model for Tanzania, LANDSAT analysis of urbanization and demographic change in Egypt, and the future impact of third world multinational corporations in the global economy.

In sum, operations research developed quickly during and after World War II, transforming itself on the one hand into a highly professional discipline and on the other hand spinning off into or merging with a variety of related perspectives, methods, and approaches, including making important contributions to futures studies.

Helmer (1983: 18, 376), however, exaggerates the importance of operations research in the origin and development of futures studies when he says that it is the "parent discipline" of futures research or that futuristics is simply a "branch" of operations research. That may be a fair assessment of his own career as a former RAND operations researcher, pioneer futurist, co-inventor of Delphi, cofounder of an early futures research institute, and author of a basic futures research book, *Looking Forward*. But it neglects many other important contributions to the futures field, some, as we shall see, going back to utopian writers such as Thomas More who published his *Utopia* in 1516.

What cannot be exaggerated, however, is the impact that futures thinking has today on people everywhere, whether it comes in the form of operations research, national planning, futures research, systems analysis, decision analysis, or policy research. For example, when Robert S. McNamara became secretary of defense in the Kennedy Administration, RAND techniques moved with him to the Pentagon. He introduced program budgeting, systems analysis, gaming, cost-effectiveness, and new procurement policies. By 1966, some ninety RAND staff members alone held 269 advisory posts, not only in the Department of Defense, but also in many other places from the National Science Foundation to the Department of Commerce to the White House. The purpose was to require thinking beyond the short term and to make decisions fit in with long-range goals (Dickson 1972: 64, 86).

Today, there are hundreds—perhaps thousands—of think tanks dealing with some aspect of the future throughout the world. Some are focused on particular policy areas and others do futures research per se. Some, including in the former Soviet Union and in the United States, continue to deal with war research and strategies of international relations. Increasingly, the focus has shifted to economic development, social change, and protection of the environment.

Among futurists, of course, there are many who have always spent their time doing peace research from nonmilitary perspectives, among other things, focusing on methods of conflict resolution, or simply doing research on civilian, rather than military, problems. For example, people such as Kenneth Boulding (1964), formerly of the Center for the Study of Conflict Resolution at the University of Michigan, and Donald N. Michael (1963), then of the Peace Research Institute in Washington, D.C., have written such books and articles (Cornish 1977: 88).

Johan Galtung of the International Peace Research Institute in Oslo, Norway, among others, organized the "First International Future Research Conference" in Oslo in 1967. Under the leadership of futurists such as Galtung, Igor Bestuzhev-Lada, Jouvenel, Robert Jungk, and John McHale and after other meetings in Kyoto, Japan in 1970 and Bucharest, Romania in 1972, the Oslo Conference eventually led to the official founding of the World Futures Studies Federation in Paris in 1973. Aware of the fact that almost all of the work of the think-tank industry up to that time had been "financed...directly or indirectly by the armament effort" and, therefore, "served military and related industrial goals," the organizers dedicated the conference to peace and development and focused their futures research efforts on "such enemies as urban sprawl, hunger, lack of education and growing alienation" (Jungk 1971: 10).

These people, and others following in their footsteps, imagined futures studies on a global level and began a "struggle for a better and more human future" that has continued to this day (Jungk 1971: 10). The peace researchers argued that people who care about their future and the future of humanity ought not to leave the future to those futurists *extraordinaires*, the modern war managers, whose scenarios of possibilities included hundreds of nuclear wars through the year 2024. Claiming to be realistic, hardheaded, rational, and technologically most advanced—all of which except perhaps the last can be successfully challenged, they held their fingers poised above "the red buttons of the missiles with nuclear warheads." "For the first time in its history, humankind has no objective guarantee against its own annihilation." The paradox is this: These full-time, professional futurist war managers are people "capable of depriving humanity of any possible future" (Ferrarotti 1986: 27).

The imbalance in futures research, however, continued through the 1980s, especially in the case of large-scale, highly organized, well-financed research projects where the clients' goals determined the definition of the problem and the focus of research, especially where those goals involved war, counterinsurgency, and cold war tactics and strategies. With the apparent end of the cold war in 1990 and the rapid changes occurring in both Western and Eastern Europe and the former Soviet Union itself, topics began to shift more rapidly and fully to the problems of establishing democratic political systems and of economic development in what many people believe will be a postsocialist world.

The Commission on the Year 2000

During the 1960s, a sense of turmoil swept the American intellectual community, especially the colleges and universities. The universities had become deeply linked with governmental, commercial, industrial, and military interests, giving information, advice, and ideas. In exchange the universities received research grants and contracts, some of which were classified and not open to public scrutiny, on which they became partially dependent financially. More than a few professors received handsome consulting fees for contributing to a variety of efforts that were linked generally to military goals and sometimes specifically to the cold war.

Ironically, at the same time many universities became centers of protest against the war in Vietnam and spawned self-styled revolutionaries who denounced the established order, what was called the "Welfare-Warfare state." The universities themselves were polarized and were a part of both the military-industrial-intellectual complex and the vocal and organized protest against it (Heineman 1993; Leslie 1993). Some students, admittedly encouraged by examples of some of their mentors on the faculty, systematically withdrew their respect from institutions of higher learning, vilifying faculty members with military connections. Some students not only clamorously protested against the war in Vietnam, but also demonstrated against the nuclear arms buildup, inequalities and injustices based on racial discrimination, poverty, air and water pollution, and urban decay. By mid-1968, sit-ins and police busts on campus were commonplace.

Some people, both within and outside of the universities at the time, could see justice in many of the social criticisms and goals of the protestors, for example, in the civil rights and the antiwar movements. Yet some radical extremists (as well as some conservative pro-war counterprotestors) were guilty of violating the norms of human dignity and civility in their treatment of others and appeared to have made rationality and authority themselves the targets of their sometimes destructive and reckless attacks. Many observers, including some students, many parents, alumni, and ordinary American citizens, regarded the breakdown of order on so many campuses as a shocking display of ingratitude in view of the opportunity for a college education the rebellious students had been given.

The turmoil and revolutionary spirit, of course, were not confined to the United States, but erupted in many other countries as well, especially in Europe where in France it culminated in the events of May, 1968.

In view of the swirling protests at the time, the screaming charges and countercharges, in-your-face confrontations, sit-ins and obnoxious expectorations, Daniel Bell's 1967 statement introducing the report of the Commission on the Year 2000, of which he was chairman, was sober and restrained: "The real need in American society was for some systematic efforts to anticipate social problems, to design new institutions, and to propose alternative programs for choice." We have to wonder if he would have been thinking about designing "new institutions" at all if the students had not been rioting on the campuses.

The Commission held its first plenary session during October 22–24, 1965 and its second February 11–12, 1966. The American Academy of Arts and Sciences established the Commission, whose purposes included considering hypothetical futures and methodological problems in forecasting and a variety of social problems expected for the year 2000. Bell, of course, was well aware of the surrounding social turmoil: "In the last decade we have been overwhelmed by a number of fractious problems (Negro rights, poverty, pollution, urban sprawl, and so on)" (Bell 1968: 17). No one had predicted the protests, civil disobediences, riots, and campus revolutions—and no one was prepared for them when they occurred. It was past time for some futures thinking, foresight, and anticipatory action.

After much discussion, about sixty papers were prepared and read, and integrative reviews and summaries were made by Bell. Then, working parties tackled a number of topics during the summer of 1967: values and rights (e.g., the relationship of social choice to individual values and problems of privacy), the life cycle of the individual (e.g., the psychological readiness of individuals for change and generational differences), the international system (e.g., the growing gap between the rich and poor nations and the waxing and waning of existing ideologies), the structure of government (e.g., problems of size of government in relationship to the scope of some problems governments face), intellectual institutions (e.g., the relationship of the university to industry and other sectors of the society), science and society (e.g., the codification and institutionalization of knowledge as increasingly important bases of innovation), the social impact of the computer (e.g., the increase in human intellectual power and comprehension), and biomedical sciences and technology (e.g., predictions of possible innovations and their implications for ideas about the nature of man and society) (Bell 1968: 372–77).

Most of these topics were civilian, far removed from the core issues of anticipating war technology and combating communism that originally demanded the attention and at the time still dominated the activities of the RAND researchers. Yet the RAND crowd had made innovations in futures research and their work influenced the Commission's discussions. For example, of fourteen studies sent to the members of the Commission before their first meeting, one was a RAND report by Gordon and Helmer. Also, ex-RAND researcher Kahn was a member of the Commission and one of the major works arising out of the work of the Commission was Kahn and Anthony J. Wiener's book, *The Year 2000: A Framework for Speculation on the Next Thirty-Three Years*, for which Bell wrote the introduction.

The Commission's activities marked a turning point in the development of futures studies. What began to jell was an intellectual network of people interested in futures thinking that went well beyond the American military researchers. Moreover, because Bell and other participants had ties with other futurists both in the United States and other countries, such as with Jouvenel's *Futuribles* group in France, it was now, by the mid-1960s, accurate to speak of an international futures research movement.

Also, the Commission's activities added respectability to futures thinking as a professional activity. The participants were a sampling of influential, mainstream, establishment intellectuals. Universities such as Brown, Chicago, Columbia, Harvard, Massachusetts Institute of Technology, Rockefeller, and Yale were represented. Foundations, such as the Carnegie Corporation of New York, Ford, and Russell Sage also were represented; so, too, were Bell Telephone Laboratories, IBM, and Time, Inc. from the private sector and the Department of State, the Department of Health, Education and Welfare, and the Department of Housing and Urban Development from the U.S. public sector.

Additionally, several of the Commission's participants went on to write their own articles and books on some aspect of the future, including Bell himself, whose *The Coming of Post-Industrial Society: A Venture in Social Forecasting*, published in 1973, was to become one of the most influential books ever published. By now, an emergent futures field could be identified.

The Limits to Growth and the Club of Rome

The Limits to Growth. As well-known and oft-cited as Bell's *The Coming of Post-Industrial Society* became, it never reached the world-wide fame that *The Limits to Growth* (1972) had already achieved. Nor did it create the same public furor. Even before *Limits* was published in 1972, media reports about its findings and recommendations had appeared and stirred up public reaction (Moll 1991). After publication, it became a center of international controversy and debate, selling over 9 million copies in twenty-nine languages (Cole 1993: 814). It was both acclaimed and condemned throughout the world, making its authors, Donnella H. Meadows, Dennis L. Meadows, Jorgen Randers, and William W. Behrens III, instantly, both famous and infamous. For a group of academic researchers more used to having their work gather dust in the library stacks than being in the media spotlights, the sensational public exposure must have been awesome.

The basic model used in *Limits* was an expanded version of a world model constructed by Jay W. Forrester (1971) of the Massachusetts Institute of Technology. Using computer simulations and Forrester's system-dynamics perspective, the authors of *Limits* reached a startling prediction for the time, that, unless changes are made, in the next century at the latest both population and industrial growth will stop (Meadows et al. 1972: 126).

Despite the authors' later protests that they were *not* predicting (Meadows et al. 1992), they obviously were. Some critics did not—or chose not to—understand, though, that their prediction was clearly both contingent and corrigible. It was contingent or conditional upon the assumption of no future changes of significance in either human values or how the global population-capital system has operated for the last one hundred years. It was also contingent, as are other predictions in *Limits*, on the assumptions of no war, no labor strikes, no corruption, and no trade barriers—all of which may make the results reported in *Limits* more optimistic than they otherwise would have been since these things probably can be expected to occur for some time into the future. Additionally, their prediction was corrigible in that more accurate and comprehensive data than were available to them, could result in different, more correct results.

The authors reached this conditional prediction by analyzing world-wide data on key variables, including total population, industrial output per capita, food per capita, pollution, and the amount of nonrenewable resources remaining at any given time. In their analysis, data begin with values for 1900 and, the authors claim, generally agree with their actual values, to the extent to which measurements are available, to 1970. After 1970, the authors use their computer model to simulate additional changes in the variables until the year 2100, based on exponential growth, finite limits, and feedback delays between causes and effects in the system.

The "standard" model, which is how the authors refer to their simulation of the future based on the assumption of no major change, *contingently* predicts overshoot and collapse of the world system. Growth in food, industrial output, and population occur exponentially until the depletion of nonrenewable resources forces a slowdown. The industrial base collapses, "taking with it the service and agricultural systems" (Meadows et al. 1972: 125). The death rate rises because of lack of food and health services. As a result, the population declines. Human life becomes shorter and more brutish.

Meadows et al. (1972), of course, are quite right in their disclaimer that the standard model is not necessarily the one that will predict the future as it will actually come to be. Their assumptions about no change in the behaviors that have historically governed the development of the world system may be incorrect, that is, they may turn out to be contrary to fact as the future unfolds. People's behaviors and social policies, as the authors hope and recommend, may change. Additionally, the conditional prediction of the standard model may err because general assumptions about no war, no strikes, no trade barriers or no corruption may be wrong; or it may err because the measurements of the variables and their assumed relationships which were used in the computer simulation may be incorrect.

Limits, however, does not conclude with the computation of the standard model. Rather, its authors make other contingent predictions of the future of the world system based on different assumptions about both policy and measurements. Then, they run their computer model into the future again and again to see what the possible consequences would be. Each time, even with very favorable assumptions, overshoot and collapse still occur. They may occur later than in the standard model, but they nonetheless occur as the world system hits one limit or another and then recoils into decline.

But Meadows et al. (1972) were able to find a combination of assumptions about conditions that results in a stabilized world system, one that was—or would have been—in equilibrium and sustainable far into the future. I won't give the details here, because the updated results are discussed in chapter 6 where I discuss their twenty-year followup study, *Beyond the Limits*. For present purposes it is enough to say that the authors show that a sustainable world system would have been possible *If certain conditions had been met*.

Meadows et al. (1972), thus, were not predicting which of their alternative scenarios of the future would in fact occur. They did rule out one alternative as impossible: continued unrestricted growth. Beyond that, they offered a choice. The future depended, they said, on what we do. If the global human society restricts industrial growth, stops population growth, recycles and finds substitutes for nonrenewable resources, gives priority to food production, and makes both production and consumption environmentally sound, then a good and long life for all human beings can be achieved within a world system that is sustainable far into the future. If it does not do these things, then there will be disastrous consequences: human needs will not be adequately met, death rates will go up, and life expectancies will go down. Human suffering will increase.

Criticisms of The Limits to Growth. Critics assailed *Limits* from every side, as Peter Moll (1991) documents. From the Left, critics attacked the environmentalism in *Limits* as a middle-class concern that ignores the basic needs of food, clothing, and shelter for the mass of poor people in the world. They argued further that this concern is hypocritical because, in their view, poverty and environmental damage are significantly created by the high levels of living of the well-to-do classes themselves. To their minds, *Limits* leaves out most of what needs to be changed: the market forces of advanced capitalism.

From the Right, critics claimed that *Limits* fails to take adequate account of future technological possibilities that would find more resources and produce new materials. They said, additionally, that *Limits* ignores how the free market economy functions, producing adjustments in usage and research priorities through price mechanisms. As some resources became scarce, according to this view, their price would go up, both reducing use and increasing research to find substitutes (Moll 1991: 117).

From the third world, critics charged *Limits* with being yet another example of the first world trying to freeze the existing order of things and to stop the poor nations of the world from developing. Policies advocated by *Limits*, such critics argued, would put additional burdens on countries that already "had difficulties enough in trying to produce enough goods and wealth simply to prevent their peoples from starving" (Moll 1991: 118). So *Limits* was partly dismissed as another first world attempt to get the biggest possible share of the world's wealth, "an instrument of persuasion or at best, of rationalization" (Cole 1983: 415).

And there were other criticisms of *Limits* too, including the following:

• The data base is insufficient, since adequate data to construct such a world model simply do not yet exist. Therefore, the results are not sound.

• To be meaningful, data ought to have been disaggregated and analyzed by different regions not by the world as a whole, because different regions are different with respect to growth in the key variables and will hit different limits at different future times.

• Pollution is considered only as an abstract single class of pollutants, lumping together all kinds of diverse things that might have quite different futures.

• Likewise in the case of resources; only a generalized resource measure is used. For example, energy, simply considered as a part of "nonrenewable resources," is important enough to require separate treatment (Freeman 1975).

• The model does not adequately take into account serious changes in the world system that could totally alter the parameters.

• The model fails to include human adaptability, ingenuity and changes in human values (Cole et al. 1975).

• The model gives a "spurious appearance of precise knowledge of quantities and relationships which are unknown and in many cases unknowable" (Freeman 1975: 12).

There is some truth, perhaps, to each of these and other criticisms of *Limits*. But the *Limits* authors have tried to answer them, more successfully than less, I think, and to explain in more detail their assumptions and their implications in a variety of places (Meadows and Meadows 1973; Meadows et al. 1973 and 1975; Meadows et al. 1974). Moreover, they have now published a restudy bringing their alternative scenarios of possible futures up to date, taking into account the actual behavior of their key variables for the world system during the twenty years since they wrote *Limits* (Meadows et al. 1992) (see chapter 6).

Other writers, of course, had written about various problems of growth, from population to environmental degradation, at least since Thomas Robert Malthus published his well-known essay on population in 1798. But no one before the authors of *Limits* had dealt simultaneously with so many key variables relevant to the survival of humankind into the far future, their interrelationships, and their consequences, holistically, comprehensively yet simply, and so persuasively. Moreover, no one had ever done all this before with such mathematical explicitness. *Limits* got the world's attention—and it deserved to. It may not be perfectly correct, as the authors themselves point out. Its many scenarios may be contingent and corrigible. But its message that the Earth's human population and economies cannot continue growing as they have been without collapse of the world system got through. Also, its recommendations about what must be done to prevent such a collapse are the foundations of much subsequent research and policy debate. It has helped shape, too, both the futures field and the field of global modeling. Most important, perhaps, although environmentalism as a modern social movement may have emerged with the publication of Rachel Carson's *Silent Spring* in 1960, *Limits* made environmental concerns a permanent part of the human agenda.

Aurelio Peccei and the Club of Rome. Limits was the first commissioned study and, perhaps, the most influential single product of the Club of Rome, which is another important element in the recent development of futures studies. As intended by the Club of Rome, *Limits* affected the hearts and minds of millions of people (Cole 1983: 410).

The Club was founded in 1968 by the late Aurelio Peccei, an Italian industrialist, and Alexander King, originally a physical chemist who was then director for Science, Technology and Education at the Organization for Economic Cooperation and Development. It was to have a limit of a hundred members and a small executive committee that met three to five times a year and made most of the decisions (Moll 1991: 26).

Peccei, who had been a top executive with Fiat, Olivetti, and Italconsult, an international investment firm, was a key actor in the Club. His business travels had taken him to many different countries and he was shocked by the hunger, deprivation, ignorance, and humiliation that he had seen, by the unjust political and economic systems, by land erosion and pollution, by the slums, by the mindless depletion of natural resources, by the enormous military expenditures and by the "risks of new technologies such as nuclear power and genetic manipulations" (Moll 1991: 54).

Peccei believed that something had to be—and could be—done to solve such global problems and he set about, through the Club of Rome and his own efforts of private diplomacy, trying to create an organizational network to begin doing so. He believed in cooperation, the oneness of humankind, the importance of a holistic understanding of the problems, self-determination and self-reliance, free enterprise, the public good, democratic governance, and every person's right for self-respect. In 1969, he published *The Chasm Ahead* in which he anticipated many of the results of *Limits* about environmental disasters, overpopulation, and a world food shortage, subjects that he had been lecturing about since the early 1960s. In a final chapter, Peccei called for action, including a multinational feasibility study aimed at creating worldwide, cooperative planning addressing such global problems (Moll 1991: 58–60).

At a planning meeting held by Peccei and King in Rome in April, 1968, about thirty European scholars, scientists and others attended, including Jouvenel and Dennis Gabor. Erich Jantsch wrote a background report for the meeting which was published in 1967 as *Technological Forecasting in Perspective*, which was widely read during the 1970s by futures scholars and planners (Moll 1991: 64). This meeting, according to Moll (1991: 64), was not successful in sparking the participants' interest. Rather, the Club of Rome was born later, after the formal meeting during an evening in Peccei's Rome apartment where six of the participants got together, four of whom became the nucleus of the Club: Peccei, King, Jantsch, and Hugo Thiemann who was from Switzerland and then director of the Battelle Institute in Geneva.

The purpose of the Club was to alert the world citizenship to what they termed the "global *problematique*," a cluster of interrelated world problems including hunger, environmental degradation, violence, overpopulation, and increasing alienation of the working classes. It included a sense of fear and urgency about such problems and the need to deal with them holistically over the long term. Industrialization itself was considered to be part of the problem (Moll 1991: 26, 65). The founders of the Club intended it to be an intellectual catalyst, sparking interest and concern among other people and organizations and creating a worldwide effort to act to solve the problems.

In 1970, Forrester convinced Peccei and others of the Club of Rome that his world modeling technique could do what they wanted done. It was agreed that his assistant, Dennis L. Meadows, would get started as soon as possible with support from a team of young researchers at MIT (Moll 1991: 79). As reported above, they published *The Limits to Growth* in 1972 and captured the world's attention.

Since the publication of *Limits*, the Club in the 1970s and 1980s sponsored sixteen additional reports, ranging from *Reshaping the International Order* and *Goals for Mankind* to *Microelectronics and Society* and *The Future of the Oceans*, but none achieved the success of *Limits*. These publications garnered neither a large audience nor many sales, selling only between 2,000 and 12,000 copies, a small fraction of the copies of *Limits* that sold (Moll 1991: 200). One deserves special mention, however. Although reaffirming the critical problems facing the human future that *Limits* described, Botkin et al. (1979) emphasize how human learning could help solve them. Arguing that there are *no limits to learning*, these authors show how innovative, anticipatory, and participatory learning of new methodologies, new skills, new attitudes, and new values could prepare humans to deal adequately with a world of change.

Since Peccei's death in March, 1984, the Club, under the leadership of King and Bertrand Schneider, has concentrated on governability. Their aim is to reorganize existing political institutions so that they will be better able to cope with the changing conditions of global problems in the future (Moll 1991: 229).

In 1991, King and Schneider published a report by the Council of the Club of Rome. In it they outline a possible *resolutique* or resolution to the global problems enunciated in the *problematique*. They argue that decline is not inevitable, that effective human action can increase efficiency and slow population and industrial growth. Although the limits to growth may be properties of the planet, they try to show that the limits to sustainability are properties of the human mind, a somewhat more optimistic view than that put forward in *Limits*.

In sum, the Club of Rome influenced futures studies in many ways. Its emphasis on a holistic, global, and multidisciplinary approach has become characteristic of the field. It advanced environmental thinking and activism and gave both an intellectual grounding beyond what had existed earlier. It also advanced the methodology of simulation and modeling and their use in futures studies. Also, its idea of the global *problematique* and its research, especially *Limits* and its images of future overshoot and world collapse, were important correctives to the rosy views of the future put forward by cornucopian technocrats such as Kahn (Moll 1991:

251). Moreover, the "private diplomacy of the Club worked much better after the publicity generated by *Limits*" (Moll 1993: 804). People throughout the world, even top national leaders, began paying attention.

H.D. Lasswell and the Policy Sciences

Another person who deserves to be listed among contemporary pioneers of the futures field is the late American political scientist, Harold D. Lasswell, because for more than four decades, beginning in the 1930s, he struggled to create what we now call futures studies, contributing important insights, concepts, methodologies, and exemplars. As a futurist, Lasswell was ahead of his time and some of the fertile seeds of his imagination fell on the barren ground of the dogmas of the established social sciences of his day. However, we need shed no tears for Lasswell as an obscure and unnoticed scholar, such as Nathan Israeli, because he was both nationally and internationally acclaimed during his lifetime as an intellectual superstar. Yet his contributions to futures studies, which were only a small part of his many works, deserve to be reassessed and given greater recognition than they have been.

Unfortunately, Lasswell did not live long enough to synthesize his work on futures studies himself, leaving that partly to others (Eulau 1958). Also, his concerns with social policy and, possibly, the indifference of most of his colleagues to his appeals for a new futures orientation in political science led him to merge his work on futures thinking with his efforts to invent the policy sciences. Since many social scientists were receptive to Lasswell's proposals for the policy sciences, Lasswell responded by devoting some of his time and energy to guiding and shaping their development while neglecting his visions for a separate futures field. Thus, some of his achievements have been overlooked by futurists.

The developmental construct. Lasswell formulated the developmental construct as an analytic tool as early as 1935, elaborated it more fully in 1937 and especially in 1941 as an aspect of the method of futures thinking that he created. It reappeared in many of his later publications. One example of the several developmental constructs that he formulated is "the garrison state."

What he foresaw was a coming world of garrison states, a world in which specialists on violence would be the most powerful group in society. He observed a trend away from the dominance of the specialist on bargaining toward the supremacy of the soldier and he extrapolated that trend into the future. The coming garrison state, according to Lasswell, was no dogmatic forecast. It was not inevitable. Rather, it was a probable "picture of the future," not necessarily the most probable one, but a possibility. Although it is frankly speculative and imaginative, a developmental construct, such as the garrison state, is based upon disciplined and careful consideration of the past and of conditions that contribute toward its formation.

What Lasswell envisioned was not simply a military state based on force. History already had given us many examples of these. Rather, what he foresaw was a marriage of the military state with modern technology, not simply a temporary marriage as in times of emergency and war, but a permanent one. The leaders of a garrison state would promote virtues such as the duty to obey, to serve the state, and to work, using their control of the distribution of respect and material goods, official propaganda, and coercive instruments of violence, including such institutions as compulsory labor groups and, eventually, the use of drugs to deaden the critical function of people.

Writing several years before George Orwell published his chilling image of "Big Brother" in *Nineteen Eighty-Four*, Lasswell envisioned the erosion of democracy and the rise of dictatorship in the garrison state. Lasswell saw the end of competitive elections and the coming of government by plebiscite, the spread of one-party or no-party states, the elimination of free speech and other civil liberties, the suppression of political opposition, the end of majority rule, the abandonment of legislative assemblies except as a means for the ceremonial ratification of the decisions of a supreme authority, and the control of the rate of production by the state with priority given to military production. As in the yet-to-be written *Nineteen Eighty-Four*, the symbols, but not the substance, of democracy would be retained.

All of these things were possibly to take place within a military state preoccupied with danger; within a society perpetually honing its fighting effectiveness, translating social change into battle potential, and maintaining popular support through war scares, which to be credible sometimes had to be followed by war.

Related to the development of garrison-police states were two other developmental constructs, a bipolar structure of world politics (the two giant powers of the United States and the then Soviet Union) on the one hand and an opposing image of a homogeneous world culture combining science and democracy on the other.

He saw the specialists on violence becoming dominant in either of two ways, either through forcible imposition of world unipolarity or through belligerent bipolarity with weapon parity. In the first case, in order to protect itself from recurring sedition, the world elite probably would establish a garrison-prison state, in which control would be in the hands of a relatively few specialists in violence. In the second case, chronic fear of nuclear war would result in the emergence of two garrison states, each becoming like prison states, again because of fear of sedition.

The alternative—his second image of future world culture based on democracy—was contingent on deferring major war between the two powers. If that could be done, then openings would occur between them. There would be exchanges—of news, knowledge, people, and products— and there would be resulting trends toward cultural homogeneity and political unification (Lasswell 1951b: 116–17).

Lasswell thought globally. Policy-oriented science was to be worldwide. He saw the peoples of the world as a single community in many ways, where different localities and countries were interdependent and the future welfare of the one depended on the future welfare of the many. Lasswell was asking for something in addition to the collection and analysis of information about particular policies in some specified time and place. He was calling for world-encompassing hypotheses concerning the major problems of our epoch (Lasswell 1951a: 11).

Five tasks of futures studies. For Lasswell, the developmental construct was part of a larger process of intellectual investigation that he called "developmental analysis." In the late 1960s, I joined with him and a few other colleagues in establishing the Yale Collegium on the Future (a pretentious name for what was, in fact, a modest faculty seminar that met once or twice a month and managed to hire a part-time secretary/ research assistant). At a meeting of the Collegium on September 25, 1967, Lasswell circulated a mimeographed paper entitled "Projecting the future" that enumerated five specific tasks for the study of the future. They are:

1. The clarification of goals and values.
2. The description of trends.
3. The explanation of conditions.

4. The projection of possible and probable futures if current policies are continued.
5. The invention, evaluation, and selection of policy alternatives (in order to achieve preferred goals).

These five tasks must be considered together, as working in conjunction with one another in an interactive way. Lasswell listed goals and values first because of their overriding importance. Knowing what trends to study of the countless number that might be studied is a matter of judging what is important, and what is important depends upon the goals and values that are to be explicitly served by action, whether that action is carrying out research or participating in politics. Lasswell saw the need for stating values clearly and for using them as a basis of scientific activity. Why? Because research that is relevant to basic human values is useful in preventing undesirable futures or in bringing into existence desirable futures.

In the particular case of the developmental construct of the garrison state, Lasswell was concerned with the values of democracy, both the existence of human rights and civil liberties on the one hand and of competitive political systems based on free and fair elections on the other. Scientists, he believed, knowing about the possibility of the rise of the authoritarian garrison state could focus their work on research aimed at understanding the trend and at explanations of the conditions producing it, possibly learning how to prevent the trend from continuing and how to preserve democratic institutions. Moreover, the developmental construct of the garrison state could alert citizens who valued democracy to the threats to it and, armed with further knowledge, they could work to reverse the trend toward the garrison state.

Thus, carrying out Lasswell's five tasks, a futures researcher would describe trends, explain conditions, and, considering both, construct a picture of a future to which they might be leading. He or she would do these things within a context of explicitly stated human values that can be used as standards of judgment to decide whether or not the constructed future is desirable. Then, combining value judgments with explanations of conditions and with predictions of the probable consequences of possible actions, the futurist would invent and select possible courses of action aimed at creating a future that is judged to be most desirable. Note that the five tasks exist as threads woven together into a single fabric, constituting a unified whole. Also, note that Lasswell's five tasks are incorporated into the purposes of the futures field today as I describe them in chapter 2.

Decision making: Facts, expectations, and values. Lasswell understood that decision making was inevitably future-oriented. Deciding how to act, for example, inventing and choosing among policy alternatives, involves a process of constructing images of alternative futures and selecting from among them. Implicit in his thinking was the assumption that humans, individually and collectively, are value-driven decisional systems acting in and moving through time toward a hoped-for or feared future as defined by their goals and values. Lasswell assumed, further, that people would act, according to their cognitive maps of reality, so as to achieve their goals and fulfill their values, avoiding their feared futures as much as possible given relevant constraining conditions. He believed that facts, expectations, and values are all necessarily involved in deliberate decision making (Eulau 1958).

Lasswell believed that making conscious decisions involves having a worldview, that is, having beliefs about how the world really was and is and having a model of how the world really works and changes, including understanding cause-and-effect relationships and identifying trends. Aside from luck or chance, making one's way effectively and successfully in the world requires that an actor have cognitive maps of physical and social realities that are reasonably accurate concerning key relevant matters. Further, manipulating the world, as distinct from merely navigating in it or adapting to it, requires some knowledge of how the world works and what causes have what effects.

In principle, these things are all matters of fact and can be known within the limits of scientific knowledge, including the limits imposed by the measurement errors of the methodologies of the day. Thus, developmental analysis rests in part on a factual base. Yet people, including scientists, may have only partly accurate or even totally inaccurate maps of reality. Presumed "facts" may be in error. One source of inadequate decision making, clearly, is simply not getting the facts straight.

Lasswell believed, further, that making decisions intelligently means having expectations about the future consequences of present developments and of taking or not taking certain actions. This, of course, is a prediction problem. It means, in other words, having anticipatory beliefs, foresight, or images of future outcomes and developments.

Lasswell didn't say "knowing the future" was part of decision making, because he was well aware of the uncertainty of the future. Predictions may not come true. Yet he realized that various alternative present possibilities for the future are indeed real and that deliberate decision making is inconceivable without having beliefs about at least some of them (i.e., without people acting *as if* they had at least some knowledge of the future).

Certainly, Lasswell did not want his readers to think that his picture of the garrison state was an inevitable future outcome, as if they could do nothing about preventing it. Quite the opposite was the case, as we saw above. It was, rather, a statement of an undesirable possibility, a warning, and a call to preventive action to those people who valued democracy. Its chances of becoming a future reality, that is, were dependent upon future circumstances and actions, some of which people had the power to control.

Constructing alternative images of the future, grounding them in logic and in past and present facts, and specifying the different courses of action that may lead to them are matters of knowledge and prediction. But alone they are not enough to design and implement intelligent human action. Lasswell also understood that decision making and policymaking involve having goals and making value judgments. Having good guesses about the consequences of action are of no use unless we have some way of judging the relative goodness or badness of the different consequences. Thus, decision makers must have answers to such questions as: Which possible futures are most desirable in the sense of promoting the achievement of positive human values? Which are intolerable in that they violate and prevent the fulfillment of such values? Without goals and values, actors have no basis on which to select one future rather than another, no guidelines by which to make choices regarding which future to work for.

By goals and values, Lasswell did not mean merely individual preferences or the whims and desires of the moment. Rather, he was concerned with those values that are socially organized. In many different works, Lasswell elaborated his conceptions of the human values that he believed to be basic and could be properly used as general, even universal, standards of judgment (Lasswell 1965, 1977; Snyder et al., 1976). He lists eight broad values, all of which derive from his conception of human dignity (see volume 2, chapter 4).

Lasswell viewed his list of values as general enough to subsume most other lists of values and he was on the right track, I think, when he turned to the long list of rights enumerated in the Universal Declaration of Human Rights of the United Nations as a specific example of the sort of thing he meant. Doing so, he was searching for some means of grounding or justifying his list of values. I continue Lasswell's search in volume 2 of this book where I discuss the ethical foundations of futures studies.

Futures studies and the policy sciences. In 1951, Lasswell, Daniel Lerner, and other writers published *The Policy Sciences.* They are generally credited with laying out one of the first proposals and earliest frameworks for policy analysis (Brewer 1974; Brewer and deLeon, 1983). This is not to say, of course, that policy advisers did not exist earlier; no doubt they have existed as long as politics has existed. In the United States, the writings of John Dewey and other pragmatists are precursors, and Lasswell himself was influenced in the 1920s and 1930s by the work of Charles E. Merriam at the University of Chicago. But *The Policy Sciences* marked the beginning of a program of scholarly work that directly resulted in the emergence of the policy sciences in the 1970s "as an identifiable, respectable, even desirable professional activity" (Brewer and deLeon 1983: 1). The new field emerged largely following the framework earlier set down by Lasswell.

The policy sciences have flourished. After a slow start, they have materialized, with what became dizzying speed in the 1960s, as full-fledged professional activities. Policy analysts—sometimes referred to as "policy wonks"—became part of the decision-making process in a wide variety of programs and areas, including those dealing with anti-poverty and economic opportunity; budgeting and taxing; civil rights; crime and violence; education; national defense; population, energy, and the environment; urban affairs and housing; and welfare and social security (Dror 1971; Dye 1978: xii).

By the mid-1970s, there were university and institute centers and Ph.D. degree programs in the policy sciences (e.g., at the RAND Corporation, Duke University, the University of Michigan, and Harvard University); professional journals and textbooks on policy analysis dealing with such topics as health, energy, welfare, national security, and transportation policy; and sources of funding, such as the Ford and Alfred P. Sloan Foundations (Brewer and deLeon 1983: 6–8). In 1983, in a book dedicated to Lasswell, Garry D. Brewer and Peter deLeon published *The Foundations of Policy Analysis* which confirmed that the policy sciences had arrived as a professional activity and which gave direction to their future development.

In the preface to his 1971 systematic summary of the major ideas of the policy sciences, Lasswell said that in the twenty years since "the term 'policy sciences' was introduced" the "social sciences have 'turned around' far enough to look toward the future. Physicists, biologists and their colleagues are concerned about the social consequences and policy implications of knowledge" (Lasswell 1971: xiii). Lasswell's assessment about scientists looking toward the future and being concerned about social consequences and policy implications was—and still is—overly optimistic and exaggerates the facts, yet it highlights the connections in his mind between the policy sciences and futures studies.

Futures studies and the policy sciences share some common origins in the efforts to deal with "the Depression, World War II, and the period of readjustment and uneasy peace right afterwards" (Brewer and deLeon 1983: 7). Moreover, they both aim to reach across disciplinary lines; are contextual, problem-oriented, comprehensive, and multimethod; strive for synthesis and holism, avoiding fragmented and specialized views; hope to alert society's participants to impending problems and opportunities; try to improve the decision and policy processes in order to achieve desirable futures; are normative in being oriented to human values rather than to theoretical concerns of scientific disciplines in selecting research questions; are normative, further, in hoping to better humankind through their work; and, in being future-oriented, are often concerned with the freedom and welfare of future generations of people who do not yet exist.

Additionally, the five tasks of futures studies, given above, are recycled throughout Lasswell's work and the work of others as being involved in policy analysis, although with some relabeling and rewording. In Lasswell's *A Pre-View of Policy Sciences* (p. 39), for example, they are "five intellectual tasks" that make up an "adequate strategy of problem solving," and in Brewer and deLeon (1983: 12–13) they are "operational principles," guidelines that emphasize "a problem orientation, contextuality, multiple methods, and an overriding concern for human values."

The two major and general purposes of the policy sciences often given are (1) developing a science of policy forming and execution, that is, an analysis of the decision process, and (2) contributing to the decision process by creating relevant information and interpretations to specific policy issues. Lasswell (1971: 4), of course, included "anticipations of the future" in the latter. Decision makers need to know what will happen "(a) if they do nothing or (b) if they follow a given policy option."

Some futurists apparently assume that futures studies and the policy sciences overlap so much as to be nearly indistinguishable. For example, Kahn and Wiener (1967: 3) identify policy research not only with anticipating future events and making them as desirable as possible, but also with preparing policymakers to deal with whatever future actually arises, by considering a range of alternative futures. The same statement, of course, can be made about futures studies.

Up until 1970 there remained considerable overlap between the policy sciences and futures studies. For example, as Michael Marien (1992) points out, when the initial issue of the journal, *Policy Sciences*, appeared in 1970, there were twenty-one people listed as editors or advisers, six of whom were prominent futurists. By 1990, even though both were concerned with societal guidance, the two fields had grown apart.

Future history will settle the question of whether futures studies will simply become a part of the policy sciences or whether it will have an intellectual niche of its own. In my judgment, there are good reasons for maintaining a distinct futures field. For one thing the principles of futures thinking emphasized by the policy sciences tend to be applied in the public and the civic sectors. This may be an arbitrary limitation that is not inherent, since detailed and rigorous futures thinking is equally helpful to the private sector, both to large corporations and small businesses, and even to individuals in their everyday lives. But it is a limiting tendency of policy sciences.

For another thing, some images of the future call for major changes in the distant future that seem politically impractical in the present. If so, then Peter H. Rossi and William F. Whyte (1983: 17) believe that they "are unlikely to be of interest to policy makers."

DeLeon (1984), after pointing out some similarities between policy research and futures studies, goes on to say that most policy studies have a near time frame with some immediate policy in mind while futures studies typically have a longer time frame. Another difference, he believes, is that policy scientists must concern themselves with implementation analysis while futurists do not, although some futurists would take issue with the notion that futurists are not also involved in implementation.

Also, the overriding concern with immediate decision making of the policy scientist tends to focus attention on the details of given cases and on problems with apparent practical solutions. Minutiae, which of course have their rightful place in decision making, may get in the way of big futures thinking on a grand, imaginative, contrary-to-fact, and antiestablishment scale. What may get shoved aside, against the exhortations both of Lasswell and Brewer and deLeon, are the sweeping, macro, positive, idealistic images of the future that have the power to change the course of entire civilizations, as analyzed, for example, by the Dutch sociologist/futurist, Frederik L. Polak (1961).

Finally, there is a possible source of conflict in orientation between policy scientists and futurists on one crucial point. Futurists aim to open up the future, to make a virtue out of the uncertainty of the future for the purpose of empowering people to achieve futures better than the past and present. Futurists aim to teach people that the future is an open horizon that can be creatively explored and that is, for an active person, a dimension of freedom. Policy scientists to the contrary often aim to "de-futurize" the future by increasing security. Through technology, law, policy and insurance, policy scientists hope to secure the future by taking its uncertainty away (Bergmann 1992: 91). This may explain partially why the policy sciences have flourished more than has the futures field. Security is comforting to people. Change, even desirable change, has its costs because it often causes both uncertainty and stress.

In sum, Lasswell was ahead of his time, yet constrained by it. He was an innovative thinker pushing his social science colleagues to open up their minds to the future, with only limited success. He worked to transform the social sciences into effective, value-oriented tools for responsible social action, yet sometimes stopped short of carrying his own futures thinking to its logical conclusion. Thus, compared with his work directly in the policy sciences, Lasswell's contributions to futures studies, though genuine, are often unfinished and fragmentary.

Yet Lasswell gave futurists a framework for the study of the future in his developmental analysis: the investigation of goals and values, trends, conditions, projections, and alternatives. Also, he gave them a powerful analytic tool in the developmental construct and the many concrete exemplars of its use, such as the garrison state. Additionally, he gave them models of disciplined and careful analysis, grounded in logic and scientific facts. In rudimentary form, Lasswell independently invented modern futures studies (Lasswell 1948).

Evaluation Research

Another research tradition that overlaps with policy analysis, but which is yet to have its full impact on futures research, began by looking backward. This is evaluation research, which since the 1960s "has become a minor industry in the United States...with an estimated annual volume of between two and three billion dollars" by the early 1980s (Rossi and Whyte 1983: 9). Much of the money came from the U.S. Congress which demanded evaluation of its programs when authorizing various innovations. For example, the Legislative Reorganization Act of 1970 and the Congressional Budget Act of 1974 gave the General Accounting Office (GAO) sweeping powers to conduct its own evaluations.

The GAO has pushed the process along by offering a draft section that can be attached to any authorizing legislation. The section calls for a statement of detailed objectives, a report on effectiveness in achieving them, and a listing of the evidence for those conclusions (Cronbach et al. 1981: 39). In other words, the demand was for evaluations of governmental programs to see if they had achieved their goals effectively and the demand included presenting concrete evidence, that is, the results of research. "The evaluation researcher is like a person sitting on the stern of a ship," Donald T. Campbell said in a lecture at Yale University some years ago, "looking back and reporting to the captain where he has been."

Evaluation research, just as policy analysis, can be traced far back into history. It's proximate origins in the United States were definitely visible by the 1930s when several social scientists "advocated the application of rigorous social research methods to the assessment of programs." In 1935, for example, A.S. Stephen, an Arkansas sociology professor, "pleaded for evaluations of President Roosevelt's New Deal social programs" and F. Stuart Chapin empirically evaluated the effects of improved housing. During World War II, Samuel Stouffer and others of the Research Branch of the U.S. Army continually monitored the morale of American soldiers, evaluated personnel policies, and analyzed propaganda. The Office of War Information studied civilian morale using sample surveys. Many smaller studies looked at the effectiveness of price controls and evaluated various campaigns aimed at changing American eating habits. Similar social science efforts were carried out in other countries as well (Rossi and Freeman 1982: 21–22).

Originally, evaluation research aimed to look backward and to find out what had happened as a result of some policy implementation, to determine whether goals had been met, and to specify who or what were responsible for the successes or failures. Today, however, evaluation research has been largely incorporated into the policy process. It is now nearly indistinguishable from the policy sciences and has a prospective as well as a retrospective aspect.

For example, Caro (1983: 49) says that, in addition to coming after implementation of a program in order to judge if it has met its objectives, evaluation can be used to guide decisions that must be made while a program is underway, including, for example, whether a program should be continued or replicated. She thinks of program development as a succession of three-part cycles, planning, implementation, and evaluation. As defined in a leading textbook, evaluation research is "the use of social research methodologies to judge and to improve the planning, monitoring, effectiveness, and efficiency of health, education, welfare, and other human service programs" (Rossi and Freeman 1982: 20). Cronbach et al. (1981: 7) add, "An evaluation of a particular project has its greatest implications for projects that will be put in place in the future."

Moreover, preimplementation evaluations increasingly demanded research into the possible unintended consequences of proposed government programs, an assessment of their probable environmental and sociocultural impacts. Today, an evaluation researcher is not only sitting on the stern of a ship looking backward, he or she is also sitting on the bow of the ship, looking forward and reporting to the captain where the ship is (possibly and probably) going.

Signposts of a New Field of Inquiry

Establishing the Futures Field or Spreading Futures Thinking?

Is it proper to speak of "futures studies" at all? Have futurists as yet achieved the characteristic features of a profession or a discipline? Is there a sufficient body of futures work to consider it a distinctive "futures field." I believe the answers are "yes." Yet the present status of futures studies may be precarious and its future as a separate and distinct field may fail to materialize as fully as some futurists believe.

Whether a separate and distinct futures field is even desirable is debatable, because many people who are convinced of the importance of futures thinking for the well-being of humankind believe that the most important objective is to jolt people out of their primarily past- and present-time orientations, including members of the current established intellectual fields and university disciplines. Such people want "to futurize" not only the thinking of the general public, but also existing fields—from anthropology to sociology and from biology to zoology—so as to expand them to incorporate more principles of futures studies into their perspectives, theories, teaching, and research.

Working toward a separate futures field, of course, is not incompatible with such a goal. Perhaps, it is even necessary for spreading futures thinking, since fulltime futurists with their own identity, organizational bases, and sources of support may be required to help disseminate futurist ideas. For example, the emerging "transdisciplinary matrix" of futures studies, which I describe in chapter 4, can best be developed coherently through the work of full-time futurists who are willing to pull together the disparate futures-relevant works of many different disciplines.

Today, there are many indications that the first steps toward the creation of futures studies as a distinct field—or "multi-field" as Marien (1985) would prefer to call it—have already been taken. Over the last several decades, people such as those already mentioned in this chapter plus many others have devoted their efforts to the systematic exploration of alternative possibilities for the future and to the methods by which such exploration can be carried out. Some of those people have formed themselves into communities of part-time and full-time professionals held together by networks of communication and both formal and informal organizations. At the same time, futurists' activities have been institutionalized within—and legitimated by—business firms, publishing companies, government agencies, intellectual centers and institutes, university teaching and research programs, private foundations, and professional societies. The results of futurists' work constitute a significant body of knowledge published in professional journals, newsletters, magazines, monographs, books, government documents, and research reports. Let's briefly review some of these signposts of a new field of inquiry.

The Rise of the Futures Movement

Richard L. Henshel (1981) says that the modern futures field, in the sense of a self-awareness or collective self-consciousness among people doing futures studies of some kind, started sometime between 1950 and 1960. Ossip K. Flechtheim (1971: 264) coined the term "futurology" in the 1940s and proposed it as a new branch of knowledge. Jungk (1971: 9-10) refers to the "first haphazard beginnings in 1944." According to Helmer (1983: 17), although its roots go back to Europe of the 1950s, the futures movement developed over the last twenty years, that is, since 1963, Jouvenel's *The Art of Conjecture* in 1964 and Gabor's 1963 book, *Inventing the Future*, being key works. Dretha M. Phillips (1983: 1) sees the field burgeoning in many areas of the world as early as 1953 and O. W. Markley (1983: 47) picks 1963 as the year during which a definite futures field emerged. Burt Nanus (1984) sees the late 1940s to the mid-1960s as the stage in which many of the basic conceptual and methodological foundations of the futures field were laid down.

Thomas E. Jones (1980: 5) cites research done during the 1950s as providing the "fertile soil" for the seeds of futures research that led to the rapid growth of the early 1960s, giving as important landmarks the publication of Harrison Brown's book, *The Challenge of Man's Future*, in 1954, the 1950s conferences dealing with the future initiated by Brown and some of his colleagues at the California Institute of Technology, and the opening of Stanford Research Institute's Long-Range Planning Service in 1958. Futures studies, according to Magoroh Maruyama (1978: xvii), has been in formation for "two decades," placing the start in 1958.

One of the first futurists of the post-World War II period was Polak, who completed his monumental work, *The Image of the Future*, in Dutch in 1955 and saw it published in English in 1961 (Boulding 1978: 7). Edward S. Cornish (1977: 97) picks the 1960s as the decade when the futurist movement began developing rapidly. Jones (1979: 21) concludes that by "the late 1960s, a new policy-oriented discipline, dubbed 'futurism' or 'futuristics,' was definitely emerging."

Thus, whatever date they pick as its exact starting point, observers of futures studies generally agree within a decade or so about the timing of its recent takeoff. The collective activities of modern futurists, though reaching back to earlier times, became clearly visible in the second half of the 1940s and the 1950s. By the mid-1960s, they took on many of the features of a social movement, began growing rapidly, and encouraged the self-identification of participants as futurists.

Futurists and Futures Organizations

In the second half of the twentieth century, there existed a considerable number of associations of futurists, including, to name only a few, the Association Internationale Futuribles (of France), the Canadian Association for Futures Studies, Instituto Neuvas Alternatives, SA (of Mexico), International Centre of Methodology for Future and Development (of Rumania), the Japan Society of Futurology, Mankind 2000 (of Belgium), the Norwegian Society for Future Studies, the Swedish Society for Future Studies, the Swiss Society for Futures Research, the Teilhard Centre for the Future of Man (of England), the World Future Society (of the United States), and the World Future Studies Federation (with current secretariat in Australia).

The World Future Society, for example, was founded with headquarters in the United States in 1966 by Cornish and others. Starting with only a few hundred members, it grew rapidly and, by 1979, peaked at a total membership of nearly 60,000 from more than eighty different countries. Its membership has fluctuated, however, hitting a low of 22,500 in 1985 and rising again in 1994 to about 30,000 (Marien 1987; World Future Society, personal communication).

The World Future Society is, as its brochure says, an association for the study of alternative futures composed of people "interested in how social and technological developments are shaping the future." In addition to its general membership assemblies, the World Future Society also holds forums for those of its members who are engaged professionally in futures-oriented research and activities. In 1991, the World Future Society's *The Futures Research Directory: Individuals* listed 1,200 individuals professionally involved in studying the future.

Nearly all major corporations today have some formal system of long-range planning or technological forecasting, that is, some futures research capability. Allstate, American Motors (Canada), American Telephone and Telegraph, Champion International Corporation, General Electric, General Mills, Kaiser Aluminum, Nissan Corporation (Japan), and Phillip Morris, for example, all have provided homes for futurists in their research and development units. Moreover, business firms may hire futurists as consultants, such as the private business organization, Coates & Jarratt, Inc. of Washington, D. C., which has done futures work for many corporations including IBM, Du Pont, Exxon, and the Gillette Co.

The World Future Society compiled *The Futures Research Directory: Organizations and Periodicals 1993–94* (Jennings 1993). In it were listed 187 futures organizations. The facts that the number of organizations was down from 230 in 1979 and that only a third of them existed at the earlier time are not surprising in the development of a new field. There are fluctuations, periods of high excitement and prosperity as well as periods of retrenchment and reexamination, in the history of nearly every human endeavor. In the case of the futures field, the 1960s and early 1970s appear to have been a flourishing period of high hopes and optimism, a period, as Marien (1987) calls it, of "futures vogue."

It was followed by shrinking interest. For example, futures research in the early 1980s lost some ground in the private sector (Amara 1984: 401). More generally, there was some disappointment and disillusionment, and perhaps some compartmentalization of the field, in the 1980s and in early 1990s. This period included both reassessment and criticism of the futures field on the one hand, while on the other hand it involved codification and elaboration of futurist assumptions, principles, concepts and methods. As I try to show in this book, the foundations of futures studies were still being forged. Also, of course, during the last part of this period, much of the world economy, including the entire intellectual-educational complex in the United States and many other countries, faced recession, reduced budgets, and downsizing that may have contributed to cutbacks in futures work as well.

Courses and Educational Programs in Futures Studies

One of the first university courses in the United States entirely devoted to the future was taught by Toffler in the fall of 1966 at the New School for Social Research in New York and was entitled, "Social Change and the Future." A year later, in 1967, James A. Dator taught two futures courses, specifically designed and labeled as such, at the Virginia Polytechnic Institute (personal communication, 30 March 1994) and I also introduced one, "The Sociology of the Future," at Yale University. Of course, like other pioneering futurists, Dator some years earlier, around 1964, had begun teaching primarily futures-oriented courses—in his case in his regular political science courses at Rikkyo University in Japan (which may have led Audrey Clayton [1988] erroneously to report Dator's first official futures course as being taught in 1966). When Toffler, Dator, and I first began introducing futurist ideas into our courses, by the way, we were not aware of the fact that Flechtheim had first proposed such courses two decades earlier.

These courses were soon followed by others and, by 1973, there were between 350 and 400 such courses on the future being taught in colleges and universities in the United States and Canada alone (Rojas and Eldredge 1974: 347–48). In 1991, Howard F. Didsbury, Jr. produced a twenty-six-part television course, "Visions, Nightmares, and Forecasts: Humanity's Attempt to Know the Future from Ancient Times to the Present," making a futures course available to the public at large through Kean College of New Jersey.

Today, there are special futures studies courses at the elementary and high school levels, and, at universities, there are entire programs in future studies, such as the Alternative Futures Option Master Program in Political Science at the University of Hawaii at Manoa; the University of Massachusetts, Amherst; the University of Minnesota; and Portland State University in Oregon. According to John Naisbitt (1984: 9–10), the number of universities or colleges offering some type of futures-oriented degree increased from two in 1969 to over forty-five in 1978.

The Studies of the Future Program of the University of Houston at Clear Lake City, Texas is among the largest and most comprehensive graduate programs in futures studies. By 1983, it had ten core faculty members and about forty students. It was designed "to permit completion of the masters degree in 12 months of full-time study" and offers a Master of Science in Studies of the Future or a Master of Science in Education with a specialization in Futures Studies. The program aims to prepare students for further graduate work leading to the Ph.D. degree involving futures studies; for careers as futurists in business, government, or nonprofit organizations, as journalists involving futures research, or as teachers, educational planners, or administrators; or simply for intellectual or aesthetic appreciation of futures studies (Markley 1983: 51–53).

Futurist Publications

Popular and professional journals devoted solely or primarily to the future, according to Naisbitt (1984: 9–10), jumped from twelve in 1965 to more than 122 in 1978. Among the 124 periodicals listed by Lane Jennings in 1993, are *Alternatives, Future Survey, Future Trends Newsletter, Futures Research Quarterly, Futuribles* (in French), *Futurics, The Futurist, Technological Forecasting and Social Change, Technology Review, World Futures Studies Federation Newsletter, World Watch,* the slick and attractively packaged *Omni,* and the academic and scholarly British journal, *Futures.*

Although few achieve the status of runaway bestsellers as did Toffler's immensely popular and influential *Future Shock* in 1970 and Naisbitt's *Megatrends* in 1984, literally hundreds of books dealing with some aspect of the future are now published each year. For example, even though it is limited to English-language publications, and mostly American at that, the 1981-82 *Future Survey Annual* (Marien 1983) lists 1,366 books and articles from 172 publishers and 146 journals. By the appearance of the 1994 *Annual,* a total of 12,746 items had been abstracted in a total of fourteen annual volumes since publication began in 1979.

In the 1994 *Annual,* 604 books, reports and articles were summarized. Marien (1994: v) points out that the smaller number of items when compared with some earlier annual volumes "does not reflect any decline in worthy futures-relevant books and articles. Rather, it reflects greater attention to depth over breadth, and thus longer abstracts filling limited space." He says, further, that the "quantity of relevant books appears to be increasing slightly. But futures-relevant journals, magazines, and newsletters appear to be proliferating."

Notably, on the international level UNESCO (United Nations Educational, Scientific and Cultural Organization) has begun a "futures clearinghouse" and publishes a variety of futures-oriented documents, including *UNESCO FUTURE SCAN.* In 1966 with the publication of Encyclopedia of the Future (Kurian and Molitor 1996), a new Series of volumes, The Knowledge Base of Futures Studies (Slaughter 1996), and other major works, futures studies moved to a new level of intellectual achievement.

At the end of the twentieth century, there has been a heightened interest in the future as hundreds of otherwise nonfuturist journals bring out special articles or entire issues exploring the twenty-first century and otherwise nonfuturist authors publish books preparing readers for the next century. Futurists ought to applaud this explosion of interest and use it as an opportunity, as the year 2001 comes and goes, to explain why people ought to maintain such an interest in the future continually, year in and year out, not only at the beginning of a new millennium.

Futurist Conferences and Meetings

Earlier, I pointed out that the First International Future Research Conference was held in Oslo in September, 1967. It had seventy participants from over a dozen countries (Jungk 1971: 9). Since then, there have been hundreds—perhaps thousands—of meetings of futurists to exchange ideas, to establish networks of communication, to discuss the nature of the new field of futures studies, and, most important, to study and design action to help solve the pressing problems that face humankind.

Meetings of futurists have included both national and international assemblies and conferences, meeting in a wide range of places, from Kyoto to Bucharest to Dubrovnik (Hayashi 1978; Masini 1978), from Beijing to Budapest, from Barcelona to Turku, Finland, and from Washington, D.C. to Rio de Janeiro, to mention only a few.

Among the larger meetings of futurists was the First Global Conference on the Future which was held in Toronto in July, 1980. (It was not, of course, the first such conference, but "First" was in its official title.) It was cosponsored by the World Future Society and the Canadian Association for Futures Studies. Over 5,000 people from more than thirty countries attended. So many showed up that at one point the Metropolitan Toronto Police had to set up barricades to control the crowd and some plenary sessions had to be videotaped and shown by closed-circuit television in hotel rooms so that people who couldn't get into the auditorium could watch.

Many, if not most, futurist organizations, such as those listed earlier, hold occasional workshops, seminars, courses, or special-issue meetings. Some, such as the World Future Society and the World Futures Studies Federation, organize large international conferences and assemblies at regular intervals, annually or biannually. Such meetings, of course, advance the futurist agenda and aim to carry out the manifest purposes of intellectual exchange, learning, and criticism, but they also help to achieve the latent purposes of creating a community of futurists, establishing the abstract identity of what a "futurist" is, and encouraging people to think of themselves as futurists.

Too Fragmented to be a Field?

Some writers, such as Marien (1987), see futures studies as too fragmented to be a field. It is certainly true that futures studies covers a wide
range of subject matters. After all, everything in the world might have a
future, thus we might explore the future of everything from the aardvark,
the family, and politics to physics, the universe, and zymology. Moreover, the people most qualified to write about these things may be experts in aardvarks, the family, politics, and so on, not futurists.

Yet, as I will try to show in the following chapters, there is a place for
the integrative and holistic work of futurists that, although often based
on specialized knowledge, also transcends it. Specialists sometimes fail
to see how events, processes, and actions that occur outside their special-
ties can affect the phenomena that they are trying to understand just as
they sometimes fail to explore the consequences of what they are studying for other things. Specializations may advance our knowledge in a
microscopic way but they may also give specialists blinders that narrow
their vision.

Additionally, some other fields that have managed to maintain some
degree of intellectual unity are not all that different from futures studies. For example, compare the futures field with area studies programs.
Scholars of many disciplines—from history and economics to literature, art history, and even irrigation engineering—are unified mainly
by the fact that they work in a particular country or region. Yet they
manage to organize themselves into professional associations and create a loose community of scholars. The same can be said of fields such
as African American studies and women's studies. There are unifying
themes to each, but members come from a variety of traditional disciplinary backgrounds and often have joint appointments in different
university departments.

But without going so far from traditional departments, look at anthropology, embracing the disparate interests and skills of physical anthropology, archaeology, linguistics, and ethnography. Or take history
and English. Both are divided into compartmentalized specializations
by time and geographical space. The historians, particularly, are fragmented according to different periods of past time and different regions
of the globe, often working literally in different languages and social
worlds than their colleagues.

Also, the historians—just like futurists—are further split by topic. Since anything that happened in the past—just like anything that might happen in the future—may invite investigation, historians are fragmented by the subjects they choose to investigate, from the history of naval battles, great leaders, medicine, science, technology, corporations, economies, and polities to ordinary family life and child-raising, sexual life, eating habits, typical clothing, and the role of women to farming techniques, land ownership patterns, animal husbandry, irrigation methods, the use of energy—you name it. History as a field, thus, may be a good analogy for futures studies in that it ranges over a diverse subject matter, though focused on the past rather than the future, yet has a clear identity, is well established in colleges and universities, and is well organized into professional societies.

Another factor that is sometimes mentioned as an indicator of the fragmentation of futures studies is the lack of a common background among futurists. Again, it is certainly true that futurists have been trained in a variety of different fields from physics, mathematics, and engineering to journalism, sociology, and history. Yet this fact is not convincing as an argument to support the contention that futures studies is not a field. No new field could ever be created if common background in its own nonexistent past were an essential criterion of its being a field. In the beginning, practitioners of new fields necessarily must have been trained and educated in something else.

Additionally, common background is not as important as common fate in making such a judgment, especially in the case of a relatively new field. No field is born fully mature. It is what happens to it after it is born that matters. If many people become and continue to function as futurists and if they create the other accoutrements of a field of study, then it will be reasonable to say that a futures field exists.

Finally, the diversity of backgrounds of futurists may be a strength for a field that attempts to be holistic and integrative, to deal with the reality of interrelationships among things in order to inform human decision and action. We need a mix of expertise on the one hand and experts who are willing to inform themselves about other disciplines on the other. Thus, even as the field grows to include individuals trained specifically as futurists in the various graduate programs that have come into existence and comes to have a more cohesive common profesionnal culture than it now does, the futures field would be well advised to continue to recruit into the futurist community people with diverse knowledges and capacities.

In sum, the transdiscipline of futures studies and the profession of futurist may not define populations of people whose boundaries match perfectly. For the transdisciplinary boundary, following Stephen Toulman (1972), we look to such things as shared concepts, procedures and methods, problems worked on, and goals striven for. For the professional boundary, however, we look to other things such as academic degrees, positions, memberships in organizations, publications, and participation in professional meetings and conferences. For a developing field especially, these two sets of boundaries may only more or less overlap.

What Should the New Field be Called?

What should the systematic exploration of the future be called? I have used several different names in this book so far and others have been proposed by various writers. In the end, of course, it will be called what people call it, common usage being the ultimate arbiter, and no amount of advocacy is likely to be of much influence. As I write, "futures studies" appears to be the label that most American futurists are comfortable with and it may win out as the most used term in the future, while "prospective studies" or "*prospective*" is generally accepted in Europe.

"Futures field" has been suggested by Amara (1981a) and it is also widely used. Some writers object to it, as does Marien (1985: 13), for example, who rejects it on several grounds in addition to fragmentation of the enterprise, including that "its practitioners do not share a common academic background" and "it is not recognized as a field." Neither argument seems convincing to me for reasons that I have already mentioned. Moreover, my dictionary defines a "field" in this sense merely as an intellectual endeavor or a sphere of activity or interest, especially within a business or profession. The definition does not specify that a field must be old or new, well established or emerging, unified or fragmented, have practitioners of the same background, or whatever.

"Futures research" is also widely used by futurists to describe what they do. It has the disadvantage, however, that not everything written in the futures field is adequately described by "research," as we have seen. The plural, "futures," is today generally used rather than the singular, "future," whether combined with "research," "studies" or "field" in order "to avoid possible mix-ups with studies done in the future, and also to emphasize that we should always study different alternative futures" (Dror 1971: 45).

"Futurology" is used by some futurists, particularly in Europe, but has not caught on among American futurists, perhaps because non-futurist critics often use the term in their sometimes uninformed attacks on the field. "Futuribles" never caught on in the United States and is limited to Western Europe. "Prognostics" is the most used name in Russia and Eastern Europe (McHale and McHale n.d.: 3).

Some futurists refer to the "futures movement" which captures an important aspect of futures studies, certainly in the late 1960s and to some extent continuing up to the present. For the dedicated fervor, commitment and activist behavior aimed at changing the world exhibited by some futurists—from peace advocates to environmentalists—are quite similar to the defining characteristics of political and social movements.

Other terms either are in very limited use or have been proposed by various writers without success in gaining acceptance, such as "alleotics," "mellology" (another version of the Greek, "mellontology," that GilFillan proposed), "stochology," "posthistory," and "the science of the future" (Wescott 1978: 524).

"Futurism," as a name for the future field, has the disadvantage that it remains associated with the art movement of the early 1900s. It began in February, 1909 with the publication of a manifesto by Filippo Tommaso Marinetti in *Le Figaro*, a Parisian newspaper. Leading a group of Italian artists, Marinetti created and publicized the movement. It enshrined the dynamism, speed, motion, and the power of the rapidly developing industrial civilization surrounding it, modernity and technological progress, and the presumed perfections of the new world of mechanical machines. Although futurism, as an artistic movement, drew on both neoimpressionism and cubism, it aimed to be a "violent departure from traditional forms" and "to express movement and growth in objects, not their appearance at some particular moment." It also incorporated similar ideas of change and movement in literature and music. By 1916, it had reached its zenith, although it continued in Italy through the 1930s (Burchfield 1972: 1182). Apparently, there is no direct connection of descent between this art movement and today's futures field.

Although "futurism" is seldom used to refer to modern futures studies, its practitioners are known in the English-speaking world as "futurists," even though this term was also applied to Marinetti and his followers.

The term, "futurist," according to the Oxford English Dictonary, was first used to refer to a person "who believes that the Scripture prophecies, especially those in the Book of Revelation, are still to be fulfilled in the future." The term, "futurist," appears to have won out over "futurologist" among practitioners of futures studies.

About what to call the field itself, however, there remains, as we have seen, less consensus. In a 1975 poll of the members of the World Future Society, Cornish (1977: 256–57) reports that only "two terms received a net positive response: *future studies* and *futures research*. The other terms, in order of preference (that is, from least disliked to most disliked), were: *futures analysis, futuristics, forecasting, futurology, prognostics, futurics,* and *futuribles*."

In this book, I'll mostly use the terms "futures studies," "futures field" or "futures research" with an occasional nod to "futures movement" or other terms on the grounds that, now as I write, in the English-speaking world futurists are more likely to use them than other labels to designate this new field of inquiry.

Conclusion

In this chapter, I've said that time perspectives and futures thinking are fundamental aspects of human cognition and can be traced back to the beginnings of humankind. They are universals in human experience. This is not to say, of course, that every person and every culture give the same time and effort to thinking about the future. Nor is it to say that every person and every culture search for alternative future possibilities as well as others do or consciously use futures thinking to make good decisions and take effective actions. In fact, one of the reasons that futures studies is needed is precisely for the purpose of improving the wisdom of human decision making and the effectiveness of human action by means of more self-conscious and adequate futures thinking.

The attempt to achieve this can be seen in the various strands in the recent origins of futures studies that I've discussed. Ogburn aimed to discover the future directions of social change in order to improve planning; Israeli explored future possibilities and people's evaluations of alternatives in order to increase their awareness of the future; national planners developed methods of futures thinking to do their jobs better as they managed World War I, coped with the Great Depression, tried to direct the course of social change in Communist Russia, Fascist Italy, and Nazi Germany, and planned strategies for conducting World War II and for reconstructing the postwar world.

Since the colonial empires crumbled after World War II, about 120 new states have been created, most recently by the collapse of the former Soviet Union and Yugoslavia. The leaders in the new states set about making "the decisions of nationhood" as they attempted to shape the present realities in their new states to conform with the aspirations and images of a better future held by their peoples. Also emerging out of World War II were the operations researchers and think tankers, including those at RAND, who created some of the early methods and exemplars of then-emergent futures research as aids to decision makers.

The Commission on the Year 2000 gave respectability, intellectual leadership and exemplars—such as Bell's *The Coming of Post-Industrial Society* and Kahn and Weiner's *The Year 2000*—to futures studies. *The Limits to Growth* and the Club of Rome enlarged futures studies to encompass a concern with the long-term, life-sustaining capacities of the Earth itself and a global perspective incorporating all of humanity. Individual scholars, especially Lasswell, spelled out the purposes of futures studies and added conceptual tools.

By 1997, futures studies had most of the characteristics of a separate field of inquiry. It had full-time professionals, networks of communication and formal professional associations, university futures courses and a few entire educational programs, conferences and meetings, hundreds of publications annually, shared purposes, a set of identifiable futurist methods, underlying assumptions, and shared exemplars that stand within a growing body of knowledge. Moreover, it had a thirty-plus-year recent history of development, a sense of a futurist community, and important individual and organizational pioneers who were increasingly recognized for their contributions.

Because of its diversity, it may be more correct to call futures studies a multifield or transdisciplinary field, rather than a field. But whatever it ends up being called, futures studies today not only exists but also has an important role to play in helping to shape the coming human future.

This overview of some of the strands in the recent development of futures studies is a mere sketch. No full and proper history of the emerging field as yet exists. When—or if—such a history is ever written, it surely would include what I have included here. But it would also include more. It would include, for example, a description of how at least parts of the popular culture of the mid-twentieth century were ready to receive futurist ideas and to provide positive feedback for the futures movement. It was a period when science fiction climbed out of the literary gutter and when future-situated books appeared on both the fiction and nonfiction bestseller lists and at supermarket checkout stations. It was a time when *Star Trek* and then its several clones found mass audiences on television and when motion pictures and magazines on the future attracted widespread popular attention.

A proper history of futures studies also would include a much fuller treatment than I have been able to give of the work of pioneering futurists such as Gro Harlem Brundtland, Ervin Laszlo, John and Magda McHale, Eleonora B. Masini, and Donald Michael. Also, it would include more works of the speculative, intuitive, and integrative side of modern futures studies, such as those of Elise Boulding, Duane Elgin, Marilyn Ferguson, Willis Harman, Hazel Henderson, Robert Theobald, Allen Tough, and W. Warren Wager (Tough 1991a).

It would include, too, a greater range of futurists from parts of the world other than the United States and Western Europe than I have been able to discuss. The origins of futures studies are worldwide and include important contributions by such futurists as Yujiro Hayashi, Hidetoshi Kato, Yoneji Masuda, and Mitsuko Saito-Fukunaga of Japan; Tae-Chang Kim of Korea; Linzheng Qin of China; Rajni Kothari, Ashis Nandy, and Satish Seth of India; Uvais Ahamed of Sri Lanka; Sartaj Aziz and Ziauddin Sardar of Pakistan; Ibrahim Abdel-Rahman of Egypt; Mahdi Elmandjra of Morocco; Margarita M. de Botero of Colombia; and Pentti Malaska and Mika Mannermaa of Finland.

From the former Soviet bloc, some futurists whose works would be included are Mihai Horia Botez, Mircea Malitza, Ionitza Olteanu, and Ana Maria Sandi from Romania; Erzsébet Gidai, Maria Koszegi-Kalas, and Erzébet Nováky from Hungary; Jan Danecki, Andrzej Sicinski, and Bogdan Suchodolski from Poland; Ota Sulc and Milos Zeman from the Czech Republic; and Mihailo Markovic and Radmila Nakarada from Yugoslavia (Dator 1994a). The list could go on and on.

Until a proper history of futures studies is written, I hope that the reader will acquire from this chapter some appreciation of the scope and diversity of its recent origins. Moreover, I describe additional futurists and their works, sometimes at considerable length, in the chapters that follow.

Let's now turn to the purposes of futures studies which I explain and illustrate in chapter 2.

2

The Purposes of Futures Studies

The most general purpose of futures studies is to maintain or improve the freedom and welfare of humankind, and some futurists would add the welfare of all living beings, plants, and the Earth's biosphere for their own sakes even beyond what is required for human well-being. Thus, at the most general level, the goals of futurists are to contribute toward making the world a better place in which to live, benefitting people and the life-sustaining capacities of the Earth. Of course, something similar may be said of the manifest aims of any number of other occupational groups, from scientists, physicians, and religious leaders to artists, carpenters, farmers, and garbage collectors. Most members of such groups like to believe that they are contributing something to human well-being, whether it is knowledge, health, peace of mind, a beautiful or useful object, food, or the removal and storage of waste.

A distinctive contribution of futurists is *prospective thinking*. Through prospective thinking, futurists aim to contribute to the well-being both of presently living people and of the as-yet-voiceless people of future generations. Futurists explore alternative futures—the possible, the probable and the preferable. These are the key ideas emphasized by many futurists, including Amara (1981a, 1981b), Markley (1983), Masini (1993) and Toffler (1978), and they recur again and again in two summary collections of prominent futurists' ideas (Marien and Jennings 1987; Coates and Jarratt 1989).

The purposes of futures studies are to discover or invent, examine and evaluate, and propose possible, probable and preferable futures. Futurists seek to know: what can or could be (the possible), what is likely to be (the probable), and what ought to be (the preferable).

In the words of Toffler (1978: x), futurists try to create "new, alternative images of the future—visionary explorations of the possible,

73

systematic investigation of the probable, and moral evaluation of the preferable."

In this chapter, I describe the major purposes of futures studies. Although I have organized them into nine major tasks, the reader should keep in mind that such a ninefold classification is simply a convenient way of communicating the multipurposes of futures studies. Other futurists might divide the purposes differently, into fewer or more categories. For example, I have incorporated Lasswell's five tasks described in chapter 1 into my nine.

The nine tasks or purposes overlap a great deal and some of them can be carried out adequately only in conjunction with some of the others. Additionally, not all purposes of futures studies—or of futurists *qua* futurists—are adequately described by the nine purposes discussed below. For example, a variety of additional purposes derive in part from the existence and development of futurists as a professional community and from futures studies as a set of professional activities. Such derivative purposes include working to obtain resources with which to carry out futures studies and to establish futures studies in the institutions of society, including in universities.

They also include trying to create and maintain the kind of society in which people are free to do futures research and to discuss openly alternative possibilities and preferences for the future. Envisioning alternative future political, economic, or social arrangements almost always involves questioning the existing arrangements of society. In authoritarian or totalitarian societies, questioning the status quo is often considered to be subversive and can be dangerous, resulting in harassment, physical abuse, banishment or imprisonment. Even in democratic societies, questioning the status-quo can be unpopular and sometimes efforts are made to suppress it. Futurists clearly have a stake in keeping societies open to free inquiry and to the civil exchange of ideas.

Thus, the futurist *qua* futurist has a right, if not an obligation, to work for a political and social order within which futures studies can be carried out, within the limits of respect for human dignity and the protection of the rights of human subjects of research. This may mean entering the political process as a partisan in support of freedom of speech, inquiry, and information. An Orwellian society, governed by Big Brother and dedicated to official lies and ideology, would mean an end to a great deal of scholarly and scientific activity—including much of futures studies.

Also, the purpose of educating people about the principles of futures studies is not adequately incorporated in the nine purposes given below. For any field of inquiry to flourish it must have cadres of new recruits, new personnel to carry on its activities. Thus, recruitment, training and some kind of certification as a futurist, however loose and informal, become essential for the future of futures studies. New futurists must be created if futures studies is to become a lasting professional enterprise.

Futurists are teachers not only of neophyte futurists, but also of members of society, both the general public and leaders. Futurists aim to teach the insights and tools of futures studies so that both ordinary people and key decision makers will make more effective decisions, thereby improving their individual lives as well as the public good. Futurists aim to educate, too, as they attempt to promote greater futures consciousness among individuals throughout society and to futurize the thinking of other people, including professionals. Thus, each of the purposes given below has an educational component to it as other people are taught and learn about it.

Not every futurist, of course, gives equal emphasis in his or her work to each of the nine purposes. Just as in any field of activity there is a division of labor, with some futurists focusing attention on only a few of the nine purposes while other futurists focus their attention on others. Some futurists, for example, are mostly researchers, while some others are mostly communicators and still others are largely activists, focusing their efforts on proposing and implementing social policy.

I have not elaborated on some of the purposes given below nor have I always given detailed examples, because I do so elsewhere in this book. In fact, two of the major purposes—constructing epistemological and ethical foundations for futures studies—are among my primary objectives. Thus, I deal with them only briefly here, since I discuss them extensively elsewhere in these volumes.

Nine Major Tasks of Futures Studies

The Study of Possible Futures

The exploration of possible futures includes trying to look at the present in new and different ways, often deliberately breaking out of the straitjacket of conventional, orthodox, or traditional thinking and taking unusual, even unpopular, perspectives. It involves creative and lateral thinking in order to see realities to which others are blind. It involves thinking of present problems as opportunities and present obstacles and limitations as transcendable. It involves not only asking what is, but also asking what could be. It involves, most of all, expanding human choice.

Present possibilities for the future are real. Present capacities of individuals, groups, and society as a whole for change and development, no matter how suppressed and unrecognized they may be, are factual. The potential for future development and growth exists in the present and, thus, can be investigated. Even if a dam isn't being built or even intended, it might be possible that one *could* be built. We can document that the knowledge, human will, resources, and other conditions to build a dam in some particular place either do or do not exist, even though we can only envision it imaginatively before the fact. To reach societal state B from state A, it has to be possible to get from A to B. That is, the possibility of going from A to B must exist in A, just as a small seed must contain within it the possibilities for a giant tree if it is to become such in the future.

Another way to think about some possibilities is to think of them as "dispositionals." Consider a fragile glass held at shoulder height over a tile floor. If dropped it would break. Before it is dropped it is breakable. If not dropped, it remains breakable. It may never be dropped and may never actually be broken. Yet it is part of the glass's dispositional present that it could have a future in which it could be dropped and broken. The fact that it could be broken remains a real present possibility for its future.

Science is full of examples of the study of dispositionals, such as the study of things that are soluble, heatable, combustible, expandable, or shrinkable and the study of people who are observable, treatable, curable, or teachable—in fact all words ending in "able," "ible" or "uble" generally describe possibilities. Other terms such as "magnetic," "elastic," "reflex," and "habit," for example, although less obviously, are also dispositional predicates. All—or even none—of their possibilities may ever come to be, but the existence of their present capacities is a fact. As Walter L. Wallace (1988: 27) says "scientists want to know not only how the world *has* been, and *is*, and *will* be. We also want to know how the world *could* be...what unrealized but realistic *possibilities* lie dormant in the world." Richard S. Rudner (1966: 22) concludes that dispositionals play an "essential role" in "theories of contemporary science."

Many human capacities in any society remain undeveloped and unrealized, that is, most people never develop more than a small fraction of their potential for learning and innovation. They generally fail to see the possibilities for change within themselves. As adults, people tend to trudge through life chained to the routines of everyday behavior that they have learned, oblivious to the more challenging and desirable alternatives open to them. This is at least partly because most of them have not been taught to look at the world as it could be. They have not been taught to search beyond the cultural conventions and manners of their own groups for possibilities either for their own personal futures or for their society's future.

General Matthew B. Ridgway who served as the U. S. Army Chief of Staff in the mid-1950s was an exceptional top leader who understood the importance of going beyond conventional thinking and exploring possibilities. When asked what he thought was his most important role as the nation's top soldier, he answered, "To protect the mavericks." What Ridgway meant was that a future war might be completely different from the currently dominant beliefs on which plans were being made. Would there be alternative plans ready to meet unforeseen challenges? He was counting on the mavericks to be looking at the future in ways different from the dominant views, thinking beyond the orthodox beliefs and school solutions. "Since the Army, Navy, and Air Force were powerful and rigid institutions, such maverick officers were not always popular and their careers were usually at risk." Hence, they needed protection (Hadley 1986: 165–66).

Conventional thinking prevents efforts to explore unorthodox possibilities for the future not only in the military. Rather, it is a general problem that also operates in every organizational and institutional setting. For example, scientists value creatively exploring the unknown, yet they often do so in conventional ways within the orthodox thought patterns of members of their scientific communities. In 1992, the then-director of the U. S. National Institutes of Health asked American scientists what they would do if they had a billion dollars for biomedical research. Her request was made to about 500 scientists attending a conference in June and was repeated in a July issue of *Science* magazine which most of the nation's 98,000 biomedical scientists read. Guess how many responses she received. Fifty thousand? Ten thousand? One thousand? No, only thirty-six answered. Moreover, even of the thirty-six, not one "recommended how research priorities might be set to serve the agency's mission: improving the health of Americans. Instead the emphasis was on supporting the research of individual scientists" (Kolata 1992).

Alfred North Whitehead said, that "almost all new ideas have a certain aspect of foolishness when they are first produced." Thus, futurists are wary about rejecting any idea out of hand. The world is too full of things that intelligent and well-educated people at one time believed to be impossible for that. One example is the response to the Kremer Prize. Henry Kremer in 1959 offered 5,000 pounds sterling for the first human-powered flight around a one-mile figure-eight course. In 1973, he raised the prize to £50,000. In Germany, a prestigious professor of aeronautics gave a series of talks on why the Kremer Prize could never be won, saying that it was impossible to make such a flight. As a result, the Germans didn't try.

American Paul MacCready, Jr., however, accepted the challenge and won the prize in August, 1977. Pedaled by Bryan Allen, MacCready's Gossamer Condor flew a figure-eight course in seven minutes, twenty-seven-and-a-half seconds. In June, 1979, his Gossamer Albatross, again pedaled by Allen, achieved another "impossibility," staying aloft for two hours and forty-nine minutes as it flew over the English Channel (Smith 1983).

When MacCready was asked why he was able to succeed where others had failed or had not even tried, he said that "it came down to a question of attitude." People are hemmed in by their preconceptions, can't break free of their own training, and limit their own viewpoints. They prevent their own success, because they don't look at as many sides of an issue as they should (Smith 1983: 42).

This is not to say that everything and anything is possible. Quite the contrary is the case. It is vital to human individual and collective well-being that we recognize what is impossible as well as what is possible. Jumping off a high cliff and flapping our arms in an effort to fly like a bird is, obviously, potentially disastrous for human beings. There *are* real limitations to human capacities and to the technological enhancements of those capacities as well. Thus, people can err in two different ways, on the one hand believing that things are possible when they are really impossible or, on the other hand, believing that things are impossible when they are really possible.

Both errors can lead to tragic consequences, but there is an important difference between getting the world wrong one way versus getting it wrong the other way. When we try the impossible and fail, we often learn. For the way the world really is necessarily impinges on our consciousness and invites us to change our beliefs. When we don't try because we believe something is impossible when, in fact, it is possible, we do not test our false beliefs. Thus, false beliefs about the impossibility of things—events, actions, processes, whatever—become self-limiting and impede learning.

Examples of beliefs about the impossibility of future occurrences that turned out to be wrong, such as those about human-powered flight, are legion. To mention just a few: The astronomer Simon Newcomb stated that no known material, machinery, or force could be united into a machine in which a man could fly any distance through the air. People didn't make the correct calculations regarding thrust, lift, and drag from known physical laws until after 1903 when the Wright brothers actually demonstrated that powered flight was possible. In the 1930s, R.v.d. R. Woolley publicly shared his thoughts that space flight was not possible. As late as 1956, just a year before the Russians successfully launched their first manned spaceship, Dr. Woolley on being appointed Astronomer Royal, "confirmed his earlier opinion with the remark: 'Space travel is utter bilge.'" In 1945, Professor F.A. Lindemann (Lord Cherwell) told the House of Lords in Britain and Vannevar Bush told the U.S. Senate "that intercontinental ballistic missiles could not be competitive with bombers in the foreseeable future" (Ayres 1969: 5–7). For better or worse, airplanes, space travel, and missiles, of course, have become modern realities and, certainly for the better, "the foreseeable future" has become a professional purpose of the new field of futures studies.

To take two additional examples, the physiologist J.S. Haldane said in 1930 that a chemical compound having exactly the properties since shown to be possessed by deoxyribonucleic acid (DNA) could not exist. And one of the greatest surgeons of the period, Berkeley George Moynihan, also in 1930, believed that the craft of surgery of his day had achieved near perfection and could not be made much better, thus closing his mind to the future possibilities of the surgical advances that, in fact, have been made in the last half century (Medawar 1984: 306–7).

When studying possible futures, futurists may become creatively involved in constructing images of the future themselves, images—for individuals, organizations, or entire societies—that interpret the world in new ways and direct human effort in new directions. Futurists have been deeply involved, for example, in creating and exploring images of a possible future for our time that incorporate concepts such as a sustainable society, a just society, an experimenting society, the global village, a world at peace, the biosphere as a living entity, and, as I describe in volume 2, chapter 6, all the world's people as a single moral community.

Thus, one futurist task is to study possibilities for the future, no matter how unrecognized or improbable they may be. By communicating the findings to clients and the public generally, futurists, thereby, aim to expand the range of individual and social choice.

The Study of Probable Futures

Another futurist task is the study of probable futures. It focuses on the question of what the most likely future of some specified phenomenon would be within some stated time period and under specified contingencies.

The phenomenon so specified may be nearly anything imaginable. It might be a personal concern such as getting into college, being fired from a job, being happy with a particular person if you married him or her, or having a sufficient pension when you retire. It might be a social concern such as the amount of street crime and violence, the quality of health care in a country, the adequacy of the educational system, the size of the human population for the Earth as a whole, the spread of nuclear capabilities to additional countries, or the pollution of particular lands and seas. What might be studied includes the entire range of natural phenonema from the weather, rain forests and watersheds to designated plant and animal species.

One question is, what would the most probable future of some specified phenomenon be if things simply continue as they are?

If the phenonenon whose future is under consideration is influenced by human actions, then the question can be rephrased, what would the most probable future be if we humans continue to behave as we do?

Obviously, both questions invite the study of the present in order to have a base from which to forecast. They also invite the study of past trends up to the present, in order to get some idea of what has been happening in the recent past. We need to know how things have been going up until now to help us understand where they would be if they kept going the same way in the future.

Of course, things may not—in most instances almost surely will not—simply continue as they are and we humans may not continue to behave as we recently have or are now behaving. Thus, answering the question of what would be the most probable future is itself contingent. Other futures might be more probable, *If* conditions change. There is, then, a fan of probable futures contingent on different assumptions about whether conditions will stay the same or change in specified ways. Thus, futurists also ask, what would the most probable future be, *If* this or that condition changed or *If* people behaved in this or that way differently from how they are now behaving?

The various scenarios given in *The Limits to Growth* briefly discussed in chapter 1 and others from *Beyond the Limits* more fully discussed in chapter 6 are excellent examples of making different assumptions about conditions and then calculating probable consequences for the future.

Searching for the most probable future under a variety of assumptions about conditions raises the question of which assumptions are most likely to turn out to be correct. Attempting to answer this question makes the search for the most probable future both more complex and more plausible as more and more information is brought into consideration in deciding which assumptions are the most sound.

Constructing probable futures contingent on different assumed conditions clearly requires knowledge. Thus, futurists are interested in the study of cause-and-effect relationships of the phenomena under consideration. Explanatory theories, including theories of social change, and the research on which they are based are useful. But carrying out such research and constructing such explanatory theories are not the major tasks of futurists. Futurists such as Daniel Bell, Amitai Etzioni, and Alvin Toffler, to give only a few examples, have proposed such theories, but the special calling of futurists is the integrated exploration of the future not the research of causes and effects. For the most part, as they explore possible, probable, and preferable futures, futurists are consumers and synthesizers of knowledge created by other scientists and scholars.

The Study of Images of the Future

A basic concern of futurists. Yet there are several topics or areas of inquiry in which futurists play a role as basic researchers. One such topic is the study of images of the future. Here is where Lasswell's five tasks fail, because they are too narrow, too focused on the applied-science side of futures studies, too limited to the managerial and instrumental aspects of policy analysis. As important as such applied and instrumental aims are, there are other, more basic, aspects to some futures research as well. Lasswell's fourth task of projecting alternative futures does not adequately incorporate what may be the most important basic research questions of futurists: What is the nature of images of the future? What are their causes and consequences?

Although they use a variety of theories of social change in their work, most futurists share several important conceptual and theoretical commitments. One is to the concept of "image of the future"—or some nearly equivalent idea such as "developmental construct," "expectations," "anticipations," "hopes," and "fears." Another is to the theoretical proposition that images of the future help shape the historical actions that people take. That is, futurists see images of the future as being among the causes of present behavior, as people either try to adapt to what they see coming or try to act in ways to create the future they want. Because present behavior partly produces the emergent future itself, futurists see images of the future as being among the causes of the future as it becomes the present.

Thus, some futurists study the ways in which images of the future influence human behavior and how that behavior in turn contributes toward making the future. To carry out this task, of course, the content of images of the future themselves must be analyzed. Such content may be investigated using a variety of sources of data ranging from literature, nonfiction writing, and current theories of history to a nation's five- or ten-year plan, the content of social legislation, interviews with leaders or ordinary people, or the observation of behavior.

One link between images of the future and present behavior has to do with an individual's ability to balance present and future gratification. There appears to be a relationship between a futures perspective and "the capacity for deferred gratification, which is in turn seen as a prerequisite for social success and advancement" (Bergmann 1992: 86). Some writers go so far as to claim that all forms of deviant, criminal, and reckless behavior have the same fundamental cause: the tendency to pursue immediate benefits without concern for long-term costs, a disregard for inevitable and undesirable future consequences (Hirschi and Gottfredson 1994). Successful self-management from this perspective involves understanding and giving appropriate value to the future consequences of action.

Additionally, because images of the future do not simply appear out of thin air, some futurists also study how images of the future themselves are formed, the conditioning and determining psychological, political, economic, social, and cultural factors that produce particular images of the future among particular peoples or cultures at given times and places. Beyond this, some futurists want to know if there are general laws that explain the rise and fall of certain classes of images of the future, such as positive, idealistic vs. negative, pessimistic images.

A study of images of the future in Jamaica. For example, during 1961–62, James A. Mau (1967, 1968) carried out in Jamaica one of the earliest studies of the causes of images of the future. Then, Jamaica was part of the great transformation of nation-founding and nation-building discussed in chapter 1. The period of Mau's field work was marked by three important political events: the West Indies Federation of which Jamaica had been a part collapsed after a referendum in Jamaica, general elections were held to determine which political party would form the first government of independent Jamaica, and Jamaica, on August 6, 1962, changed from a British colony to a politically independent nation-state. As Mau (1967: 200) says, it was an "opportune time to study attitudes toward the trend of change, past and future," because, "as independence approached, the Jamaican citizenry was preoccupied with speculations about the island's potential, its problems, and the likelihood of making independence a success."

One of the "problems" centered on an urban slum of 100,000 or so people in the western part of Kingston, Jamaica's largest city. Many of the slum dwellers, or squatters, lived in shelters "built of refuse and scavenged material," in "hovels of board, canvas, cardboard, and tin sheets." Mau focused his study on this area, interviewing both slum dwellers themselves and the local and national leaders who had anything to do with "the formulation and implementation of policy concerning West Kingston and its people" (Mau 1967: 202, 203).

To take just one example from Mau's study, he constructed an Index of Belief in Progress based on leaders' images of the future as being positive and optimistic at one extreme or negative and pessimistic at the other. As classified by Mau, leaders with positive, optimistic images of the future were those who believed in progress, believed that generally the people of Jamaica would be better off in the future, that independence itself would have a favorable effect, that the people of West Kingston would be better off in the future, and that civil disorder and violence were not likely. Leaders who did not believe in progress had contrary images of the future.

What caused some leaders to believe in progress—even including the future of the 100,000 West Kingston squatters? Mau found that three factors provided an answer. Leaders who had positive, optimistic images of the future were people (1) whose knowledge about the West Kingston squatters was most complete and accurate; (2) who held the most influence and power to make decisions about the future of West Kingston; and (3) who were morally committed to the proposition that a just Jamaican society ought to be based on social equality and exclude no Jamaican. Contrariwise, leaders who were unknowledgeable, less powerful, and inegalitarian and exclusivist in their definitions of justice had negative, pessimistic images of the future.

Mau's use of knowledge is of particular interest. He defined it specifically as knowing about "the complaints and discontents of the lower-class people who reside in West Kingston" (Mau 1967: 219). Mau scored each leader by comparing leaders' responses to the question about such complaints and discontents with the results of a question which he had asked a sample of West Kingston residents themselves. Thus, Mau could compare what the leaders believed about the slum dwellers with what the slum dwellers told him about themselves, their conditions and their attitudes.

Mau found, as you probably would expect, that some leaders were very knowledgeable about the detailed complaints, discontents, and hopes of the slum dwellers while other leaders were much less knowledgeable, some having totally inaccurate views. But would you have expected the relationship between knowledge and the belief in progress that he found? *Knowledgeable leaders were more likely to believe in progress than unknowledgeable leaders.* Mau (1967: 222) concludes that to believe in progress is not necessarily to be unrealistic and unknowingly idealistic. Just the reverse is the case in his study: pessimistic images of the future tend to be based on ignorance.

Mau's study was relevant to public debate and policy at the time in Jamaica. Imagine the repressive, police-state policies toward the people of West Kingston that would have been—and in fact were—advocated by unknowledgeable leaders who saw no hope for a positive, optimistic future of peaceful change for them, but instead had only nightmares about hostile and angry black West Kingston masses charging out of their slums, swinging machetes, and attacking the mostly brown- and light-skinned middle and upper classes living in the Kingston suburbs. Contrast the constructive policies that would have been and were advocated by knowledgeable leaders who rightly understood the squatters' modest hopes for a better life, fundamental decency, willingness to work, and, yes, even ingenuity—because it takes both brains and a certain amount of entrepreneurial zip to survive in a slum.

A study of images of the future in Western Civilization. Another, though quite different, example can be found in the work of Frederik L. Polak (1961). Unlike Mau's study which was based on a research problem limited in time and place and on the methodologies of survey research and statistical analysis, Polak's study was on a sweeping historical scale, covered a vast period of time, and stretched geographically over many areas and different societies. Mau wanted to explain how different leaders came to have certain images of the future. That is, he wanted to discover the *causes* of their images of the future. Polak was primarily interested in the *consequences* of images of the future, how such images contributed to the rise and fall of different societies and cultures throughout the development of Western civilization.

Ranging from Sumer, ancient Greece, and those areas where the Judeo-Christian religions arose to the Renaissance, the Enlightenment, and the early industrial era, Polak shows what achievements were possible as a result of positive, idealistic images of the future. In any culture, he shows that a core capacity is the ability to imagine the future. When he turned to the modern world, to mid-twentieth century humans, he concluded that this imaging capacity in the West was in decline (Boulding 1978: 7, 11).

A considerable amount of modern social research has confirmed the fact that positive images of the future can have beneficial consequences. For example, John A. Clausen (1986) in a longitudinal study of two generations of people finds that "planfully competent" adolescents become adults who are more stable in their major social roles, less beset by crises over the whole life span, more stable emotionally, and somewhat happier as they approach old age than their peers who were less planfully competent in adolesence.

Moreover, optimistic images of the future, even when seemingly illusory, help people master their environment, overcome vulnerability to adversity, and meet their long-range goals (Taylor 1989). Of course, such images cannot actually be "illusory," contrary to the author's interpretation, because, if they were, then there could not have been any observed effects on the actual future well-being of people.

Thus, a purpose of futures studies is the study of images of the future themselves, their content, causes, and consequences. No one suggests that images of the future are the sole determinants of the future as it becomes the present and, certainly, no one suggests that the future turns out exactly as envisioned. There are too many unintended and unantici- pated consequences that result from human actions for that, not to men- tion the existence of causes of change well beyond human control. But futurists agree that images of the future constitute one important factor in guiding human behavior and that, through behavior, whether it is pri- marily adapting or controlling, such images contribute to the creation of the shape of the coming future itself. Purposeful action requires the an- ticipation of future occurrences (Reichenbach 1951: 194).

The reader will find many examples of studies of images of the future throughout this book, some of which are described in considerable de- tail. Such examples include Hadley Cantril's (1965) cross-national sur- vey of hopes and fears of ordinary people, various Delphi studies of expert judgment, Robert B. Textor's (1990) study of images of Thailand's future through his anthropological probing of beliefs of a single infor- mant, and Henshel's (1976, 1978) work on the self-altering prophecies. Other pioneering empirical studies of images of the future include Steven F. Alger's (1972) research on images of the future and the two cultures, Bettina J. Huber's (1973) study of white South African elites, and C. Timothy McKeown's (1990) study of scientists.

The Study of the Knowledge Foundations of Futures Studies

Any field of knowledge faces the question of how it knows what it claims to know, of stating and justifying its epistemological foundations. Thus, another purpose of the futures field is to provide philosophical grounds for the knowledge that it produces and the research and other intellectual procedures that give rise to it.

Although the knowledge foundations of futures studies contain some of the most well developed aspects of the field, they also contain, ironi- cally, some of the least developed. On the one hand, futurists have spent a lot of their time and effort adapting or constructing methodological tools for the systematic investigation of alternative futures. Also, they have conducted numerous empirical studies in which they have used them. Thus, there are many research methods and exemplars some of which I describe in chapter 6. This is not to say that there is no room for improvement in methodology and no need for additional research. There certainly is. But it is to say that futurists have been prolific in constructing, using and criticizing methods of futures research—from Delphi to Ethnographic Futures Research and from computer simulation to participatory futures praxis.

On the other hand, futurists have accomplished much less in stating the philosophical bases of their assertions about possible, probable, and preferable futures. With a few exceptions (Helmer 1983; Slaughter 1993a), the major issues of the nature of knowledge in the futures field—as apart from the mechanics of specific methods—have received relatively little sustained attention and critical thought.

I devote chapters 4 and 5 to a discussion of the foundations of knowledge for futures studies, since a major aim of this book is to propose an epistemology for futures studies implicit in current futurist practices. In chapter 6, I summarize some of those practices by describing a variety of specific methods of futures research.

The Study of the Ethical Foundations of Futures Studies

The study of the ethical foundations of futures studies follows directly from the futurist purpose of exploring *preferable* futures. In order to assess the desirability of alternative futures, futurists must study, evaluate, and apply human goals and values. They must concern themselves with the nature of the good society and with the standards of judgment and evaluation that they use. Going beyond that, they may have to investigate human nature, the larger natural world, and even the cosmos in search of the meanings and purposes of life in order to find justifications for their value standards.

Why, for example, is a sustainable society better than an unsustainable one? Why should present generations care for the well-being of future generations? Why should people want to cooperate with others, work toward a just society, desire peace and harmony, tell the truth, work hard, be loyal to others, respect authority, be generous, and so on?

The study of the ethical foundations of futures studies incorporates Lasswell's task of clarifying values and goals, but goes well beyond it. It includes the many empirical studies of the goals and values people hold, from leaders and experts to ordinary citizens, that futurists have carried out. It includes the exploration of people's value judgments underlying their notions of the good society. It includes the construction and justification of some objective standards of value judgments by which values and goals can themselves be evaluated. It includes, also, the formulation, codification, and legitimation of professional ethics for futurists which as yet mostly exist as an informal and often unstated set of guidelines.

As in the case of knowledge foundations, I go beyond existing formulations and explore the ethical foundations of futures studies more fully than has been done before. Thus, I devote all of volume 2 to this task. Fortunately, there exists a rich tradition of utopian literature that examines the nature of the good society and that is where we begin in volume 2, chapter 1.

Let it be said here loud and clear, however, that the study and fostering of deep caring about the freedom and welfare of future generations are among the most important purposes of futures studies.

Interpreting the Past and Orientating the Present

It is well known that we use the past to guide our present behavior and to help construct our images of the future. For example, we sometimes use images of the past for rationalizing what we want to do now, for revising our ideas about effective behavior according to lessons we believe we have learned, or for reviving for the future something from the past we regard as having been good. Thus, our beliefs about the past can help shape our beliefs about the future.

Although it is less well known, the reverse is also true. Our beliefs about the future help shape our beliefs about the past. That is, different images of the future tend to invite different images of the past, even in the same social and cultural setting and even in describing the same period of past time.

For example, before Jamaica began the road to self-government in 1944, most of its histories had been written by Europeans or people of mostly European descent. Such histories dealt with a variety of topics—from European naval battles in the Caribbean to wars in Europe that resulted in Caribbean lands changing hands and from the economics of sugar to the detailed political arrangements of colonial regimes. They seldom dealt with the social and cultural histories of the majority of the people, for example, people of African and East Indian descent.

That all changed, however, with the end of colonialism and with Jamaicans facing a future of political independence. Local brown and black people began writing Caribbean history and they told a totally different story about the past. It prominently included the forced migration of Africans, the conditions of transport of indentured laborers from India and elsewhere, and the cruelties and injustices of plantation slavery and indentured labor in the Caribbean. It also included a story of a long struggle for equality and social justice on the part of the oppressed slaves and laborers themselves. This new view of the past was a direct result of a new image of Jamaicans' future: the coming independent statehood through which, it was believed, local freedom and equality would be achieved. History is often the story we tell ourselves about who we want to be.

Also, futurists do not neglect the present. Eighty-seven percent of futurists responding to McHale's 1971–72 survey (p. 19) said that they aimed their futures studies toward bringing about change in the present. Jantsch (1972: 4), who equates futures research with planning for the most part, says that, "Planning is long-range thinking affecting action *in the present*." The present is where action takes place that shapes the future. To decide what to do *now* is largely what futures thinking is all about. Thus, futures studies helps to interpret the past and provides guidance for present action.

A related function of the futures field is an orientating one. Futures thinking lets us know where we are in the present. Often the rapidity of change results in confusion about what is happening around us in the present and the immediate past. Unless we have some perspective on where we have been, where we are going, *and where we want to go in the future*, the present is unintelligible. This is illustrated by a story the American comedian, Bill Cosby, tells. When in college, Cosby faced the old question, is the glass half-empty or half-full? At home, he ran it by his father. Without hesitation, his father answered, "It depends if you're pouring or if you're drinking" (*New Haven Register* 14 May 1995: A2).

The present can be viewed as a transition or turning point, where we move from somewhere to somewhere else, from some time to some future time, from what we were to what we will be. We want to know how far we have moved toward some future. What progress, if any, have we made? What decline, if any, have we experienced? To answer these questions we compare our beliefs about the past, our beliefs about the present, and our expectations about the future.

Additionally, futures studies contributes to our ability to balance the demands of the present against those of the future. As McHale (1978: 15) says, the present may need protection from apparent future demands. We can do so much now for hoped-for future payoffs that never come that our present is shrunken and deprived. Thus, there are times when *carpe diem*—seize the day—is what we ought to do: enjoy the present as opposed to placing all our hope in the future.

But the opposite is also possible. We can deprive the future. Thus, the future may need to be conserved from the excesses of the present. By never delaying our present gratifications, we can draw too much from the future, mortgaging it beyond its capacities. If we do, then, when the future comes, it will be overburdened by past debts.

Thus, futures thinking is both indispensable to and consequential for (1) interpreting the past, (2) understanding the present, (3) deciding and acting in the present, and (4) balancing the use of present and future resources.

Integrating Knowledge and Values for Designing Social Action

On November 13, 1985 the Nevado del Ruiz volcano erupted killing 21,000 people in Armero, Colombia. Their deaths were unnecessary. More than a month before the catastrophe, geologists published in many Colombian newspapers a prediction of the probable coming eruption in great detail, even giving a map of the probable courses of the mud flows and describing the resulting hazards which, it turned out, were deadly accurate. The social action that needed to be taken in order to save people's lives was clear. Both Armero's mayor, Antonio Rodriquez, and the local Red Cross recommended that the town be evacuated. But nothing was done, because of national official inaction and public apathy—and, it should be added, because of an expert prediction that a crucial dam would hold that turned out to be fatally false.

Designing social action incorporates Lasswell's task of the invention, evaluation, and selection of policy alternatives. But in order to design social action, a futurist must organize and focus a great deal of disparate knowledge and critically examine the relevance of many different values. Action, unlike research, is not granular, reductionist, and analytic. Things cannot be held constant, separated into small bits for investigation out of their context and function, nor can "other things" assumed to be equal. Action is holistic and synthethic. It is interconnected with other things in a complex set of interrelationships. It may have a ripple effect such that an action designed to achieve one thing may influence everything else around it, having countless unanticipated and unintended results. Values other than those defining the intended goals of action may be affected, perhaps negatively, and they must be taken into account in decision making.

Most of the knowledge that futurists may need for making a particular policy decision will have been created by other people, specialists in fields about which futurists may know little—in the case of the Nevado del Ruiz volcano by geologists and volcanologists. For effective action and policy design, what determines what knowledge is relevant and needed is not the training and expertise of the policy makers and their advisers, but the nature of the problem being dealt with, the policy goals, social context, and technological means that could possibly be used.

An enormous range of expertise is often involved in designing social action. Take, for example, the decision to bury or not to bury low-level military nuclear waste in the Waste Isolation Pilot Plant (WIPP) in New Mexico. WIPP is located twenty-six miles southeast of Carlsbad, New Mexico and is a deposit site for nuclear waste that would be buried more than 2,000 feet below the surface in what is mostly salt in the Salado Formation. It was created by the U. S. Department of Energy in an effort to protect people and the accessible environment from the radiation effects of military nuclear waste. Something ought to be done soon, because some of this waste has been inadequately stored since the 1940s, some has been lost, and some is already contaminating the environment and damaging people and other living things. The range of expertise that has gone into the decision making that led to building WIPP is truly astonishing. At various stages of the policymaking process, the experts contributing their special knowledges included physicists, mining engineers, hazardous waste engineers, civil engineers, mathematicians, hydrologists, geologists, nuclear engineers, safety system engineers, geographers, economists, sociologists, political scientists, psychologists, anthropologists, linguistics experts, astronomers, historians, lawyers, librarians, and many, many others.

You may wonder what all these different experts could contribute, but the wonder disappears when you consider all the knowledge that is required to transport nuclear wastes safely from their currently scattered locations throughout the United States to WIPP, to package them and move them underground, to store them and seal the storage chambers, and to keep them inaccessible from human intrusion for 10,000 or more years. Special knowledges and skills are required to construct the site and decide how best to store the wastes safely. Also to be considered are events or processes that might disturb the site and release radioactive materials into the environment any time within the next 10,000 years. Toronados? Earthquakes? Hurricanes? Floods? Volcanoes? Climate changes, such as a new ice age? A direct hit from a meteor from outer space? Human intrusion, such as mining or drilling for something valuable—for example, potash, water, or something else that we do not now regard as worth anything—thought to be in the area? How can the site be marked for such a long period to keep people away? Would people 1,000 or 5,000 years from now even understand warning markers written in contemporary languages? Would they understand pictograms? Would warning markers have the opposite effect than that intended and attract future people to the site rather than keeping them away?

Somehow all the small bits of relevant knowledge held by many different experts must be put together to develop an overall design and policy that would achieve the desirable future of removing nuclear wastes from the accessible environment and keeping them removed so that their radioactive radiation can do no harm to living things, both for now and the long-term future.

The above is only one small example of the need of synthesizing knowledge in order to make effective decisions. Marien, as reported by Coates and Jarratt (1989: 195), believes that the "intellectual task of seeing and integrating the many pieces of knowledge we have of our complex physical, environmental, governmental, and social systems is our outstanding world problem." Such integration is certainly a necessary task for designing appropriate social action, a task that the specialization in the way we have organized existing knowledge and the departmentalized structure of colleges and universities have made more difficult.

In the design of social action, the futurist transforms herself or himself from a passive observer into an active participant in policy processes, into a synthesizer of knowledge on the one hand and a policy formulator on the other, into a social architect (van Steenbergen 1983). What actions can be taken to accelarate, retard, or prevent entirely some trend? What actions can be taken to change a stable and steady situation for the better? How can particular images of the future be achieved and brought into reality if they are desirable—or prevented from happening if they are undesirable? To answer such questions the futurist becomes a social system designer.

To create such a social system designer is the aim of the Alternative Futures Option Master Program at the University of Hawaii at Manoa. "Generally if one wants to 'change the world' legitimately," its director, James A. Dator (1984b), says, "one might either enter law school or a planning department. Neither provide the skills and orientations I feel are necessary: social science, creativity, ethical and critical concern [and] design tools." Thus, Dator includes the role of social designer, which includes such tasks as inventing, evaluating, and selecting policy alternatives, within the purposes of futurists. He does so, of course, in the context of open, democratic decision making in which the participation of ordinary citizens is fully encouraged. An early exemplar of the kind of "anticipatory democracy" that Dator proposes can be found in *Hawaii 2000* (Chaplin and Paige 1973).

Images of the future, of course, are involved in designing social action. They provide the goals and the motivation. Starting with a desirable image of the future of some person, place, organization, or phenomenon, futurists sometimes work back to specify the actions necessary to achieve it. Thus, images of the future provide guidance mechanisms for social action (Encel et al. 1975: 27). We must remember, however, that "Ideas—even images of the future—have no power in themselves to change the world. Their effectiveness depends upon the existence of people both willing to put them into action and capable of doing so" (Miles 1978: 82).

Finally, policymaking and decision making themselves always involve social science knowledge, since they are social processes. The people of Armero died, not because of any lack of scientific knowledge on the part of the volcanologists, but because the social processes involved in putting that knowledge to effective use in designing and implementing remedial social action were inadequate.

*Increasing Democratic Participation
in Imaging and Designing the Future*

Most futurists include among the purposes of futures studies the goal of increasing democratic participation in the processes of imaging and designing the future. There are some futurists who may give relatively little attention to this purpose, especially perhaps those futurists who do proprietary work for particular clients and those who work full time for establishment institutions, both government agencies and private profit-making organizations. For other futurists, however, it is a primary responsibility. The World Futures Studies Federation, for example, in a brochure published in the late 1980s includes as one of its aims "encouraging the democratization of future-oriented thinking and acting."

In chapter 6, I give a detailed example of one such futurist, the late Robert Jungk. I describe Jungk's "future workshops" by which he expanded the participation of ordinary people in proposing and evaluating alternative images of the future that affect their own lives.

Another example of increasing the participation of ordinary citizens in shaping images of the future took place in Honolulu in 1982. Aiming to increase citizen participation in governmental decision making, Ted Becker and Dator organized the first Honolulu Electronic Town Meeting (ETM). They wished both to inform the public and to allow members of the public to respond with their opinions about Reaganomics—issues involved in the role of government in the economy, for example, sources of funding for public services, the use of public versus private-sector providers, federal versus state control and responsibility, what services should be provided, and the probable impact of such things on the State of Hawaii. They also considered alternative economic futures for Hawaii (Dator 1983: 212).

The Honolulu ETM had several different parts. There was a scientific-information-gathering part through the Hawaii Televote in which 700 persons selected by random digit-dialling were given printed information about the issues and were invited to be interviewed and to vote using a specially prepared ETM ballot covering the issues under discussion. Also, two daily newspapers, three commercial radio stations, Hawaii public radio, a commerical television station, Hawaii public television, and the island's largest cable television station all participated. The radio and television time was used to provide the audience with information and a variety of opinions about the questions on the ETM ballot. There were phone-in programs, so that persons could express their views or ask questions over the air on issues presented during the program. In addition to the usual panel discussions with phone-in periods on radio and television, the melange of media offerings included three half-hour television programs of dramatic/satirical vignettes dealing with the issues, similar in style and format to the American television program, *Saturday Night Live*, and a documentary film (Dator 1983: 214). The ETM ballot was published in a newspaper and everyone was invited to vote on the issues being discussed and dramatized. The culmination of media—and interactive responses from the public—was a final hour-long live televison program over Hawaii public television (KHET-TV) during which viewers were invited to telephone in their evaluation of the project and to ask questions about the project of Becker, Dator, and then-Lieutenant Governor Jean King.

The Honolulu Electronic Town Meeting is an example of futurists attempting to increase democratic participation in proposing and evaluating both images of the future about important public issues of the time and policies designed to achieve or prevent them. It appeared to be a success. There was widespread response from the public and increased public discussion and awareness of the policy issues being discussed. Practically all of the people who telephoned during the ETM made favorable comments about it (Dator 1983: 214, 216).

In sum, futurists generally work toward increasing democratic participation in imaging and designing the future. There is considerable evidence that such democracy, despite its sometimes frustrating slowness and occasionally paralyzing public disagreements and debates, contributes toward human betterment more than authoritarianism. For example, not only do democracies not go to war with each other, they also do not let their people starve. "In no country with a modicum of political liberty and a free press has there been a famine such as those that killed millions in Bengal in 1943, and more recently in Ethiopia and Somalia" (Ryan 1995: 25).

Communicating and Advocating a Particular Image of the Future

Futurists in one way or another often go beyond encouraging democratic participation in imaging and designing the future. In the example of Jungk's futures workshops to be discussed in chapter 6, the images of the future and policy designs created in the workshops may become the basis for social action. When the workshops end, participants often organize themselves and collectively act to put their designs into practice, usually with the advice and full cooperation of Jungk or his workshop leaders. Members of the Science Policy Research Unit of the University of Sussex in England, to take another example, explicitly state their value-based objective of the reduction of worldwide inequalities in such things as levels of nutrition, housing, health, and income (Encel et al. 1975: 15–17).

Some futurists aim at overarching visions that have transcendent elements. They may include speculative and creative images of "the other" perfect society, contradictions of the present, discontinuous futures that foretell the coming of new and different worlds. They aim to surpass the limits of the now, the limits of present understanding, and the limits of past experience. Drastic social change is sometimes envisioned. The unknown is defined and portrayed as the coming reality. Sometimes, as in the case of Polak (1961), such liberation of human thought comes with hortatory pleas for rejuvenating the imagining capacity of humankind, for the creation of new, religion-like, positive, idealistic images of the future that will have the power to galvanize people into taking actions that will change the world.

Such futures thinking in its scope and power seems far removed from the more mundane and delimited policy questions that concern many futurists, such as how the Yale-New Haven Hospital should change its methods of delivering health care next year or whether the country of Guyana should spend more in agriculture or education in order to increase its gross national product within the next few years. As Elise Boulding (1978: 8) says, "Social planning, blueprinting, and the technological fix are not what Polak had in mind."

Yet futurists may be involved in all three—as well as imaginative visioning—as they disseminate particular images of the future, evaluate them as desirable or undesirable, and then advocate, more or less explicitly, practical social actions either to bring them into reality or to prevent them from occurring. The authors of *The Limits to Growth*, for example, construct and communicate images of global overshoot and collapse, while advocating policies to prevent such images from becoming reality.

Such futurists, of course, become part of the political dialogue. It is the proper nature of futures studies that this is so. For futurists are not imprisoned in an ivory tower. Although they are committed to truth—to the creation, dissemination, and preservation of knowledge—just as other scientists and scholars are, they are also committed to creating an anticipatory and action science that has some effect on society. Mannermaa (1986: 662) goes so far as to claim that a "futures study which does not have any kind of direct or indirect impact on the development of society is totally useless, and cannot really be called a futures study." Futurists aim to contribute to human betterment by translating knowledge and values into action. Proposing action, then, is part of a futurist's job *qua* futurist. Proposing action, though, brings responsibilities. Action means that there may be consequences for people's lives, making them better or worse than they are. Futurists, thus, must examine goals and values. Not only are futurists obligated to state clearly what values are being served by their proposals for action—and whose ox is being gored—but also to test and justify their belief in those values in public debate and critical discourse, to give the reasoning and supporting evidence underlying them. As educators, the tasks of futurists, as Robert E. Carter (1984) says of education generally, are to help people develop capacities for logical analysis, critical judgment, caring, choice, and moral commitment, with the special obligations to envision and evaluate the consequences of action for the future.

Partisan politics can be a dirty business, which may be why scientists—especially social scientists—have tried to stay out of playing a direct role in politics and have adopted the protective ideology of being "value free." Futurists—given the nature of their commitment—have no choice. They can't have it both ways. That is, they can't give advice about what the future ought to be and how to get there, while at the same time claiming the immunity of the academy and the innocence of only reporting the facts.

The only right answer is to own up to the political dimensions of the futurist's role and to try to elevate the larger political discourse by bringing more reason, evidence, logic, complexity, civility, and humility, and, most important, more valid futures thinking to it. The futurist's commitment is not to partisanship in the narrow sense of the term. Rather, it is to truth and goodness, and to fairness and openmindedness.

The Role of Prediction

A Definition of Prediction

Although the term "prediction" may refer to the relationships of things at a given time (Schuessler 1968), in this book we are interested in a definition that incorporates the future: things that, under certain circumstances, will, could, or would happen, exist, or change *in the future* (Henshel 1976). Most generally, a prediction is a statement about the expected occurrence of some future event, outcome, state, or process. Some writers give different names to such statements, calling them "forecasts" if they deal with concrete and calendar-bound predictions (Schuessler 1968), "projections" if they are based on the quantitative extrapolation of time-series data, or "prophecies" if they are either broad, holistic statements about the future or if they refer to self-altering predictions.

None of these conventions, however, is followed by every writer and, sometimes, each has been contradicted. For example, Marien in a personal communication (16 February 1988) insists that "prediction" should be restricted to the concrete and calendar-bound, and that "forecast" is a more general term referring simply to "an opinion about the future." Masini (1993: 54–55) follows Jantsch (1967), defining a "forecast" as a probabilistic statement and a prediction as a non-probabilistic statement. Slaughter (1993: 293) makes distinctions among "prediction" which he regards as a "confident statement about a future state of affairs," "forecasts" which he says "are based on an 'if...then' process of reasoning, involving the careful analysis of past knowledge," and "foresight" which he sees not as a technical matter but as a basic "human capacity and skill." Among demographers, "projection" usually means a scenario of change based on any arbitrary set of assumptions, while a "forecast" refers to a projection based on a presumed realistic set of assumptions (Long and McMillen 1987: 142).

Other terms are used as well, such as "prognostication," "prevision," "anticipation," "expectation," or "futuribles." Henshel (1982:59) says that defining these terms to refer to different aspects of prediction is a fruitless task, "since everyone does it differently." Certainly futurists have not reached a consensus on their definitions, although "projection" is usually associated with extrapolation of trends and "prophecy" with sweeping pronouncements or self-altering predictions.

Thus, in this book I use "prediction" to refer to the most general and shared meaning of these terms and I define it simply as a statement or assertion about how the future might turn out to be.

I include all kinds of predictions in this definition: single or multiple, conditional or unconditional, contingent or not, likely or unlikely, absolute or probabilistic, short- or long-range, small- or large-scale, trivial or momentous, based on scientific evidence or not, accurate or inaccurate, self-altering or not, publicly announced or kept secret, about a specific thing or a range of possibilities, plausible or not, desirable or undesirable, and so on. Whatever their source or whatever they are labeled (e.g., "projection," "forecast," etc.), if a statement concerns some future outcome, event, or condition, I include it in the definition of "prediction." Sometimes, I may refer to a prediction as a "projection" or a "forecast" or a "scenario about the future" to be consistent with the author or authors of a particular work that I'm discussing or a convention in a particular subfield, such as demography. But all such terms are "predictions" in the sense that I am using the term.

The Role of Prediction in Everyday Life

Predictions are a necessary part of decision making and planning, both in the worlds of planners, architects, engineers, business executives, statesmen, and other leaders, and in the everyday world of ordinary people pursuing their individual projects. Imagining the future, even if it is limited to the very near future of minutes, hours, or days, is how people steer themselves through space and time as they make their way in the world. As Nathan Keyfitz (1987: 237) says, "One cannot act purposefully in any small respect except within a picture of what the world will be like when the action produces its effects."

People make statements—or construct mental images—about the future in order to act, even though they sometimes do so at the margins of consciousness. They predict the probable consequences of their own behavior, of other actors' behaviors, and of forces beyond their control in a contingent, complex, and often largely implicit process of choosing how to act in order to achieve their goals. This process is subject to constant revision and correction as people receive feedback from their experiences.

For example, Thomas Gabor (1986) points out that prediction "is a routine human preoccupation inherent in all social behavior.... Thus the decision to turn on a light switch is based on the expectation that the action will result in the illumination of a room." This is not to say that predictions—whether made by ordinary people or professionals—always, or even usually, turn out to be true. Yet people are all moderately good at prediction or living would be impossible (Furbank 1993).

The Role of Prediction in Science and Social Science

Many philosophers of science, of course, view prediction as among the goals of science. Prediction, it has been said, is both an indicator of

the progress of science and a measure of the validity of scientific knowledge. Karl R. Popper (1959; Weimer 1979: 52), for example, views a
theory as trustworthy if it yields reliable predictions and instrumental
utility; Otto Neurath (1959) views successful prediction as a measure of
the fruitfulness of a science; Israel Scheffler (1967: 101) says the "preva-
lent view" in science is that when a "prediction is borne out by experience, the set of beliefs in question has passed a critical test"; and Hans
Reichenbach (1951: 89), in what may be the most extreme view, asks,
"What is knowledge if it does not include the future? A mere report of
relations observed in the past cannot be called knowledge."

In the social sciences, Schuessler (1971: 302), echoing these views,
writes that "American sociologists have more or less taken for granted
that prediction is one of their main objectives." Not all social scientists
would agree, perhaps especially not those with a humanistic bent. Yet
demographers project population changes, economists have created an
economic forecasting industry, a social-indicators movement which is
partly aimed at prediction has been around at least since the late 1960s,
and a recently published set of papers on forecasting in the social and
natural sciences has reasserted a social scientific commitment to prediction (Land and Schneider 1987).

Such a commitment, of course, goes back at least to a founder of
sociology, Auguste Comte, who asserted *savoir pour prévoir*. It continued with nineteenth-century social analysts such as Marx, Weber,
Marshall, Spencer, and Durkheim, all of whom carried out their comparative and historical studies with predictive intent. "They were concerned with where their own societies were going and with what might be
done to facilitate or block these tendencies" (Hopkins and Wallerstein
1967: 35).

Even anthropology, "whose forte," as Reed D. Riner (1987: 311–12)
says, "has been disciplined hindsight," has had its face rubbed in the
future in recent decades. It is no longer limited to describing and analyzing the exotic, past, and often dying cultures of the world, primarily for
the benefit of other anthropologists. Today, it increasingly includes applied anthropology, that is, active, client-oriented research aimed at the
solution of contemporary organizational and community problems.

In addition to being a goal of science and an indicator of the validity
of scientific knowledge, prediction is linked to science in another way
too. As Richard S. Rudner (1966: 60) has pointed out, the logical struc-

ture of a scientific *explanation* is, with one important exception, the same as that of a scientific *prediction*. The exception, of course, is "the purely pragmatic one of the temporal vantage point."

From past observation, for example, we may conclude that x causes y. But we can add a speculative leap and say that if we want to produce y in the future, then we should do x. Logic such as this, whether qualitative, implicit, impressionistic, abysmally incomplete, and informal on the one hand or quantitatively modeled and based on precisely measured variables and clearly stated explanations on the other, is often involved in transforming hindsight into foresight. Knowledge, or partial knowledge, of the past, often expressed in theoretical or quasi-theoretical statements about relationships, can be applied inferentially to the future, although, obviously, not with certainty.

Hempel (1965: 81–96), while clinging to the positivist acceptance of prediction as an aim of science, acknowledges the tentativeness of knowledge confronting an as-yet-unknown future. He says that Galileo's law refers both to past instances of free fall near the Earth and to future ones. But he recognizes that future instances are not covered by present evidence. Hence, the evidence for Galileo's law, and, he adds, for any other scientific law, is incomplete. Thus, there is always an "inductive risk" that a presumptive law may not generally hold and "future evidence may lead scientists to modify or abandon it." Although Hempel states it less dogmatically than an extreme logical positivist would and pulls back from the claim of empirical certainty, he offers only a caution. It may be risky going, but the scientific road to prediction, he says, remains open.

Caution is warranted, because, even if an explanation has been found by all past experience to be true under all conditions and with a probability of 1.00 of y occurring when x occurs and never occurring when x does not occur, some future test of the explanation could show that there are conditions under which it would not hold (Hempel 1965). Even at the most simple level, of course, a prediction based on an explanation is contingent by its very nature, being conditional on the initial "If..."

Although most, if not all, social scientists working within the science model may be willing to accept prediction as a part of the social scientific enterprise as a *general principle*, we must acknowledge that many of these same social scientists actually fail to make prediction a part of the specific work that they do. They tend to be content with assessing the "systematic associations between events observed in the past" rather than with "testing out the prognostic value" of research findings (Miles 1975: 2). They tend to deal with relationships among variables at one point in time, not through time. When time is introduced as a variable, with some important exceptions in economic forecasting and population projections, prediction often is from some past times to more recent past times, not to the future of time present.

The Role of Prediction in Futures Studies

Since the purposes of the futures field are to explore possible, probable, and preferable futures, it seems obvious that futurists must agree that prediction is one of their goals. After all, predictions are statements about the future and futurists talk about the future. The fact is that there is disagreement among futurists about the role of prediction in the futures field. The disagreement, however, is more apparent than real.

On the one hand, some futurists embrace prediction. Joseph P. Martino (1987), for example, focuses much of his work on the central task of forecasting. Panos D. Bardis (1986: 117), to take another example, defines the field as follows: "The science of futurology deals with all types of forecasts of future social events." Toffler's *The Third Wave* (1981) is filled with what is coming or what will happen in the future, however much he hedges his predictions by contingencies. F. A. van Vught (1987: 188) notes that "Futures researchers claim that they can predict the future scientifically." Finally, the futurist literature is filled with sweeping, though contingent, predictions of doom, like the one made by two leading futurists: "If we fail to learn this lesson, we will march like lemmings to extinction" (Henderson and Theobald 1988: 35).

On the other hand, some futurists minimize or deny prediction as a purpose of futures studies or are ambiguous or contradictory in their views. Daniel Bell (1976: 3–4) says that we cannot predict the future. He adds that it is difficult to predict. Then, he claims that forecasting (as distinct from prediction—a distinction that he recognizes is arbitrary) is possible under some circumstances, subtitling his book, "A Venture in Social Forecasting." Although it's not entirely clear whether he's for or against, one thing that he apparently means is that we cannot know that predictions are accurate before the fact and another is that accurate prediction—or "forecasting" as he calls it—is difficult to do.

Masini (1988), former president of the World Futures Studies Federation, recently affirmed that *exact* prediction "is by no means the task" of futures research. She argues that predicting only *one* future is not the purpose. As she says further, futures research "should rather reveal the alternative possibilities, and analyze the risks concomitant of these possibilities and their consequences." Such a view, of course, is not new. Speaking of the ancients, Martin Buber (1957) said that it was not prediction but rather confronting people with the alternatives of decision that was the task of the genuine prophet.

Yet prediction is clearly involved in specifying both the "risks," which is an inherently future-oriented concept, and the future consequences of various possibilities. Discussing what "could happen" or "might happen" involves making predictions that are conditional, contingent, chancy, or multiple.

Jouvenel (1967: 54–55) makes useful distinctions among what he calls "primary forecasts" (which are predictions "*conditional* on the absence of any corrective action"), "secondary forecasts" (which are predictions "*conditional* on the taking of a certain definite action" or a series of conditional predictions each based on specified actions), and "tertiary forecasts" (which are predictions of what definite actions will be taken and of outcomes that will follow from them).

Amara (1981a: 25), says that the "future is not predictable." Rather, he stresses the futurist's role in raising people's consciousness of time and in increasing their awareness of the openness of the future. Like Masini, he highlights the possibilities and probabilities of *alternative* futures rather than predicting what one future will happen. He aims to evaluate consequences of different actions for the future, to involve other people in thinking about the future, to structure communications about possibilities for the future, and to create different images of the future.

Lasswell, as reported earlier, makes the projection of *alternative* futures one of his five tasks of studying the future. Coates (1985: 21) bluntly says, "The central principle of futures research is *alternative futures*: there is no single predictable future." McHale and McHale (n.d.: 3) say, futurists do not aim "to prophesy what a specific future will be, but rather more to explore the plurality of future(s) states, which may be contingent upon our actions or accessible to our choice." Dator (1994b) embraces "foresight" as a purpose of futures studies, but claims that it is not prediction which for him implies certainty and determinism.

These purposes, as we have seen, certainly are part of the futurist's role, but most—if not all—of them *do* involve prediction as I have defined it. For example, to the extent to which each of them ultimately is aimed at increasing the effectiveness of human action, each comes down to some statement about the possible, probable, or preferable *future* consequences of some action, event, or process. Each comes down to a statement of some future outcome, however uncertain, contingent, and conditional it might be or however inaccurate it turns out to be. Predictions, sometimes of a specific future, but most often of some possible and more or less probable alternative futures under specified assumptions and conditions is what the futurist enterprise is all about.

There may be instances, though, when dealing with the possible futures of certain phenomena when there *is* a single predictable future. It all depends on what kinds of futures we are talking about. Do not put the barrel of a loaded and functioning gun into your mouth and pull the trigger. The chances are close to 100 percent that you will kill or severely damage yourself. There are many things about which we know so much that outcomes can be predicted with great certainty. Thus, how certain a prediction is depends on what the prediction is about, because every action, event, process, and phenomenon may have its own particular range of possibilities and degrees of error. This is a factual question.

Another apparently antiprediction view among futurists is expressed by Jungk (Gordon, Gerjuoy, and Jungk 1987: 25). He says that he does not see himself as a forecaster. Yet he clearly believes that people need to look ahead and become more responsible for the possible consequences of their actions by considering them in advance. Obviously, people can't consider the future consequences of their actions without predicting what they will be. Thus, his view is contradictory.

There seem to be two aspects of Jungk's resistance to the purpose of prediction in futures research. The first appears to be a general antiscience and antitechnology attitude. For him, prediction reeks of dehumanized methodology, impersonal techniques, and abstract quantification, while he wants to emphasize humanistic values. No doubt it sometimes does reek of all of these things. We can add that quantitative prediction, just as any other kind of prediction, is sometimes incompetently carried out, methodologically obscure, or totally unrelated to the real world. Yet some technical reports are obviously useful for many purposes and are sometimes of potentially great human consequence (e.g., the quantitative scenarios of *Beyond the Limits* or *On Thermonuclear War*).

A second source of resistance flows from Jungk's emphasis on the exploration of preferable futures, especially what people—ordinary people, not leaders, intellectuals, bureaucrats, or technicians—desire. He is less concerned with what the future *will* be if things continue as they are than with what the future *ought* to be according to some set of valued goals and *could* be if people acted so as to achieve such goals.

Jungk is right to caution futurists against mechanical (and often arrogantly wrongheaded) applications of techniques, limited focus on only the possible and probable, and deemphasis of the preferable. But predictions of various kinds are necessarily involved in his own efforts to "give the public and all professionals access to futures information," "to futurize public opinion," and "to give people a chance to look ahead" all of which he says are among the purposes for which he established the International Library for Futures Research in Salzburg. Moreover, as I try to show in volume 2, chapter 2, evaluating alternative futures according to desirability—and, more important, according to some objective standards of moral rightness—also involves predictions because it involves making assumptions about the probable future outcomes of choice and action.

For example, picturing what "ought to be," that is, what is morally right, is, in part, a matter of predicting what social consequences will exemplify some value, as in "justice will be served by giving equal pay to women doing the same work as men." In part, it is a matter of predicting what means will achieve what ends, at least to the extent that it is assumed that human action can lead to the preferred futures toward which it is directed. Thus, although futurists may be more interested in *constructing* the future than in merely predicting it, the predictive and constructive aspects of futures studies are not mutually exclusive. For predicting the probable consequences of particular actions is necessary for construction. Finally, picturing what "ought to be" is, in part, a matter of predicting what conditions or goals people will regard as preferable when some particular future becomes the present, that is, will they still value the same things then as now?

One explanation of why some futurists have become wary about making predictions is because they've been embarrassed by being wrong too often themselves or by irresponsible predictions proclaimed with little or no grounding whatsoever by others. Wildly speculative predictions stated as fact are easy targets for critics of the futures field and have been used to discredit futurists. Thus, such critics are disarmed when futurists say that they do not predict.

Additionally, the renunciation of prediction on the part of some futurists has been an understandable attempt to legitimate futures research as a careful and respectable intellectual endeavor. An easy, though misguided, way of doing this is simply to deny that futurists make predictions at all. The correct way is to deemphasize making predictions of *the* future, to focus on alternative future*s*, to acknowledge that the future resists efforts to know it before the fact, and to point out that predictions, even sound ones, don't necessarily turn out to be true.

Finally, a flight from the task of prediction is encouraged by a fixation on exactly those future events that are the most difficult to predict. If we predict that the horses will run at the racecourse at Aqueduct next Saturday, then we have said nothing exceptional even if it is true. The schedule has been announced and the list of races has appeared in the newspaper. That is, we can predict thousands of things, most of which will turn out to be true if we stick to routine events. Such predictions may appear uninteresting and unexceptional precisely because they are so certain (even though they may sometimes not come true: for example, some freak weather conditions or the outbreak of war might lead officials to cancel the scheduled horse races at Aqueduct).

But what is exceptional, profitable, and interesting is to make predictions about events that are by their very nature uncertain. Although people do need to anticipate that the horses will run next Saturday if they intend to go to the racecourse and place some bets, what they would regard as most valuable and useful, obviously, is accurate predictions about which horses will come in first, second, and third in each race. Alas, picking the winners, as we all know, is fraught with uncertainty.

Thus, if we overlook the countless number of human and other events that are reasonably certain and appear to be obvious and focus instead only on the most uncertain and chancy events, then the conclusion *seems* to follow that we cannot predict. Yet, obviously, we can and do make predictions all the time, although not always accurately. Futurists who have been most interested in the least obvious of future events, of course, have good reason to be the most frustrated by the entire effort to predict.

In sum, constructing alternative futures, that is, writing scenarios depicting different futures, always involves the process and act of prediction. It does so because such construction necessarily includes evaluating the effects of different conditions, assumptions, or models on the phenomenon whose possible futures are under consideration. Prediction, thus, plays an important role in the futurist enterprise, even though the predictions (forecasts, projections, etc.) may be multiple, conditional, contingent, corrigible, uncertain, and, as we shall see, presumptively or terminally true or false, or self-altering.

Thus, we can conclude: (1) Futurists can and do make predictive statements, but they cannot be certain that their predictions will be accurate. (2) They can—and in order to increase the effectiveness of decision making often must—consider not just one future outcome but a range of alternative possibilities for the future, assuming different conditions. And (3) they can—and must in order to encourage responsible action—evaluate possible and probable futures according to some scale of values so as to judge how desirable various alternative futures will be.

Will Futures Studies Lead to Fascism?

Some people believe that planning, particularly the collective planning of economies or societies, somehow is "unnatural" and leads to authoritarian or totalitarian control that violates individual freedoms. Futures studies, too, has been criticized in this way, largely because, like planning, it involves control, deliberate decision making and conscious intervention in the direction of social change. Bending the future to human will, if it involves the future of a group, means that at least some individual members of the group must be guided and their efforts coordinated with the efforts of other people. Clearly, some group authority and decision-making structure is necessary for producing coherent, integrated individual behaviors in any kind of collective activity, from team events in sports to running a business to governing a nation.

Individual freedom, indeed, may be limited by such social arrangements, both because of the prohibition against interfering with the legitimate pursuits of other people and because of obligations to contribute to group projects. But individual freedom can also be enhanced through group formation and collective effort. For, as sociologists know, social order permits members of society to count on the routine cooperation of other people, for example, as contained in the division of labor. Cooperative efforts—whether based on personal relationships in the family or

impersonal relationships in the market, for example—dramatically increase total social power, production, and wealth. Thus, they also dramatically increase an individual's success in the pursuit of his or her own
projects. They are liberating because they allow individuals to break out
of the cage of limitations imposed by the relatively meager capacities
and resources of an individual acting solely alone.

Despite the obvious payoffs of group living, there are always critics
who seem blind to them and who emphasize only the negative. Two such
critics are Kenneth M. and Patricia Dolbeare (1976: 226–27). They claim
that national planning as advocated by reform liberals and democratic socialists as well as by conservatives is leading to fascism in the United
States. Then, they proceed to attack the modern futures field as even more
of a threat than mere planning. They say that futurism may be a "radical
ideology that most fully serves the fascist cause" (p. 228). They see futurism as serving existing decision-making elites and as increasing the control
of such elites over their societies (pp. 211–12). They add that the futures
field legitimates "massive centralization and totalistic control, inevitably
on behalf of the major units of the world capitalist economy" (p. 214).

There is possibly some merit in their view as a criticism of *some* examples of futures research. Every field of inquiry, including futures studies, may produce works that might be used—or misused—to support one
politically partisan view or another, including authoritarianism. Moreover, trying to be of assistance to existing decision makers does tend to
carry with it support for the powers that be.

As I pointed out in the discussion of think tanks, there is a danger of
the cooptation of futurists by established elites, of the monopoly of futures thinking by powerful interests in society, and of the merging of
futurists into the military-industrial-intellectual complex. Much of the
research in futures studies to date has been carried out for specific clients and on topics of interest to them. In his 1971–72 study, for example,
McHale concluded that 80 percent of the financing for futures research
came from either government (50 percent) and corporations (30 percent).
Only 10 percent came from academia and another 10 percent from foundations (p. 17). Thus, if futures studies is a tool that helps control the
future, there is a question of whose interests futurists serve. Who will
control futures studies and futurists and to what ends?

The Dolbeares's extreme view, however, flies in the face of the widespread commitment to democratic participation among futurists. Some

futurists, as we have seen, have made the democratization of control over the future a major purpose of their work. Jungk has shown that futures thinking is useful to "followers," citizen groups, and ordinary people as well as to war managers, business leaders, and government officials. Jungk certainly agrees with the Dolbeares that undemocratic control of futures research is undesirable and a potential problem, but that does not make him any less a futurist or any less favorable to the further development of futures studies. Rather, it leads him to do what he can, since he understands the importance of futures thinking, to keep futures studies open to all sections of the population.

Galtung and Jungk (1971: 368), for example, point out that a similar problem exists on the international level in that some countries have more futures research capabilities and resources than others. This means differences in power, because "he who has insight into the future also controls some of the present." Thus, they propose the internationalization of futures research as soon as possible. They want futures research to be democratized, with its intellectual tools shared and more equally distributed "The future," they say, "belongs to all of us, not only to small oligarchic groups and interests." The point is not to cut down the tree of futures knowledge, but to share its fruit.

Planning collective futures, of course, is already part of our daily lives. It is involved in such commonplace things as the postal service, water supply systems, waste disposal systems, social security, fire protection, delivering electricity and telephone services to households, General Motors' five-year plans, the U.S. Federal budget for next year, the space shuttle flights, or a family picnic. Sometime automobile executive, Lee Iacocca (1984: 330), asks whether or not planning is un-American. His answer is no. He points out that Chrysler, like every other successful corporation, plans all the time. "Football teams plan. Universities plan. Unions plan. Banks plan. Governments all over the world plan—except ours." Of course, he is wrong about the U. S. government not planning, but it was not doing enough national economic planning at the time to suit him.

This is not to say that any of these things—and many others—can't be done better than they are. That's exactly the point. They probably can be done better. If so, then futures thinking has a role to play.

As Australian writer, Keith Suter (WFSF 1985: 34–35) says, there are three routes to the future, disaster, drift, and design. Major social changes can come about in each of these ways, through calamities such as war, through aimless and thoughtless stumbling and backing into the future, or through conscious design. "Failing to plan," he says, "is planning to fail."

Futurists dealing with broad cultural goals and the analysis of whole societies are just as likely, perhaps more likely, to be subversive of existing arrangements than supportive of them, as I pointed out earlier. The alerting functions of futures studies have included warnings of everything from nuclear arms races and overpopulation to environmental pollution and resource depletion. Toffler (1981) goes so far as to say that the dominant thinking in modern societies is that of the outmoded "Second Wave" and that we must change completely to "Third-Wave" thinking to prepare for the emerging future. We have seen Lasswell's concerns with the values of democracy and human dignity in his writing about the possible rise of the authoritarian and totalitarian garrison-police state and the world revolution of our time.

In sum, futures studies, planning, policy analysis and related fields by no means necessarily imply increasing authoritarianism or totalitarianism. That depends on how they are carried out and whose values they serve. They often do imply, however, some form of increased social control, some enlargement of group or societal or even global coordinated efforts. Such efforts can be consequential for shaping both the futures of individuals and of societies. They require that some people become organized into hierarchies of power and influence. That is the nature of society, organized social life. From any one individual's point of view, the state or some other agency of society may be coercive and it may curb individual freedom for the sake of the public good. But that is true in any kind of society, no matter how simple or complex, or how democratic or authoritarian. It is true whether it results from the seemingly spontaneous evolution of society or from a conscious and public decision process where goals, values, and means are made explicit and subjected to scrutiny and debate.

Increasing democratic participation, as we have seen, is part of the futurist agenda. But making sure that polities and societies are democratically organized and democratically directed is a task that goes beyond the boundaries of futures studies or any other form of knowledge or tool of administration. Making sure that public liberties are maintained is the job of everyone who values them. Yes, there is always the danger that knowledge of technologies—including futures studies—will be used by particular groups to further their own ends. To prevent this, to echo Lasswell's earlier warning, all people who value democracy should do what they can to make sure that such abuses do not occur and that all people have an equal opportunity to participate in the decisions that will shape their futures.

Conclusion

The overriding purpose of futures studies is to maintain or improve human well-being and the life-sustaining capacities of the Earth, the futurist's distinctive contribution being *prospective thinking*. The purposes of futures studies are to discover or invent, examine and evaluate, and propose possible, probable and preferable futures. The futures field is an integrative science of reasoning, choosing, and acting.

Specifically, the aims of futures studies can be conveniently divided into nine major purposes:

1. The Study of Possible Futures
2. The Study of Probable Futures
3. The Study of Images of the Future
4. The Study of the Knowledge Foundations of Futures Studies
5. The Study of the Ethical Foundations of Futures Studies
6. Interpreting the Past and Orientating the Present
7. Integrating Knowledge and Values for Designing Social Action
8. Increasing Democratic Participation in Imaging and Designing the Future
9. Communicating and Advocating a Particular Image of the Future

Futurists seek to know what causes change, the dynamic processes underlying technological developments on the one hand and changes in the political, economic, social, and cultural orders on the other. They seek theories to explain such changes and to help people to recognize and understand them. Futurists seek to determine what anticipated changes may have to be accepted because they are, temporarily, that is, at a given time and place, or intrinsically and always, beyond human control. They also seek to determine what can be changed by human actions, what trends can be accelerated or prevented, or what phenomena are amenable to individual or collective human action, such as altering the rate of consumption of scarce resources, producing less dangerous levels of air and water pollution, or changing the birth rate. These things *could* be done, which is not to say that they will be done.

Additionally, futurists are concerned with prediction in the general sense of making assertions or statements about the future. Prediction is necessarily involved in futures studies, just as it is in human action of any kind. It is absolutely essential. Unfortunately, futurists' thinking has not been fully clear on this issue, partly because prediction itself is by no means the only purpose of futures studies, as we have seen. Rather, the purposes are more complicated and more far-reaching than mere prediction and involve the construction and evaluation of alternative futures for the purpose of increasing human control over the future.

Thus, some futurists have even disavowed that prediction is one of their goals, although those same futurists use a variety of different euphemisms for the same idea, such as "foresight," "forecast," or "projection," to describe their work. Also, there are different kinds of predictions and it is sometimes useful to distinguish among them. So far, the attempts to define particular forms of prediction by giving them different names has not resulted in a consensus of usage.

Although prediction is a necessary aspect of futures studies, futurists seldom predict a single, unconditional, and certain future. Rather, futurists' predictions (forecasts, projections, prophecies, etc.—whatever they are labeled) are usually multiple, conditional, contingent, corrigible, and uncertain.

Futures studies, of course, is still developing. It still remains, as John M. Richardson (1984: 384) says, a fragmented and fractious field, although more focused than it was a decade or so ago. But there are positive aspects to its youth and newness. There are "exceptional opportunities for originality and creativity."

To date, the record of the new field is mixed, as we shall see in subsequent chapters. "As in any field, but particularly in a new and popular one," Kahn (1975: 405–6) says, "there are the fashionable, banal, polemical, and sometimes even charlatanical elements. And even much of the 'legitimate' work does not attain very high levels of originality, creativity, or scholarly rigor." Yet futures studies, he adds, includes many activities that are reasonably careful, scholarly, and scientific.

Reviewing the futures research of the 1960s and 1970s, Amara (1984: 402) sees four major accomplishments: (1) alerting society to coming changes in the external environment, such as in the energy supply, employment patterns, environmental hazards, and technological developments; (2) developing judgmental and qualitative methods of forecasting, such as the Delphi technique, that reduce complete reliance on historical data; (3) describing, rather than suppressing, the uncertainty inherent in any forecast or assumption about the future; and (4) encouraging the reassessment of choices and objectives, especially the creation of new alternatives and approaches to emerging problems and opportunities.

One question posed by critics of futures studies must be met head on. It is whether futures studies, since it involves conscious decision making and an attempt to control the future, leads to fascism or some form of authoritarian and totalitarian control of the many by the few. I answered, no, not necessarily.

Yet, like any tool, futures studies might be of help to the management, administration and manipulation of people on the part of some special or sectional interest rather than for the public good. Moreover, it has been so used, as we have seen in the cases of some developments in Russia, Italy, and Germany and of some of the work of the think tanks. But many futurists are aware of this possibility and direct their work toward the democratization of futures studies, both by encouraging the participation of everyone within a society in the issues that concern their future and by internationalizing futures studies. This democratizing tendency is so strong among futurists that I have included it as one of the major purposes of the futures field. Furthermore, in volume 2, chapter 3 I suggest some preliminary considerations for the professional ethics of futurists that include concerns for the protection of the public good.

Finally, futurists have no intention of trying to monopolize futures studies. Just the opposite is the case. Every topic of study may have a future. Nearly every discipline and field of study can include, if it doesn't already, a future and future conditional tense. Each one can be cast in a prospective or future-oriented framework. Even history, if historical analogies are drawn linking past events to the future outcomes of present actions, can have a futures component. Moreover, the study of images of the future of past times, their content, development, and consequences are proper subjects for historical study. Thus, professional futurists do not want to have and probably never will have the future to themselves. Surely, there will be other sources of specific forecasts and general images of the future to inform public policies and to help guide society (McHale 1971–72: x).

Let us next turn to a discussion of some of the key assumptions that futurists make that underlie their futures work, including the crucial assumptions concerning the measurements and meanings of time.

3

Assumptions of Futures Studies

Futurists necessarily make assumptions in order to carry out their studies of the future. Many of these assumption, such as holism and a belief in the reality of present possibilities for the future, have already been briefly mentioned. In this chapter, I discuss them and some others, trying to make their meaning clear and to show that they are reasonable and plausible assumptions to make.

Of course, it is nearly impossible to state all of the assumptions that futurists—or the members of any other scientific or scholarly group— make. There are too many auxiliary assumptions that are necessary for that. We don't need to state, for example, that futurists believe that the Earth is shaped approximately like a sphere, that oxygen and water are essential for human life, that gravity exists, that human beings have ideas about the world in which they live and that they communicate with each other. That is, there are simply a lot of things that all of us, including scientists, simply take for granted. If we try to write them all down, they would make a very long—and boring—list indeed. Moreover, most of them would not be distinctive to futurists.

What I try to do here is to be selective and to pick out a few assumptions that I believe are both distinctive and important in understanding the futurist enterprise and in carrying out futures studies. Some of them are well-known and explicitly recognized by most futurists. Others are implicit, often hidden premises that have not yet been critically examined by the futurist community. I label these underlying beliefs "key assumptions."

Additionally, I state a few other assumptions that are more general and that deal with the nature of human behavior and society. Although futurists may share these general assumptions with some other scholars, they are worth making explicit because they emphasize foundational beliefs that have to do with that part of futures studies that is an action and prescriptive science. I label these "general assumptions."

Although not every futurist accepts every key and general assumption that I identify, I try to show how each assumption contributes both to the conduct of futures studies as most futurists understand the aims of the field and to the intellectual products that futurists create. If working futurists do not now make these assumptions, then I suggest that they consider doing so.

Before stating the assumptions, however, I discuss the meanings and measurements of time, because a conception of time is itself among the most basic assumptions of futurist thought and because ideas about time importantly affect some of the other assumptions that futurists make. Sohail Inayatullah (1993: 251) says that any "adequate theory of the future must be able to problematize time and negotiate the many meanings of time even as it might be committed to a particular construction of time." To the extent that Inayatullah emphasizes the importance of a conception of time to futures studies, he is right on track. I will try to show, however, that there are meanings of time that are not problematic. For no matter how many meanings people may give to time as they experience it psychologically and socially and as they interpret it in different situations, there are, nonetheless, underlying meanings and measurements of time that are immutable and universal. Time is, for example, unrepeatable and irrecoverable. Regardless of people's awareness of it and regardless of their possibly contrary beliefs about it, time inexorably goes on and on.

Time Machine Earth

Time Traveling

How would you like to be a time traveler? Would you like to get into a time machine or step into a beam of energy and travel freely into the past or the future? Then, when you were ready, would you like to travel again—and yet again—to some other time?

I have asked many people, both young and old, that question over the past three decades or so. Usually, they ponder it for awhile, a few becoming very wary and cautious, suddenly not so sure that they want to travel in time at all. A few others scoff at the idea, saying that time travel is impossible. Still others, the majority over the years, decide that indeed they would like to be time travelers, although frequently with the proviso that they want to be able to return to the present when they wish. What about you? Would you like to be a time traveler? Close your eyes a minute and think about it.

So, what did you decide? You were right if you decided that you don't really have any choice. We cannot move freely in time, back to the past and forward to the future and return to the present in a time machine, as writers such as H.G. Wells depict.

You don't have a choice, because, whether you like it or not, *you already are a time traveler*! All of us here on Earth are time travelers. We are constantly moving with the arrow of time on a one-way trip out of the past in on ongoing present toward the future. We are traveling with time because the universe in which we live *is* a Time Machine. We can think of the universe as a giant clock, a machine that gives measurement and meaning to time.

The metaphor of "Spaceship Earth" is now well known. For example, the American politician Adlai Stevenson in one of his last speeches before his death said in 1965, "We travel together on a little spaceship, dependent on its vulnerable supplies of air and soil...preserved from annihilation only by the care, the work, and...the love, we give our fragile craft" (McHale 1971–2: ii). This conception of all humanity hurtling through space on Spaceship Earth is a powerful image. But it is only part of the reality. The other, equally powerful metaphor necessary to complete the picture is of the Earth hurtling through time. The Earth itself is part of a "Time Machine."

The Measurement of Time

The astronomical clock. St. Augustine said, "What then is time? If no one asks me, I know; if I want to explain it to a questioner, I do not know" (Priestley 1968: 146). We still may not know what time is, but we do know how to measure it precisely. Our measurements of time begin with the observation of the apparent movements of the macro astronomical objects in the Earth's sky. From the beginning of human society, the concept of time was linked to the apparent rising and falling of the sun as the Earth rotated around its own axis, dividing time into day and night and the gradations within and between them. It was linked, too, to the revolutions of the moon around the Earth, giving us the time measurement of months. It was linked as well to the orbital revolutions of the Earth around the sun and, eventually, to the apparent movement of planets and stars.

Changes in many observable phenomena associated with these astronomical movements were noted early in human development: the movement of the ocean tides (the English word "time" has the same root as "tide"), the changes in the condition of the land and climate, the growth of plants, and the behavior of animals—all of which were more or less geared to the movements of nature's astronomical clock.

The association between the daily appearance and disappearance of the sun and the occurrence of light and warmth and then darkness and cold seems so obvious to us today that it is hardly worth remarking. Yet people at some point long ago in the prehistory of human development must have marveled at the connection. They must have wondered as they discovered the seemingly endless and complex interactions and influences of the motions of the objects in the sky and how they affected the changes in plant, animal and even human life on Earth.

Several different peoples in various parts of the globe discovered that they could predict the apparent movements of the sun, moon, planets, and stars and their intersections, such as eclipses, when they occurred. By observing the movements of astronomical bodies, they also discovered that they could predict the changes of the seasons—the cold, the warmth, the rain, the snows, and the dry periods. They could predict, for example, when it was the right time to hunt and to fish, to gather fruit, to plant, and to reap.

Most early peoples believed that various astronomical objects represented some supernatural powers and were gods and goddesses who could use these powers to affect human life. Thus, they believed that observed astronomical movements could be used to order human behavior and social life and from them predicted the best times to pray, to work, and to procreate, among other things, in the social ordering of human behavior. Many viewed astronomical interrelationships as determining the fate of individual people and the destinies of whole societies. In ancient times, much effort went into astrology and the casting of horoscopes.

Of course, in many ways human destiny *is* determined by the cosmos and the movements of astronomical objects. There is a fundamental truth here. For without the sun life on Earth would be impossible. Too much or too little heat and all life, as we Earthlings understand it, would die.

There really is a correct time to sow and a correct time to harvest. Plant at the wrong time and no crops will grow. Since animal behavior, too, is often time dependent—on seasons and climate, for example—there really *is* an ideal time to hunt and fish for certain game. Get it wrong and there is no food to eat.

On quite a different level of explanation, most of us respond to some natural conditions, such as the seasons and daily changes in the weather. We feel exhilarated on some days, say a brisk, clear, sunny day, and depressed on cloudy, hot and humid, or cold and dark days. Some of these conditions—for example, the amount of light our bodies receive— may work directly on our biological systems. Others may work through our conditioned, psychological responses. But each is a long way from believing that astral movements directly determine a person's personality, career, correct marriage partner, or fate in life. There appears to be no sound evidence to support these latter beliefs. They are superstitions.

The calendar. There are many gaps in our knowledge of the early development of time reckoning, but with the development of writing, the record of the past becomes readable. The use of a calendar, which is basically an effort to model some astronomical observations and keep a written account of them, occurred at least as far back as 2600 B.C. in Mesopotamia. All the early calendars were theocosmic, incorporating both religious beliefs and astronomical observations. Astronomy and astrology developed together, to some extent indistinguisably so, and so did supernatural beliefs about the nature of gods and goddesses.

The cosmic part of calendars is based on empirical observations and mathematical calculations. Objective astronomical realities, obviously, set limits to the construction of calendars based on them, whether the calendar builders were Egyptian, Greek, Mayan, or whoever. Thus, because calendar builders observed aspects of the same astronomical realities, there are many similarities in different calendars, even though they were independently invented.

For example, whatever their language and culture all calendar builders were constrained by such facts as the 365-plus days it takes the Earth to revolve around the sun (even if the revolution was perceived as the other way around, that is, the sun revolving around the Earth) and the twenty-nine-plus days it took the moon to revolve around the Earth. Also constraining was the fact that there was no simple way to coordinate the two. Thus, lunar calendars that ignored the sun got out of phase with the seasons, so that a given season such as summer would not occur at the same calendar time each year. Combining them within a single system of time measurement is not easy to do and lunisolar calendars are complex. Extra days have to be inserted in them from time to time.

Calendars had another part too. They were also magical and religious, invested with spiritual and moral significance. Social life was organized so as to conform to movements of astronomical objects as modeled by calendrical reckoning. Because these objects were believed to be deities, such social organization was, thus, sanctioned by the sacred beliefs of the group.

The Hindu calendar, for example, was and is consulted to know when to worship, when to feast and when to undertake all sorts of particular projects (Baumer 1976: 80). In medieval Europe, annual fairs were not only economic and social occasions, but also usually doubled as religious celebrations. They depended on some shared time calculations so that people from different places could gather together at specific times (Goody 1968: 35). Among the Greeks, too, the temporal and the moral orders were linked together, the notions of justice and the moral order being tied to time through both the seasons of the year and the seasons of man's life (Lloyd 1976: 121). As early as 255 B.C., the Chinese used the calendar to order both their public and private lives, so as to coordinate them with the presumed cosmic rhythm of the universe (Larre 1976: 59–60).

The Gregorian calendar is most widely used today, although often in conjunction with other calendars, such as the Hindu or Muslim calendars. It is shot through with religious significance, most obviously Christian by design, although still carrying many hidden or disguised features of the pagan origins of Christian beliefs. It's zero point, of course, is based on the presumed existence of Jesus Christ; thus, time is either "Before Christ" or "*Anno Domini*" ("in the year of our Lord," i.e., since Christ was born).

The Gregorian calendar is amazingly accurate. It measures a year with an error of less than half a minute. As we all know, it is still used to regulate religious life and also organizes political, economic, and social life as well, not only locally, but worldwide. Named after Pope Gregory XIII, who had requested its construction, it was first introduced in 1582, replacing the Julian calendar. The latter had served Western Europe for more than 1,500 years, having been invented by the Egyptians and introduced by Julius Caesar in 46 B.C.

All the great civilizations had a calendar, whether lunar, solar, or lunisolar, more or less well developed and reasonably accurate. Usually, there was a single calendar that combined and integrated the sacred and the secular. The Egyptians, Greeks, Romans, Hindus, Jews, Muslims, Chinese, Mayans, and Aztecs, among many other peoples, developed or revised and improved calendars. All constructed units such as the day, week, month, year, and longer periods of repetitive units, although with different names, of course, and sometimes—for example, the week—with different durations.

Two major efforts to eliminate the religious elements from the calendar failed. The French Revolution produced a new calendar that was adopted in 1793 and used until January 1, 1806. Known as the French Republican calendar, it attempted to remove all Christian (and, incidentally, earlier pagan) associations contained in the Gregorian calendar. The second attempt to expunge the religious elements was an aftermath of the Bolshevik Revolution in Russia. A new calendar was introduced and used between 1929 and 1940. It first had a five-day week and then was changed to a six-day week.

The failure of these two attempts to remove religious elements from the calendar ought not to be seen as a victory of religion over secularism. The calendars failed not so much as a result of their nonreligious character, but rather because of the consequences of the lack of social coordination they produced. In France, the French Republican calendar put the French out of synchronization with the people in other countries who still used the Gregorian calendar. In Russia, city dwellers who lived by the new calendar were out of synchronization with the country dwellers who stubbornly stuck to the traditional seven-day week (Zerubavel 1985: 42).

The mechanical clock and beyond. The calendar is an instrument of time reckoning that brings time orderliness to nearly all societies, social institutions, and social groups. It does so "by regulating the temporal location and rate of recurrence of socially significant collective events" (Zerubavel 1981: 31). But the calendar is of little help for ordering schedules within the day, that is, for micro-times like hours, minutes, and seconds. For such micro-temporal ordering, the continuum of time somehow has to be divided into smaller and smaller units, measured ever more precisely. Only then, is it possible to have ever more complicated and intricate time coordination of societal activities and ever more punctuality in the behavior of individuals.

Early technolgies for such micro-division of time were unsatisfactory. Sundials, sand clocks (e.g., the hour glass), and fire clocks had been long used, but each had drawbacks. Sundials, for example, were of no use at night or days with heavy cloud cover. Sand and fire clocks did not work for extended periods of time without frequent periodic human attention.

What led to a breakthrough was an unanticipated consequence of the endeavor to model astronomical movements more precisely by mechanical means, a breakthrough that allowed the measurement of particular durations of time, even if quite short, to be dove-tailed with the perpetual movements of astronomical bodies. As early as 979 A.D., the Chinese had water clocks (clepsydras) that were, in fact, more calendar than clock, their purposes being to coordinate actions and events with calendrical calculations and to perform astrological divination. Usually, they were constructed as physical models of moving parts imitating the movements of astronomical bodies, and, as an unintended side effect, they also gave the hours of the day. Europe, too, had such clocks.

This is a case in which an unintended consequence (the micro-measurement of time) of an action (the effort to physically model the movements of astronomical bodies) was beneficial to human purposes. The mechanical clock was invented and its human uses understood, at least in Europe. It spread in medieval Europe, the first unmistakable reports being in the fourteenth century. There were, among others, the tower clock of the Norwich Cathedral built by Roger Stoke about 1321–1325 A.D. and an astronomical clock at St. Albans started around 1330 A.D. and completed in 1364 (Landes 1983: 53).

The clock has been called "the key machine of the modern world" (Goody 1968: 33). It may not be quite as important as was the invention of fire and the wheel, but it brought far-reaching changes in technologies, the polity, economy, society, and culture, and even in the personalities of individuals (Landes 1983: 6).

The mechanical clock has obvious advantages compared with either the water clock or the sundial. Being weight-driven rather than water-driven, it does not freeze and stop in the winter. Also, unlike the sundial, it keeps working through the night and on cloudy days. It worked because of the energy of a falling weight through "a *gear train* (wheels and pinions)—what later generations were to call clockwork." The greatest invention of all was "the use of oscillatory (continuous back-and-forth, to-and-fro) motion to track the flow of time." The "oscillatory timekeeper was combined with that part of the mechanism known as the escape-ment" that blocked and released "the wheel train at a rhythm dictated by the regulator" (Landes 1983: 7–9).

In the West, the earliest clocks were used in monasteries. They made it possible to keep to a schedule of prayer, for example, eight times a day for Benedictine monks. Watching the clock, a timekeeper would strike the monastery bells at the appropriate times, signaling the monks that it was time to assemble for their Divine Offices. The word "clock" itself derives from the medieval Latin word for bell—"*clocca*" (Zerubavel 1981: 37, 38). Eventually, bells or other alarm systems were put on the clocks themselves (Landes 1983: 63, 69).

In England, from about 1100 to 1300 church bells were the dominant source of micro-time marking. But church bells were often within ear-shot of people in nearby communities and they increasingly were relied upon to schedule secular as well as religious activities, particularly in the towns. By the fourteenth century in England, some bells and clocks became totally secular devices and functioned as a part of solely commerical and civic activities (Thrift 1988: 74). In cities, bells marked the start and end of everything from sleep and work to meal breaks, drink service and curfews (Landes 1983: 59, 72). Eventually, every town had or aspired to its own clock tower.

Mechanical clocks permitted equal hours in all seasons and, eventually, as we all know, they became miniaturized and portable on our person, first as pocket watches and then as wristwatches. The inexpensive standardized watch became widely used starting about the middle of the nineteenth century, with production first in Geneva and then in the United States. The watch made possible individual time scheduling, "as distinct from the com-munal (and less flexible) timekeeping of the muezzin's call, the village drum, the church bell, and the town-hall clock" (Goody 1968: 33). In soci-eties where mechanical clocks and watches were widely used, time disci-pline became internalized in individuals as an ordinary and expected part of their personalities and being on time became a social norm.

Accuracy of time measurement was actively and somehat urgently sought by two groups of people: astronomers and people engaged in sea commerce. Astronomers came to realize that accurate time measurement was an indispensable tool of their observation of astronomical bodies. What we mean by motion is that something "changes position in space in a certain time interval" (Adam 1988: 207). Ship's officers needed accurate time measurement in order to navigate. Without it, they knew only their latitude (easily measured in the northern hemisphere by the altitude of the "North Star," Polaris) and could not calculate the longitudinal position of their ships at sea to get an exact fix on their position (Landes 1983: 104–5). This was considered to be so important that the British Parliament in 1713–14 offered an award of £20,000 to anyone who could devise an instrument to measure longitude. John Harrison eventually won the prize, with demonstrations both on a voyage to Jamaica in 1761–62 and another to Barbados in 1764. His chronometer was accurate to one tenth of a second per day (Hoghen 1968: 83; Landes 1983: 146–56).

With the coming of the railroads, the speed of travel over settled land was so increased that the variation in local times interfered with the construction of uniform schedules. Thus, on November 18, 1883 in the United States, for example, the nation's fifty regional time zones were converted into four. That day was known as "the day with two noons," one when the sun was directly overhead and a second according to "the new, uniform noon in the Eastern, Central, Mountain and Pacific zones of 'Standard Railway Time.'" The then mayor of Bangor, Maine, apparently unaware of the fact that his old regional time was no less a human construction than the new time zones, decried the change as "'an attempt to change the immutable laws of God Almighty'" (*New Haven Register* 18 November 1983: 40).

By the late nineteenth century, the accuracy in the measurement of time had greatly increased, error having been reduced to a hundredth and even a thousandth of a second a day as a result of the use of electricity. "At this point man's clocks were more accurate than nature's, and it was possible for the first time to detect, measure, and explain irregularities in the earth's rotation." By 1955, using atomic vibrations as controller (the first cesium resonator), time could be measured with an accuracy of only one second of error in a thousand years. In 1967, the General Conference on Weights and Measures did away with the old definition of the second as one 86,400th part of a solar day and redefined it as 9,192,631,770 periods of the radiation of the cesium-133 atom (Landes 1983: 186–87).

Finally, in 1984 researchers at the University of South Australia developed the world's most accurate clock—up to that time. The technology is called sapphire-loaded super conducting cavity and uses microwave signals resonating inside a two-inch sapphire crystal. The clock loses only one-100,000 millionths of a second a year and is 100,000 times more accurate than an atomic clock (*Australia News* 27 September 1984).

Meanings of Time

Events and processes. All normal adults have conceptions of time. We distinguish *now* from past and future, and we distinguish the past and future from each other too (Rappaport 1986: 4–5). We speak of physical time, biological time, psychological time, social time—and, sometimes, many other kinds of time as well. We speak, for example, of a time to play, a time to work, a time to eat, a time to sleep, a time of war, a time of peace, a time to live, and a time to die—a time to do this or that. We speak of time standing still or speeding up, of a few minutes seeming to last forever and of several hours flying by as if only a minute or two had passed.

None of these rhetorical devices, nor the perceived or subjective reality underlying them, is in any way incompatible with the fact that time has a reality and characteristics of its own—such as, for example, its relentless march on and on. Time is not, as the French sociologist Emile Durkheim would have us believe, merely a category of the human mind (Gell 1992).

Sometimes, when we speak of social time, we are referring to particular social events or processes which happen at a given time. For example, devout Christians observe Sundays as a special day for participating in religious services. But this is not to say that there is "religious *time*" apart from the underlying system of time measurement itself, but it is to say simply that a certain designated time, as measured by some agreed-upon way of time reckoning, is set aside for particular human activities that are religious.

Sorokin and Merton's (1990: 57) classic argument that astronomical time is only one of several concepts of time and that social collaboration is at the root of all social systems of time is fundamentally misleading, if it encourages the view that time is somehow "influenced by the number and importance of concrete events occurring in the particular period under observation." Such a view confounds time itself with the time choreography of social activities and with the human perception of time. Indeed, human activites, events and processes must take place at or during some specific time or times. But to label time by the human uses of it is to identify the activity, event or process, not time itself. Thus, when systems of time calculation are used for the purpose of coordinating human activity and making collaboration possible, the nature of time does not change.

When we speak of our human sense of time or how we perceive or experience time, we are talking about our subjective judgments about it. How much time certain activities subjectively *seem* to fill is notoriously dependent on what those activities are, how absorbed we are in them, how boring they are or how pleasurable or painful. As Alfred Gell (1992: 94) says, "our organic sense of durations is relatively unreliable." Sitting in a dentist's chair undergoing a painful procedure can seem interminably long, even if only a few minutes have elapsed. Contrariwise, engaging in a pleasurable experience in which a person can lose him- or herself can result in the perception that little time has passed even if hours, in fact, have gone by.

Research has also shown that time seems to fly by as we get older, while time passes slowly for young people. We do not know why this is so, although it has been proposed that each year that we live is a smaller percentage of our total time lived. Thus, for a ten-year old, a year is a full 10 percent of his or her total life so far. For a sixty-year old, a year is much less of his or her total life, less than 1.7 percent. This fact may affect perception of the speed of passing time. Each year we live is a smaller percentage of the total years of our lives.

Subjective experiences of time can be—and have been—objectively measured by asking standard questions of people. As a measure of clock time, however, subjective experiences are often inaccurate. The experience is an illusion compared to clock or astronomical time and people know that it is (Gell 1992: 316). Psychological experiments on time perception, for example, have focused on how different people differ on their estimates of the duration of time under different conditions, for example, when they are placed in different settings, asked to carry out different tasks, given different drugs, and so on. But we can— in fact in order to carry out such studies we *must*—"distinguish how long an interval 'seemed' to last (our perception of its duration) and how long it actually did last (the cognitive judgement we arrive at on the basis of all the information at our disposal, including, for example, making use of a clock)."

There are at least four temporal aspects of situations, events, processes and other phenomena that can be distinguished. There are "sequence," that is, in what order they take place, before or after; "duration," that is, how long they last; "temporal location," that is, when they take place according to some time scale, such as a calendar date; and "rate of recurrence," that is, how often they take place, as in the case, for example, of events that are repetitive (Zerubavel 1981: 1).

Every event and process may have its own appropriate time scale or typical duration. Some, such as the evolution of navigational capacities of birds, may be measured in millions of years. Others, such as changes in the human immune system, may be measured in hours or days. Neither define time as such. Both, rather, describe the rapidity of changes in the event or process under consideration. Time itself remains the same.

Time and social convention—the week. Human beings construct time measurements. However, with one possible exception, they do not do so arbitrarily, but incorporate constraints derived from observations of nature. The year, the month and the day—for that matter, the hour, the minute and the second—"are linked to observed relative movements of the sun, the moon and the earth" and in recent decades to the microscopic movement of atoms. The week is different. It may be a cultural artifact, resting mostly on social convention. Eviatar Zerubavel (1985: 4, 61) claims that its "invention was one of the first major attempts by humans to break away from being prisoners of nature and create an artificial world of their own." Note, for example, that a week differs from other units of time in that a new week does not usually begin at the beginning of a new year as does a new month, a new day, and a new hour.

Also, the week has had a variable number of days at different times and places. For example, it had three days in ancient Colombia and New Guinea, four in West Africa, five in ancient Mesoamerica and Indochina, eight in ancient Rome, ten in ancient Peru and Egypt, and twelve in ancient southern China (Zerubavel 1985). Thus, the week seems to be an arbitrary and primarily social convention as to the number of days it has contained.

For example, the week clearly has an important social function in filling a gap in the scheduling of human events left between the day and the month. The week appears to have been invented in connection with the regulation of economic transactions through periodic markets characteristic of peasant economies that occurred less frequently than daily but more frequently than monthly (Zerubavel 1985: 45).

But that conclusion raises a puzzle. If in fact it has no basis in nature, then how has it come about that today most of the peoples of the world use the seven-day week? The answer may be the convergence and rein-forcement of several fortuitous factors, including a partially mistaken—at least incomplete—notion of the universe.

First, there are two connections between the seven-day week and natural phenomena. One is the link to the various phases of the moon. As Zerubavel (1985: 8–9) notes, the religious Assyrian calendar as early as the seventh century B.C. recognized as "evil days" the seventh, fourteenth, nineteenth, twenty-first, and twenty-eighth days of a lunar month. More-over, using the quarter, half, three-quarters, and full moon as markers divides the moon cycle into four, and four divided into 29.5306—the length of elapsed time between any two new moons—comes to 7.38 days, or closer to seven than to any other whole number.

Another source of the seven-day week comes from the association with the early, though mistaken, belief that there were only seven astro-nomical objects in the sky that moved. They were called "planets," which derived from the Greek verb "planasthai" and which literally means "to wander" (Zerubavel 1985: 12).

Each astronomical "deity" had his or her own day: the day of the sun (Sunday), the moon (Monday), Mars (Tuesday, *Mardi* in French), Mer-cury (Wednesday, *Mercredi* in French), Jupiter (Thursday, *Jeudi* in French), Venus (Friday, *Vendredi* in French) and Saturn (Saturday). The reason that some of the English names don't appear to relate to the deity is that in the Germanic languages the names of the Nordic counterparts to the original Roman gods and goddesses were used. Mars was Tyr, Mercury was Woden-Odin, Jupiter was Thor-Donar-Thunar, and Venus was Fria-Frigg (Zerubavel 1985: 12).

Finally, there was the Jewish contribution of observance of the Sab-bath, which established a regular pattern of a seven-day cycle from one Sabbath to the next. This regular observance of seven-day intervals ap-pears to be a Jewish innovation. This seven-day system was preserved in Judaism, Christianity, and Islam, the most holy day in each case being Saturday, Sunday, and Friday. The biblical account of Creation, for ex-ample, gives a powerful justification for it: "And on the seventh day God finished His work which He had made; and He rested on the seventh day.... And God blessed the seventh day, and hallowed it" (Zerubavel 1985: 6–8).

But this religious belief may be derivative, itself a product of much older beliefs whose source was observation of the phases of the moon and movements of the "seven" planets. For example, the origins of the special significance of the number seven in liturgy, ritual, magic, and art goes back at least to the ancient civilizations of Mesopotamia.

With the spread of Islam and the European colonizations, of course, the seven-day cycle spread around the globe (Zerubavel 1985: 25–26).

Continuous time. The question of whether or not time is a continuum or is composed of discrete units goes back at least to the Greeks (Lloyd 1976: 145). The alternative hypotheses are still with us. Some modern physicists, for example, believe that time, like light, is composed of quanta, that is, tiny particles, rather than being a continuous flow. If so, one second would contain more quanta than all the seconds since the birth of the Earth, but this is sheer speculation (Landes 1983: 391).

If we can assume that time shares many of the properties of space, such as continuity, connectedness, orientability and a metric structure, then we can also adopt the assumption of its infinite divisibility. That is, we can subdivide it again and again without limit. Although at this stage of our knowledge it is a provisional proposition, we can envisage both space and time as "an infinite collection of points packed so closely together that they are *continuous*" (Davies 1977: 5).

Time is unidirectional. Is time unidirectional, circular or cyclical? Much has been written about time appearing to be circular, but it is not entirely clear what writers mean by the term. Do they really mean that time goes around in a circle or do they simply mean that there are cycles in which certain events recur again and again? Repetitions of similar events may be perceived—or, more accurately, misperceived—by people as circular. In peasant communities, for example, cyclical, if not circular, conceptions of time dominated—and still dominate—many aspects of social life. The future tended to be similar to the past in important ways. Thus, future activity could be anticipated as a recurrence of past experience—or at least the occurrence of markedly similar and familiar events that could be more or less predicted and often met quite adequately with past coping methods.

Menno Boldt, a sociologist who has spent decades working with Native Canadians (i.e., "Indians"), in a personal communication (19 October 1989) says that in some sense Canadian Indians do believe in circular time, that, for example, each summer is the same summer as previous summers. But, he adds, that this belief is similar to their view that individuals may age, change in various ways, and display different moods, etc. even though they remain constantly themselves. The spirit of a season, so Canadian Indians believe, "is subject to similar observable transformations but constantly remains itself." They seem also to have a conception of time moving on and on, however, even though they express it as distance. An elder, for example, has "covered a long distance."

By the time of the ancient cities, if not before, circular or cyclical conceptions of time were supplemented by conceptions of time as unidirectional. The increasing scale of society and centralization of political systems required planning, especially for such activities as "the storage of grain against famine, the construction of capital works, the build-up of supplies for military campaigns, and the organization of long-distance trade" (Goody 1968: 41). Such planning required notions of a manipulated or controlled future and of a time that moved, allowing for the occurrence of new and different events and states of the world.

The ancient Greeks are often cited as having a cyclical conception of time, but they also had linear conceptions. Although they recognized the cycles of daily, weekly, seasonal, and yearly movement as repetitive, they understood that youth is transient, aging is irreversible, and that death inevitably comes (Lloyd 1976).

A. J. Gurevich (1976: 239) argues that different conceptions of time coexist in most cultures, cyclical vs. linear, seasonal vs. directional. It is a matter both of a difference in what aspect of life is being considered and of emphasis. For example, in the Middle Ages, somewhat different conceptions of time were applied respectively to agrarian life, constructing genealogies of ancestors and descendants, understanding the Bible and liturgy, or writing history. In nonliterate societies where change is slow, most events may appear to be repetitive, and, where primitive religious beliefs include the present existence and possible future influence and intervention of dead ancestors, past, present, and future merge into one in the time perspectives of the people. In such societies, the emphasis is on cyclical conceptions of time.

Although cycles were still recognized, the coming of the Judeo- Christian religions highlighted conceptions of unidirectional time. The Jews, for example, have been called the "builders of time." For them, time was set in motion in the beginning, "and ever since then history has been moving forward irresistibly" (Neher 1976: 149, 152). The Christians, too, broke with the cyclical vision of the world. They believe that time "unfolds between the two poles of the two Advents of Christ, the one which took place at the time of Palestine and the one which will take place with the 'close' of the 'last day of the world'" (Pattaro 1976: 189). Thus, for Christians, there are "three decisive moments: the beginning, the apogee and the end of the human race. Time becomes definitively vectorial, linear and irreversible" (Gurevich 1976: 235).

Time, unlike the hands of a clock, is not going around in a circle or an ellipse or anything like them. Time itself is not cyclical. Rather, time is constantly moving on or out or beyond into new beginnings, whether things change or whether they stay the same (Gell 1992: 36). Even experts on the measurement and meaning of time have been confused on this point. For example, Zerubavel (1985: 83) mistakenly believes that the week as a measurement of time presupposes a circular conception of time. Not so.

Yet cycles are a basic part of nature, biology, and human societies. The seasons come and go and return. Women in the fertile ages normally menstruate cyclically. Patterns of daily traffic to and from work on week days, weekend variations in work behavior, the annual vacation, and periodic religious celebrations are all examples of cyclical behavior. *But cycles are not a type of time.* They are changes in events, activities or processes that occur in or with or during time. They are repetitive temporal sequences of phenomena. The *identical* time is never lived over again and the same *identical* event never occurs again.

Presidential elections in the United States, for example, may be held every four years, but each of the presidential elections is a different election. This summer is not the same summer as last summer. The sun rises every morning, but it is not the same sunrise as yesterday. Cycles in everyday life—whether based on natural or social rhythms—do not result in a system returning to the *exact* state it was in before.

In figure 3.1 I try to illustrate the difference between the false conception of time as circular, appearing to repeat itself endlessly in a recurring duplication of the past and the correct interpretation of cycles as events that move in or with time. Both the upper and lower parts of the figure show the seasons, the upper with time as a circle and the lower with events as cycles moving through or with time. Abstracting out from the seasons only the temperature in the lower part for the sake of illustration, we can see that the temperature varies in time as the seasons change, getting hotter in the summer and colder in the winter. Although the events (i.e., temperature changes) that are correlated with the seasons go through patterns of repetitive movements, time itself is neither circular nor cyclical. Time simply continues to move forward.

Time may be the "distance" between the beginning and the end of an event or a happening. That is, events and happenings take place in time or for a time. But we must distinguish between the temporal characteristics of events and happenings, however recurrent they may be, and time itself. Time keeps moving on.

Time is irreversible. Time is unidirectional and it is also irreversible. Time goes only in one direction, forward. It does not go backward.

Yet some physicists, at least until recently, believed that time could move either forward or backward. Newtonian physics, for example, allows for time-reversibility. The mathematics of mass, space and time give no privilege to the idea of the forward march of time. Dynamic equations such as those used to describe the motions of astronomical bodies in classical mechanics imply that the system can work in reverse and, therefore, go backward in time (Prigogine and Stengers 1984: 61).

But, as Barbara Adam (1988: 208) shows, this reversibility of time is achieved only by simplifying assumptions that are contrary-to-fact. One has to exclude friction, gravity, radiation and electro-magnetism, and interaction and complexity. Only then is one left with a universe of directionless time.

Even Einstein said that the irreversibility of time is an illusion. Thus, for him, there was no distinction between the past, the present, and the future (Prigogine and Stengers 1984: 15, 294), and his theory of relativity gives no privileged direction to time (Davies 1977: 70). Yet this conclusion again involves making a contrary-to-fact assumption: that speeds approaching the speed of light could in fact be achieved.

Einstein changed physicists' conception of time by making it relative to the framework of the observer, defining time as local and contextual. According to his special theory of relativity, "if anything could travel faster than light, it would go backward in time." But according to the same theory, this is impossible since the theory implies that nothing can travel faster than light (Morris 1984: 160). Of course, Einstein's special theory of relativity may not be a theory about the nature of time at all, but rather a theory about events and happenings, especially the behavior of light rays.

Beyond Einstein, some views of quantum physics that are concerned with the motions of individual microscopic particles include the idea

FIGURE 3.1
Representations of Circular Time Showing Seasons (above) and
Cyclical Events Showing Temperature (below)

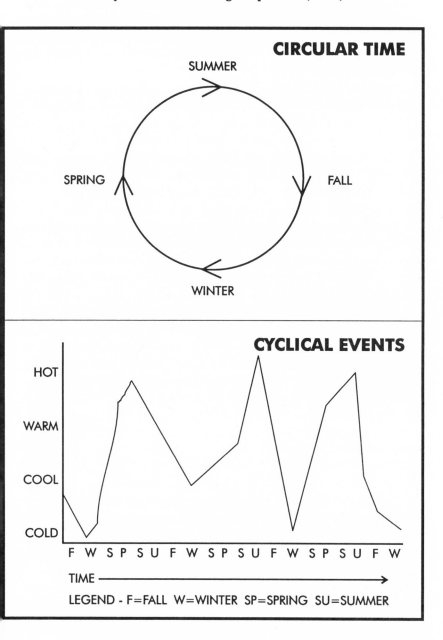

that "interesting objects could in theory move forward or backward through time" (Calder 1983: 16). Sequential time and causality—both of which are central to human experience—"cease to be meaningful at the sub-atomic level" (Adam 1988: 212). Thus, there is a sense in which Newtonian mechanics, the theory of relativity, and quantum physics all describe "the world as reversible, as static" (Prigogine and Stengers 1984: xxix).

Yet Adam (1988: 213) concludes that conceptions of past, present, and future are relevant for quantum theorists. She refers to the theory that matter cannot be separated from its activities and to the current belief in physics that "this activity-matter does not resemble the conventional conceptualisation of inert matter. Quantum physicists describe a 'virtual state' where electrons, for example, seem to try out all their possible futures instantaneously before settling down to exist, and later to disappear again."

There are other aspects to physics, however, where the physical world is shown unambiguously to be asymmetrical and characterized by unidirectional and irreversible time (Adam 1988: 213). Additionally, there are other good reasons to believe that time is irreversible and moves in a single direction. Let's briefly consider them.

The second law of thermodynamics. First, there is the second law of thermodynamics, the scientific equivalent, according to C.P. Snow, of Shakespeare's works in literature. It gives a one-way direction to time. It states that for "all isolated systems, the future is the direction of increasing entropy" (Prigogine and Stenders 1984: 119). "Energy may be conserved but not re-used for the same work" (Adam 1988: 213). It cannot be reversed, which confirms the experience of our everyday lives.

One way of expressing the second law is to refer to the everyday experiences of heat conduction. We know that heat flows from hot things to cool things, dissipating away. It is a time-asymmetric phenonemon because the reverse process of heat passing from cold to hot never occurs without some outside mechanism being used (Davies 1977: 64). The textbook example is that of a cold saucepan of water on a hot brick. Soon both saucepan and brick will be tepid. The tepid saucepan never sends its heat energy back into the tepid brick resulting in a cold saucepan and a hot brick. The second law concerns not only heat flow, but is of general applicability and describes irreversible time-asymmetric phenomena of many types (Davies 1977: 64).

Such an irreversible process, however, should be thought of as an underlying principle applying not to an individual system but to a total environment when thinking of everyday life, because there are few isolated systems in nature or society. Living things, whether plants or animals, are not isolated systems and they draw energy from other systems in their environment. For example, the sun provides the energy for biological activity but only at the cost of the increase in the entropy of sunlight (Davies 1977: 67). Similarly, social organizations and entire human societies are not isolated systems. They, too, draw energy from their environments. In fact, a fundamental measure of modernization and development of human societies, at least in the course of human history up to the present, is the amount of energy from inanimate sources per capita that they use. Thus, the evolution of living things and human organization tends toward increasing structure and complexity, because energy is drawn from exogenous sources.

Nonetheless, for the system as a whole, including all the sources of energy being drawn on in the surrounding environment, the trend toward entropy appears to be true. The total "universe tends toward a lukewarm uniformity" (Calder 1983: 17). If this is correct, then time's arrow points in only one direction, toward the future.

Although some of them also derive, ultimately, from the second law of thermodynamics, additional reasons for believing that time is irreversible should be stated explicitly. They are listed below.

Biological development. There are at least two ways that biological development contributes to the conclusion that time moves only in one direction. The first is biological evolution itself in that it is an example of asymmetric change. For example, natural selection results in changes in existing species as well as the elimination of some species and the creation of entirely new species. Thus, belief in the irreversibility of time is necessary to the belief that nature produced living beings through evolutionary processes, including humans (Prigogine and Stengers 1984: 96).

The second is the biological development of particular organisms and other forms of life during maturation in their own lifetimes. For example, we humans start life by emerging from our mother's wombs as babies. With time, we slowly grow taller, heavier, and more capable as we become young adults. If we are lucky enough to have a long life, we continue to change through time into middle and then old age. Finally, at a later time, we die. We never see a decayed body emerge from the grave, become an old person, grow younger with the passing of time, become a baby, and, then, disappear into a mother's womb (Davies 1977: 60–61). Every living entity has a biological clock or clocks, including circadian rhythms that order the repetitive events of life's biology. In mammals, chemical changes create signals preparing the organism to wake, to eat, to make hormones, to sleep, and so on. They order, too, nonrepetitive changes and determine chemical and biological aspects of an organism's life's course, resulting in aging and ending ultimately in death.

Wave motion. Another asymmetric process that does not depend directly on thermodynamics is wave motion. When we throw a stone into a pond, for example, we see the immediate impact and then a resulting disturbance in the water, a set of circular waves that spread outward from the point of impact toward the edges of the pond. We never see the reverse happening. That is, we never see waves spontaneously starting at the edges of the pond, converging toward the center, and then disappearing (Davies 1977: 83).

Similarly, radio waves are another example of time asymmetry. They are broadcast from a transmitter and disperse in all directions out into space at the speed of light. The reverse process has never been observed, that is, where radio waves spontaneously arrive from outer space, coming in from all directions and converging into a radio transmitter. "A radio message is never received *before* it is sent" (Davies 1977: 62).

The history of the universe. According to the well-known calculations of Archbishop James Ussher and the Bible scholar John Lightfoot, the Earth was created at 9 A.M. on October 23, 4004 B.C. Today, we know that they were off by quite a few billion years.

Most twentieth-century physicists have come to believe in what has become known as the "Big Bang" theory of the creation of the universe. It includes the belief that, about 13.5 billion years ago, the universe was in a hot dense state and could not have been in a steady-state (Davies 1977: 189). The universe, according to this theory, had a beginning, is expanding and, therefore, changing, and may have an end. This means that the universe has a history and it requires the belief in the irreversibility of time. The discovery of a residual black-body radiation empirically supports the Big Bang theory.

Whether an end will come in a Big Crunch or whether there will be an endless cosmological expansion is debatable, but in principle the Big Bang theory of creation and the current expansion of the universe seem reasonably well established today. The universe has a history and, just as the totality of space is the universe, the "totality of time is the history of the universe" (Davies 1977: 141). The time dimension in that history has sequence, duration, and direction. (See table 3.1.)

Other traces of the past. There are other traces of the past in the present, from footprints embedded in rock and growth rings of trees to recorded history, archaeological remains—both fossils and relics, and geological layers of the Earth. There are no such traces of the future. Thus, we have present evidence that there was a past. In our own lives, we have memories of it. Although we have anticipations, expectations and images of the future, we have no memories of the future.

Finally, research that relates to the meaning of time is continuing. Ilya Prigogine and Isabelle Stengers (1984: xxviii, 177, 232) believe that the physical sciences are in the process of rediscovering time, that quantum mechanics occupies a kind of intermediate position in the reconceptualization of physics now going on, and that the representations of physical realities in physics are "moving from deterministic, reversible processes to stochastic [probabilistic] and irreversible ones." Eventually, mainstream physics may be brought into line with the facts of life, and the one-way reality of cosmic time may be related to the fundamental character of cosmic forces (Calder 1983: 17).

Two conceptions of time. For some decades there was a debate concerning the relative merits and reality of two different conceptions of time. One was called the A-series which involves a past-present-future perspective of time. According to this view, events are classified as being in the future at some time, in the present at another, later, time, and in the past at still another, even later, time. Although all events are one or another of these, which of them they are changes. For example, any event that actually happens was a future event before its occurrence, then, as it occurs, it is a present event, and, finally, it becomes a past event (Gell 1992: 151).

A second conception of time emphasizes not this past-present-future aspect of time, but a before-after perspective. According to this view, known as the B-series, events are classified as being before or after one another in time and this classification does not change. If an event is after another event in time, then it remains after it at all times, forever.

Earlier, some writers believed that A-series time (past-present-future) was how individuals experienced time, but that it was largely a subjec-

TABLE 3.1
History of the Universe: Above—Looking Back in Time from Now.
Below—Looking Forward from the Beginning of Time.

Feature	Age in Years
Technological culture	100
Civilization	10,000
Humankind	5 million
Mammals	200 million
Terrestrial life	3 billion
Earth	4½ billion
Universe	13½ billion

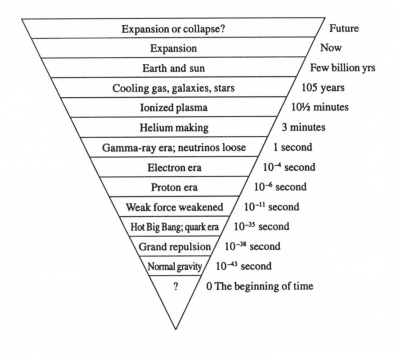

Expansion or collapse?	Future
Expansion	Now
Earth and sun	Few billion yrs
Cooling gas, galaxies, stars	105 years
Ionized plasma	10½ minutes
Helium making	3 minutes
Gamma-ray era; neutrinos loose	1 second
Electron era	10^{-4} second
Proton era	10^{-6} second
Weak force weakened	10^{-11} second
Hot Big Bang; quark era	10^{-35} second
Grand repulsion	10^{-38} second
Normal gravity	10^{-43} second
?	0 The beginning of time

Sources: Adapted from Davies (1977: 156) and Calder (1983: 98–100)

tive definition of time with little basis in scientific fact. B-series time, these same writers believed, was objective and scientific (Gell 1992).

Contrary to these views, both A-series and B-series conceptions of time are equally valid as objective conceptions of time and Prigogine has established both as laws of nature (Adam 1988). Both define a sense of time that is experienced and understood by human actors, each is compatible with the other, and both are founded on empirical observations.

Thus, the A-series and the B-series conceptions of time are not two different kinds of time. They are, rather, simply two ways of looking at the same phenomenon of time. Events can be dated and the date informs us which events occurred before or after others. The same dates also tell us, from the vantage point of the present, which dates refer to the past and have already happened and which dates refer to the future and to possible events that have not yet happened and which may never happen at all.

The concept of the "extended present." Astronomical time keeps relentlessly moving on. Clock time keeps ticking or pulsating away. Both invite a conception of the present that is only the briefest of moments, a knife-edge between the past and the future, between the last tick and the next tock. There is a reality to this conception and for some purposes— in dealing with some machines or electronics, for example, or in setting a micro-date and micro-time of some occurrence, as in astronomical navigation—it may be useful.

Yet living beings—at least human beings—seldom actually experience the present as knife-edged. Rather, we tend to experience an "extended present" that includes not only the knife-edge of the present, but also the immediate future and immediate past, some duration of time on either side of the passing moment as well. Most of our experiences of the present include a length or duration of time, a kind of freezing and prolonging of our sensations of the present. We take some portion of the continuous flow of time "as being a minute or a year, even a millennium, and in a particular context seize on that as the present" (Young and Schuller 1988: 4).

Edmund Husserl (1966) as early as 1887 proposed the concept of the extended present that includes not only the now, that is, the present moment, but also both memories of the immediate past, that he labeled retention, and anticipations of the immediate future, that he labeled protention. He speaks of the horizons of a temporally extended present.

This gives "thickness" and "temporal spread" to the present. Retentions and protentions are past and future phases of current events, that fuse with them to create consciousness of the present (Gell 1992: 223–24).

Ulric Neisser (1976) speaks of the three phases of the "perceptual cycle" each of which corresponds to a faculty of the mind: memory of the immediate past, perception of the immediate present, and imagination or foresight of the immediate future. These three faculties are not viewed as independent, but as interacting parts of a single process. Perception, thus, is not viewed as passive reception of external signals. Rather, it is a dynamic process in which memory functions as a frame of reference giving meaning to perceptual inputs and in which foresight functions to direct search and scanning movements that produce the perceptual inputs (Gell 1992: 232).

Modern futurists have also proposed the idea of the extended present. For example, Richard A. Slaughter (1993c: 294), following the lead of Elise Boulding, suggests the notion of a 200-year present. Such a time span connects us directly with our parents and grandparents, our aunts and uncles and grand-aunts and grand-uncles, and even with people of the earlier generation of our great grandparents. It also connects us with our children, grandchildren, and great grandchildren and the other kin of their generations too. Such a time-chain of family relationships when used to define a relevant present for social action and decision making might help reduce the undesirable consequences of short-term thinking.

Of course, every event or process in the world may have its own appropriate extended present in the sense of providing a distinctive frame of reference for thinking about what the proper duration of "the present" ought to be. To say this is not to deny the reality of the knife-edge of the now, but rather to add the human need to understand appropriate time durations for different phenomena in order to design effective action. Remember that even a knife-edge has *some* thickness and is itself an appropriate extended present for some purposes.

Key Assumptions

Key Assumption 1: The Meaning of Time

The first key assumption involves the meaning of time. To summarize: *Time is continuous, linear, unidirectional and irreversible. Events*

*occur in time before or after other events and the continuum of time
defines the past, present and future.*

One final cautionary note about the measurement and meaning of time: Human conceptions of time rest in part on cosmology which is, after all, about the secrets of the universe. We humans do not yet know all—or even most—of those secrets. Although we have discovered the sun's revolution around the galaxy, cosmic expansion, black holes, dark matter that remains unseen and that may constitute up to 99 percent of the universe, thousands of galaxies other than our Milky Way, quasars, and the microwave echo of the Big Bang, many mysteries remain. We are learning more all the time, especially as a result of continued research and data from such sources as the New Technology Telescope in Chile, the Keck Telescope in Hawaii, unmanned observational probes into space, and the Earth-orbiting observatories, such as the Roentgen Satellite, the Hubble Space Telescope and the Cosmic Background Explorer. In a decade or two, new theories and understandings will surely be formulated and accepted that may affect human conceptions of time. Yet from what we now know it is reasonable to believe that they will be consistent with the assumption that time is continuous, linear, unidirectional and irreversible.

Key Assumption 2: The Possible Singularity of the Future

The second assumption is: *Not everything that will exist has existed or does exist.*

Past time may not be a good sample of all time. If we are not continually reliving past time, nor always facing again the exact events that have occurred before, and if we are not moving in circular, repetitive time, then the future as it becomes the present contains some events, happenings, processes, structures—perhaps an unlimited number of things—that have never occurred or existed before. The future, as it becomes the present, contains new time. It, thus, may contain some new, unprecedented things as well.

Without social change, of course, societies and the repetitive patterns of social behavior that define them remain nearly the same, subject only to random variations and errors in behavior that more or less average out around a norm and, perhaps, leave no permanent mark. Even when there is change, if it is slow enough, people may not notice the minor adaptations that they make and how their present customs slowly drift away from their past behavior. Moreover, without written records the past may be continually redefined as similar to the present, even if, in fact, it wasn't. Thus, "living memory" and oral history are themselves partly shaped by present practices and may be inaccurate accounts of past ways.

But when change is rapid, then the disjunction between the past and the future is great and people, especially where written records exist, become aware of the social transformations, both great and small, that have taken place. The past then can be used to confront the present.

Without rapid social change, our cognitive maps of the past and present work reasonably well for the future, since the future is similar to the past— or at least is similar enough that the maps, perhaps with minor adjustments, still get us where we want to go. But with rapid change, our cognitive maps of the past and present may be so out of date that they no longer are accurate or even recognizable representations of the future as it becomes present reality (Jouvenel 1967: 10). When that happens, we need to construct new maps that correspond to the changed world, if we wish to act effectively. The awareness of the disjunction and the process of such construction obviously raise doubts about the moral bases of society and invite both social criticism and the effort to justify traditions in the face of new ways of doing things. In this way, crises of legitimacy of authority occur.

Some cultural traits may become extinct while others have a tendency to persist. Our knowledge of past traditions, even if not their present acceptance and practice, continues to expand as archaeological researchers push our conscious time-scales into the past. Much of the past is still living today. Thus, the present is a diverse and wondrous kaleidoscope of ideas, artifacts, technologies, and behavior patterns, some of which, though modified and adapted, originated in many different ages of the past, whether or not we are conscious of it. They coexist with the new, which, no matter how standardized in particular cases, occurs in such prolific heterogeneity overall as to add to the variety. The unprecedented new mix of beliefs, practices, and products itself fosters still more creation and renewal, speeding up the rate of change and creating new possibilities and alternatives of choice for human life.

Key Assumption 3: Futures Thinking and Action

The third assumption is: *Futures thinking is essential for human action.*

We cannot consciously act without thinking about the future. Although we may be able to "react" without images of the future, we cannot "act" without them. Action requires anticipation. People have reasons to act. In this sense, action "is explained by its final cause, its goal" (Jouvenal 1967: 26–7).

Of course, there are also causes—that is, explanations—of human behavior other than the goals or purposes of behavior. We can—and some social scientists do—ask how people acquire their goals and how actions are shaped by structural factors such as social class and organizational forms.

Yet any human response that can be dignified with the description "action" rather than mere behavioral ticks, reflexes, or habitual mannerisms requires some futures thinking, because people necessarily "anticipate the effects of their actions" (Ferkiss 1977: 5). "We cannot be viable human beings," Peter Medawar (1984: 298) says, "without taking some view of what will happen in the future." This has been known at least since Aristotle who believed that deliberate human action is performed for the sake of some end to be brought about in the future. Because of the arrow of time, the results of our actions are always in the future, at some time after the action. Thus, all normal, mentally unimpaired human beings have learned that they must look to the future for the outcomes of their actions.

In order to act we must decide what action to take. That is, we choose an action in a given situation from among a number of possible alternative actions. Thus, decision making is part of the process of action and it is explicitly future-oriented (Snyder et al. 1962: 90).

But could it be otherwise? Can we make a decision from among alternative courses of action without considering the future consequences of those actions? Generally, no. The only instance where the future isn't considered, however short-sighted such consideration is, may be in the case of a "meta-decision," when we do not consciously make a decision because we are not aware of the fact that we even have a choice. How many children from inner city schools, for example, fail to excel in school because they do not believe that they can go to college anyway and, thus, have no reason to try? And how many of them, then, cannot get accepted by any college because they did not do well in school? This, of course, is exactly where futurists want to play a role in helping people to see what the real possibilities for their futures are.

In other cases of nondecision, however, in which the choices are known and understood, avoiding making a decision is itself a decision. That is, deciding not to act is itself an action.

Although futurists believe people do—indeed, inevitably must—engage in futures thinking, they also believe that most people most of the time don't do it as well as they might and can learn to do it better. Generally, they do it with little data, no explicit methodology, no rigor and no constructive criticism by others (Jouvenel 1967: 277). Moreover, people often fail to think far enough into the future, focusing only on the short term. These criticisms apply to individual decisions in people's personal and family lives as well as to people's collective decisions.

It is not rational, of course, to spend more time, money, and effort on research in order to make the right decision than the cost would be of making the wrong decision. Thus, the costs of any research—including futures research—must be trimmed to be lower than the expected net gains from doing it.

Furthermore, there are situations in which the speed of decision is crucial. When you must decide now or else lose some opportunity, then it may be too costly—or impossible—to stop and engage in careful analysis of the costs and benefits of different alternatives. Instead, you make your best guess and decide. Ideally, you would have planned ahead for improbable contingencies and worked out prior to need what action you would take if one or another unlikely emergency occurred.

Key Assumption 4: The Most Useful Knowledge

The fourth assumption is derivable, in part, from the foregoing assumption. For example, Dror (1975: 148), acknowledging that "every conscious and goal-oriented decison-making process depends on knowledge about the future," goes on to say that, therefore, "if such decision-making is accepted as possible and desirable, the assumption about knowledge of the future being instrumentally useful is acceptable."

Underlying much futurist work is the following assumption: *In making our way in the world, both individually and collectively, the most useful knowledge is "knowledge of the future."*

It is the future that we need to know in order to make our way through our daily lives effectively. We steer ourselves through time, as well as through physical and social space, according to our goals, our expectations of future happenings, and our anticipations of the possible and probable future time trajectories of other people. The more we know about the possibilities and probabilities of the coming events of the future, the better able we are to plan the actions that create our lives.

As we saw in chapter 2, in their everyday lives people constantly make predictions and act on them. They organize their lives by train, bus, and airplane schedules; by the ebb and flow of ocean tides; by when the football game or newscast is scheduled to be shown on radio or television; by the openings and closings of business establishments, from banks and department stores to restaurants and theaters; by the schedules of schools and universities; and by the announced timing of civic events, such as in the United States the local Fourth of July parade. People act on expectations about the future behavior of the weather, the stock market, the cost of housing, and interest rates on loans. They organize their own actions according to their anticipations of the boss's reaction to their asking for a raise or a few extra days off, the day the garbage gets picked up, when church services will be held, and the chances of getting into medical school after college.

Even a common occurrence such as driving a car requires, for sensible action, short-term predictions about where other vehicles will be moving, whether, for example, an oncoming car will stay on its own side of the road. Hopes and fears, expectations for the future, and predictions of the behavior of others govern perceptions of the options for action that people have and the choices that they make among them.

Our knowledge of the physical and the natural worlds, of cause and effect, contributes to our ability to navigate through time. So does our knowledge of the social world. For example, our understanding of social norms, social roles, and the organizational interrelationships of roles permit us to anticipate the behavior of other people with considerable success. We couldn't make our way effectively in the social world if that were not true. If all behavior were random, it would not be safe to get out of bed in the morning—or even to stay *in* bed.

Yet, despite the basic orderliness of society, there are exceptions to the norms. That approaching car usually, almost always, stays in its own lane. Yet sometimes it comes crashing into the wrong lane and into oncoming cars. Thus, we literally bet our lives each day—or in less dangerous situations than on the highway, at least some part of our well-being—that we can know reasonably well the future behaviors of other people, for example, drivers of other cars, family members, friends, barbers, bank clerks, teachers, students, airline pilots, policemen, doctors, street vendors, grocery store customers...whomever. However, those behaviors themselves are not carved in stone. There are exceptions to them. Deviant behavior as well as conformity exists, and the deviance may be desired and welcomed innovations as well as undesirable anarchic, antisocial or self-destructive behavior.

Life, in this view, is a series of gambles in which we are forced to keep making bets about future outcomes and the likelihood of given actions producing them. Every "bet has an expectation, or expected value" (Edwards 1968: 35). Feedback from our past actions is made available to us by knowing their consequences and we use it to try to reduce the uncertainties that we face as we make our next decisions to act. Part of this process involves changing these expectations, that is, our assessments of alternative outcomes of our own and others' possible actions, where warranted by past experience.

Although we generally use our knowledge of the past as a guide, sometimes we need to forget the past, because past ways of doing things are not always adequate to the changing situations that confront us (Shils 1981). There may be more effective, efficient, and equitable ways of doing things than the old ways of the past, both because new and better solutions to old problems may be invented and because there are new problems that we have never faced before, especially in a rapidly changing society, that require new solutions. Sometimes, "the present has to engage in a deliberate struggle with older modes of thought and action" if a better future is to be achieved (Goody 1968: 40).

Clearly, we use assertions about the future in order to make our way in the world. Yet there are times when some people say that they wish they didn't know the future. Occasionally, people say this when expectations of the future contain bad news, especially seemingly inevitable bad news (Dror 1975: 148). For example, the wife of a man mentally and incurably crippled by Alzheimer's disease who has seen her husband change from an intelligent, independent, dignfied human being into a dependent, mentally disabled individual may believe that it was a blessing that she didn't foresee it.

If the future contains some dreadful thing, then is a person really better off not knowing? Most futurists would say no. It is better to know the future than not to know, even in such cases. One reason is that the dreaded future may not actually be inevitable. Perhaps something can be done to avoid it. Another reason is that, even if the terrible future event cannot be prevented, at least people can prepare themselves for it. It may help the victim and his or her loved ones to prepare psychologically, permit the victim to do some things he or she wants to do while there is still time, or allow the victim simply to get his or her affairs in order.

Finally, in most circumstances of bad news individuals ought to be told the truth as a matter of respecting their rights and dignity.

In sum, futurists agree that the past should be studied for whatever insights it may offer that help understand the present and guide action aimed at shaping the future. In fact, some methods of futures research depend on it, as I explain in chapter 6. But the past seldom contains all— or even most—of the information needed to deal effectively with the new situations of the future.

In any event, as a guide to the future, knowledge of the past cannot be accepted as it is. We must somehow transform it into sound, plausible, and relevent assertions about the future. We can begin by testing our knowledge of the past to see if it is, in fact, correct. We can critically examine our sources and evaluate our interpretations. Next, we can question the applicability of past knowledge as an analogy for planning present action in some particular situation. Then, we must transform images of the past, imaginatively and speculatively, into assertions about the future consequences of action. Finally, we must search for any additional empirical, logical, or theoretical evidence that may refute our assertions. If they survive these tests, only then should assertions about the future be accepted as warranted. It is such transformed assertions that people use to design present action.

Thus, we remember the past, if we want to have a future approximately like some aspect of the past. We remember it, too, if we want the future to be different, because we don't want to repeat past mistakes. In either case, our knowledge of the past can provide some tests of our beliefs about the future. But our assertions about the future depend both on logical deductions and on speculative leaps that transform knowledge of the past.

Sometimes, though, we must forget the past and transcend the present, if we want to create a new, different, more desirable future. In such cases, knowledge of the past may be more of an obstacle than a help, acting like blinders and narrowing our vision. If so, we need to invent and imagine alternative futures that have never existed before.

Whether we are remembering or forgetting the past, the most useful knowledge for the design of effective action is "knowledge of the future."

Key Assumption 5: Future Facts?

This leads us to key assumption five: *The future is nonevidential and cannot be observed; therefore there are no facts about the future.* Although the most useful knowledge in making our way through life is knowledge of the future, how can we have any knowledge of the future if there are no future facts? Must we conclude that there is no knowledge of the future? How can we study and make justified statements about that which does not exist? Here is a seeming paradox of futures studies.

This assumption may be one of the few things on which almost all futurists are in agreement. For example, Lasswell (1951: 11) says that events "in the future are not knowable with absolute certainty in advance." Jouvenal (1967: 5) says that "the expression 'knowledge of the future' is a contradiction in terms. Strictly speaking, only *facta* can be known." Malaska (1995: 80) says that the essence of the futures field "is the lack of futures objects to be observed. The concept of future means something 'not yet and nowhere.'" Thus, in stating assumption four I put quotation marks around "knowledge of the future" to acknowledge the correctness of the proposition that there are no future facts.

There are past facts, present options, and present possibilities for the future.

There are no past possibilities and there are no future facts.

Hans Reichenbach (1951: 241) says, "A statement about the future cannot be uttered with the claim that it is true; we can always imagine that the contrary will happen, and we have no guarantee that future experience will not present to us as real what is imagination today." Brumbaugh (1966: 649) concurs that the future differs from the past and present, but emphasizes that present options and possibilities are real: "there is a genuine ontological difference in the kind and the definiteness of being which past facts, present options, and future possibilities possess.... A 'choice now open' or a 'possibility' has *some* ontological status; it is not a pure Parmenidean nonentity. It therefore also has *some*, though indefinite, correspondence with the propositions asserted about it."

The future is a domain of uncertainty. It is problematic. As cognizant beings and as researchers of the future, we face considerable trouble from this fact. We want to make statements about the future that are as accurate as possible so that we can rely on them to design our actions. We know that such statements in the future or future conditional tenses do not constitute justified knowledge in the strict sense, because the future is not factual until it has become the past. Thus, we do the best we can, formulating assertions about the future—usually a range of alternative futures—on which we can act *as if they were true*. "It is all very well to say that the future is unknown. The fact remains that we treat many aspects of it as known, and if we did not we could never form any projects" (Jouvenal 1967: 41).

Yet there are some things about the future that appear to be so certain that for all practical purposes we can rely on them. The sun will surely appear to rise tomorrow morning. The stars will be in their places next June 11 at 2100 hours universal standard time as predicted, so that astronomical navigators can locate their ships at sea. Taxes will surely be collected in France next year. Death still looms as our individual destiny, although we don't know how, when, or where it will happen. In the year 1704 "the English astronomer and mathematician Edmund Halley predicted that the comet which now bears his name would return in 1758." He was right. Modern astronomers told us for years before the event that we would see the comet come near us again in 1986 (Medawar 1984: 298). They, too, were right. Moreover, for many people in the routine ruts of their daily lives the future may be all too predictable: every day the same grind of weekday travel to work, working at the same place with the same people, doing the same thing, week after week, month after month, and year after year.

These and many other future events appear to be sure bets. Yet, no matter how sure, they are still bets. Although it may be unimaginable to us that the sun would not appear to rise every morning for the next century and beyond, who knows what weird and unlikely astronomical event might occur? Who knows when illness or economic recession might intrude into the routines of our daily lives. We might be killed in an auto accident tomorrow or be laid off from our job because of events beyond our control. No matter how certain a predicted future event appears to be, it is not a fact—it is not absolutely certain—until it occurs. There is no way that we can empirically confirm the occurrence of a future event as long as it remains in the future, because we cannot observe an event before it occurs (Dunne 1927: viii). What we can do in order to increase the effectiveness of our actions is to increase the probabilities that our statements about future happenings are *presumptively* true at the time we assert them. We can formulate conjectures about the future and subject them to various tests of logic and fact.

In sum, we have seen that the knowledge of the past and present that we use to help us orient ourselves toward the future is not used directly. By making deductions, we transform such knowledge into assertions about the future—predictions, conditional statements, foresight—on which we can base our decisions, plans, and actions. Such assertions are not justified knowledge in the strict sense because they cannot be verified until after the future time arrives to which they refer. But they can be accepted as conjectural knowledge if our serious efforts to falsify them fail, as I explain in chapter 5. Thus, we can use them as a basis for our decisions to act.

This fifth assumption, then, serves as a constant reminder that assertions about the future can never be accepted as absolutely certain. They are not facts. But, if they pass objective and rational tests, they are conjectural knowledge, that is, presumptively true, and can be used to design action *as if* they were true.

Key Assumption 6: An Open Future

Following Amara (1981a), I state the sixth assumption as follows: *The future is not totally predetermined.*

Futurists do not think of the future as fixed and existing out there in time bearing down on us in some inevitable way. Quite the contrary is the case. The future offers opportunity and contingency. Thus, we don't "discover" the future, because there is no pre-existing future "out there" to discover. "The arrow of time is a manifestation of the fact that the future is not given." Within the limits of the possible, the future is open (Prigogine and Stengers 1984: 16).

Thus, as living, breathing, purposive individuals, the fact that the future is a domain of uncertainty and openness gives us hope. For people in their roles as active agents who can choose, decide, and act, the future is a domain of liberty and power. Liberty because people are free to "conceive that something which does not now exist will exist in the future."

And power because people have "some power to validate" their "conception" through their willful actions. (Jouvenal 1967: 5).

The future is a domain of liberty not simply because we cannot know the future in any certain sense. It is also because the future itself is contingent, not only our knowing of it. What will happen depends on many things, including what we ourselves choose to do. Obviously, some aspects of the future are more open than others. The future from this point of view is an assemblage of different possibilities, contingencies, near certainties and uncertainties, constraints and opportunities, some more probable than others.

This raises the question of free will versus determinism. Some futurists, such as Solomon Encel et al. (1975: 35) prefer to sidestep this question, simply pointing out that no one has really solved the problem over the last 5,000 years or so, but that fact has not prevented us from planning and forecasting in useful ways. Perhaps, however, there is nothing to sidestep, because this confrontation may be a false problem in that free will and determinism may be perfectly compatible views.

In some ways, human beings have choice. They make choices within limits among various alternatives for the future, sometimes creating some of the alternatives themselves. This is not to say that their choices and decisions to act, and the goals and values that they reflect, don't have explanations. They do, and in that sense they are "determined."

The fact that social scientists can predict the aggregate behavior of individuals does not mean that the individuals who make up the aggregate are not choosing and deciding on their behavior. For example, sociologists know that the chances are very high that on average an individual will marry another individual of the same religion. Catholics tend to marry Catholics, Muslims tend to marry Muslims, Hindus tend to marry Hindus, Protestants tend to marry Protestants, and so on. Although in any given case, a marriage may take place between people of different religions, it is an unlikely event considering all of the marriages taking place. But does this statistical regularity mean people aren't making choices and decisions, whether the decision is largely being made by the marriage couple itself or the families involved? No. Both the probabalistic determinism of the cultural explanation underlying the pattern of mate selection and the choice and decision exist co-jointly.

Even more likely than marrying someone of the same religion is marrying someone of the opposite sex. There are cases of homosexual relationships where people of the same sex marry—or de facto marry—but there are strong social constraints that tend to limit marriage partners to persons of the opposite sex. Choice of a marriage partner usually takes place within the limits that define a field of socially acceptable eligible partners. Still more likely—perhaps a certainty—is that if two human beings reproduce and between them create a baby, one parent must be male and the other female. In this case this biological constraint defines the possible.

It is within the class of possible events that probable and preferable events, choice and decision making take place. Thus, I've added "totally" to Amara's formulation of the sixth assumption to acknowledge the existence of constraints, some of which cannot be transcended.

To say that something is determined is not necessarily to say that it is inevitable. It is simply to say that some causes and effects can be identified with some degree of probable connection under specified conditions. Some things may be so certain as to seem inevitable, but contingent events also have causes. Cause and effect can be seen in the statement, "If you drink a glass of hemlock, then you will die." It may not be entirely up to you whether you drink it or not, but the conditions leading to your drinking it, if you do, have explanations. Such explanations allow for contingencies, both because probabilities may be attached to them which allow for exceptions and because they are often conditional, being stated as "If, then" statements, or, if they are more powerful explanations, as "If and only if, then" statements. That is, you must do x, if you want y to happen; or you must do x, and only x, if you want y to happen. We all use such conditional cause-and-effect statements implicitly in the mundane decisions of our daily lives.

The philosophical position of "compatibilism" bears on the freedom-determinism debate. As its name implies, it holds that freedom and determinism are compatible. The crucial issue, compatibilist determinists argue, is not whether or not x is caused, but whether or not the causes that bring x about include our desires and actions. Compatibilists say that we need to distinguish determinism (i.e., the thesis of universal causation) from fatalism (i.e., the thesis that everything is predetermined independently of what we want or do). In the latter sense, for example, everybody's eventual death is both determined and inevitably fated. Yet the manner of our death will be determined (i.e., caused), even though it is not fated that we will die in a particular way and at a particular place and time.

Our actual death may be caused by the coming together of many different causal agents, including people's own free choices and behavior, for example by their life styles, type of work, recklessness, and so on. Smoking cigarettes increases the risk of dying of lung cancer. Knowing that causal law, people can lower their risk of getting lung cancer by not smoking. Freedom, thus, is not independent of causal laws, but arises from the possibility of people gaining knowledge of these laws and making them work toward their own ends (Mills 1989).

Also, probability enters into our world in at least three ways. The first, derived from past experience, is the objective probability attached to a given scientific law, such as the examples given above about the chances of a person marrying someone of the same religion or opposite sex as him- or herself. These are historical facts and we can learn from them, although it is always questionable whether or not they will apply to future situations. Such objective probabilities can change over time.

A second way in which probability enters into our lives is the subjective probabilities that we attach to certain outcomes from our proposed actions before the fact. We create some—if not most—of them by noting historical probabilities, both the past experience of others and of our own, and by transforming them into probability assertions about the future when confronted with present circumstances similar to those of the past. Subjective probabilities, for example, include the estimates of our chances of being in an auto accident, getting accepted at medical school, having the light go on when we throw the switch, getting caught in the rain without an umbrella, and so on. Of course, these subjective probabilities may be only more or less accurate when applied to new situations and we can alter them when we learn new information.

Finally, the third way that probability enters into our world is in the intersection of two (or more) causal chains. For example, we might say that a man was struck by lightning on a golf course by chance. Yet there are two causal chains involved, one that explains how a man got to that particular place at that time and a second that explains the occurrence of the lightning at that same place and time. The man did not choose to be struck by lightning. Thus, the event was unintended by him. Yet it was a consequence, in part, of the choices that he made that brought him to the golf course. The choices, in turn, have their own causal explanations. They coexist with the causal chain that led to the lightning and are part of another causal chain that led to the man being there.

There are some futurists, such as Dator (1984: 63), who may disagree with the belief that the future is determined (although they agree that it is not *pre*determined). This disagreement may be partly semantic, a matter of how we have differently phrased our statements to try to express a similar belief. We all agree, I think, that the future is uncertain, that there are alternative possible futures instead of *the* future, and that basically, the future, within limits, is open, and not inevitable.

It may be true that a future event is not determined until it actually happens, that is, until it moves into the present and becomes the past. When the determinants come into play, then the event happens. Thus, we might say that the future *will have been determined* after it becomes present and past, as we look back in time from some later present. But until it actually becomes past time, future time has some degree of freedom and openness.

Key Assumption 7: Humans Make Themselves

The seventh assumption is: *To a greater or lesser degree future outcomes can be influenced by individual and collective action.*

Futurists view the contingencies of the future and the determinants of action as being at least partly under the control of individuals, groups, and human organizations. Jouvenel (1967: 52) says that for "a given person, who is at once knower and agent, the future is divided into *dominating* and *masterable* parts." The masterable future is what people can make of the future by their own acts. There may be some part of the future that I can control and another part that I cannot. But even the part I cannot control may be subject to someone else's control. In "human affairs the future is often dominating as far as I am concerned, but is masterable by a more powerful agent, an agent from a different level."

This is one of the most fundamental tenets of futures studies: we ourselves help to create the future with our own present decisions and actions (Dickson 1977: 16). We seek, in other words, not only to know the future, but also to control it (Ferkiss 1977: 5).

The sixth and seventh assumptions set two tasks for futurists. One is to determine how open the future is at any given time with respect to what; that is, what is contingent and conditional and what constraints, biological, physical, political, economic, social, cultural, or psychological, must we take into account?

Another is to determine how much of the contingent future is subject to human will and how much is not. What can be changed with what degree of likelihood and with what amount of human effort and what cannot be changed? Futurists, as we have seen, aim to make sure that we recognize everything that, in fact, can be changed—or that can be kept the same, if that is the aim—and that is subject to our will. They help people to avoid fatalism. Especially, they help prevent them from accepting futures as inevitable that in fact are contingent on their own actions.

One bias that must be guarded against is focusing too much on producing change and overcoming the status quo. This is not to say that "overcoming the status quo" is excluded from the futurist agenda. Quite the contrary is the case. But one option, especially in a world in which some change appears to be out of control, is to ask what parts of the present are of value and should be preserved. Also, planned change often must be selective and carried out within a context of maintaining and using some things in order to improve others. In other words, although futurists generally believe that it is possible to create better worlds than existing ones, they temper—and ought to temper—their general enthusiasm for change by reminding themselves and others that it is quite possible to create worse ones as well.

Key Assumption 8: Interdependence and Holism

The interdependence in the world invites a holistic perspective and a transdisciplinary approach, both in the organization of knowledge for decision making and in social action.
Futurists view the world as so interrelated that no system or unit can be viewed as totally isolated (van Steenbergen 1990). Rather, in order to understand the dynamic interrelationships in the world, they argue that every unit that is the focus of futures research should be considered as being to some degree consequential for every other unit. McHale (1969: 14) points out that today there are global systems that depend for resources on the entire planet, that they are interlocked with each other, and that each depends as well on a global knowledge pool of technical and organizational expertise. No nation today is self-sufficient.

One implication of the interrelatedness in the world is that a holistic, all-encompassing perspective is necessary for adequate decision making and social action. Futurists, thus, expand their fields of vision to make certain that they become aware of the unintended and otherwise unrecognized consequences of social actions. Dickson (1977: 19–20) gives an example of the banning of flammable sleepwear for little children by the Consumer Product Safety Commission. Afterwards, it was discovered that some of the new nonflammable fabrics cause cancer. Every solution to a problem may—probably will—cause a new problem of another kind as a second- or higher order consequence of any action. Thus, decisions made only on the basis of considering first-order consequences are usually inadequate.

Platt (1975: 156) uses the analogy of steering a bus. Every time the driver turns the wheel to avoid a pothole in the road the bus takes a new course and heads for new potential dangers. To steer adequately the driver must have knowledge of the total visual field, not simply of one possible path or a few potholes. The best driver will consider many alternative paths, so that he can see what new obstacles exist that must be avoided to pick a safe path. There are few—if any—final solutions. Rather, there are corrections, and still more corrections.

A second implication of assumption eight is that decision making and social action, to be carried out adequately, must be transdisciplinary. Thus, the futures field is developing as a transdisciplinary science, going against the trends toward disciplinary specialization and professional balkinization. The interrelatedness in the world means that formulating and implementing policy lead futurists into topics that may be the purview of a variety of different disciplines. Thus, futures research projects often have included multidisciplinary teams of researchers who synthesize disaparate bodies of knowledge to focus on a specific problem, as we saw in chapter 2 in the example on storage of nuclear wastes. Where a single futurist works alone, he or she often must learn things far removed from his or her own field of expertise.

For example, major topics of futurists such as those dealing with war and peace, environmental problems, sustainable development, technology assessment, international relations, space exploration, communications and ethnotronics, education and human potential, and healthcare, to mention just a few, have psychological, political, economic, social and cultural as well as engineering, mathematical, biological, and physical aspects. No discipline contains all of the knowledge needed to deal fully and adequately with such topics. Each must turn to other disciplines for help.

Holism, it is important to add, is not a matter of adding more and more variables to make a model more and more complex. Linstone (1977: 8), for example, says that the analysis "of a system in terms of many elements and interactions does not constitute a holistic view of the system. The familiar reductionist analytic processes are inherently unsuited to this task." He adds that the most important lesson that he has learned as a futurist "is the need to consider *simultaneously* multiple models and *Weltanschauungen*, even if they appear to be contradictory."

Futurists try to grasp the main outlines or features of the whole rather than dealing only with the details of the parts. Action, unlike research, must take the wide scope of interdependence and holism into account.

Key Assumption 9: Better Futures

Obviously, if futurists propose and critically assess preferable futures, they must believe that: *Some futures are better than others.*

Some futures—and presents and pasts too—are more desirable than others. People judge outcomes as being more or less preferable. If this were not true, if any outcome were just as good as any other, then it would hardly be necessary to be concerned about the future and to work for one kind of future rather than another. Futurists, as we saw in chapter 2, accept, as part of their tasks, helping people to assess alternatives, to explicate value judgments, and to examine critically the basis on which such value judgments are made.

In volume 2, where I discuss the ethical foundations of futures studies, I show that there are universal or near-universal human values that can be applied as a standard of critical judgment everywhere on Earth, that cultural and ethical relativism are false doctrines, and that objective methods exist by which to evaluate values themselves and to judge them as being right or wrong. I show that we know—and can demonstrate objectively to others that we do in fact know—many of the values that describe what is most desirable. Thus, we know what many of the features of the good society are and we have a basis for judging alternative possible futures as more or less desirable.

General Assumptions

There are three additional assumptions, more or less implicit in what I have already said, that can be stated explicitly. These assumptions are by no means exclusively held by futurists. Indeed, they are generally shared by a wide variety of other scholars and scientists, by some people in many different disciplines. Yet they are not shared by every observer of human behavior and, in fact, some writers disagree sharply with one or the other of them. They are, I believe, essential to the futurist enterprise.

General Assumption 1: People and Their Projects

People are creative project pursuers; they are acting, purposeful and goal-directed beings.
As Riner (1987:315) puts it, "Humans are, potentially, rational, self-directed, intentional beings capable of generating reliable knowledge about experience, and equally capable of applying that knowledge to change their circumstances." Human beings are self-conscious, value-driven decisional systems, actors with purposes, goals, and images of the future.

Most futurists agree with Encel et al. (1975: 24) who view a human being as "an actor rather than as a passive robot driven by inexorable forces to fulfill his [or her] fate." There are, of course, alternative images of human beings such as objects that are mostly—if not fully—determined by structural factors, genetic, economic or social, a mere empty receptacle into which is poured the meaning of a given culture or an unformed metal onto which is stamped the behavior patterns expected by a particular society.

Sociologists face a chicken-or-egg problem of which came first, society or the individual. Since the work of the French sociologist, Emile Durkheim, they have tended to emphasize how society shapes the individual. Certainly, the human infant is born nearly helpless and dependent on adult members of the society for its food, shelter, and general safety and care. Certainly, for each infant a language, culture, and approved and disapproved patterns of social behavior exist when it is born. The infant learns these things as it grows and develops into a cultural and social being. Certainly, language, culture, and society tend to persist long after a given individual has died, and they are seemingly indifferent to the birth and death of any particular individual member of society. The processes by which individuals become fullfledged members of society and by which the social order is maintained have captured much of the attention of social scientists—and they deserve to do so.

Yet it is a onesided and inadequate view, because the relationship between the individual and society is more reciprocal than the above view
suggests. Without individuals, human society and culture are obviously
impossible. Individual behavior in the aggregate defines the social structure and individuals are the carriers of culture—the language, beliefs,
values, and normative ideas of the human group. Just as the egg produces the chicken, individuals create society and culture. Individuals invented society and culture and they recreate them each day with their
behavior and intercommunications.

People not only learn what society intends to teach them, they also
make discoveries and learn things inadvertently. More important, they
learn what they need or want to know. That is, people become explicitly
and self-consciously instrumental about their own learning. Unlike a computer, human beings may never be washed clean of all memory, with the
possible exception of extreme cases of amnesia or mental illness. But,
also unlike a computer, human beings can reprogram themselves by their
own conscious efforts. Such self-reprograming results as human actors
purposefully, calculatingly, often quite rationally and sometimes very
emotionally, weigh the consequences of their future actions for their goals
and values and decide on the actions that they will take. Some of those
actions may change society, partly because even when people try to conform they may not do so perfectly and partly because people not only try
to conform, they also innovate and rebel.

Thus, a futurist model of social behavior centrally rests on a conception of motivated individuals pursuing their projects.

General Assumption 2: Society as Expectation and Decision

*Society consists of the persistent patterns of repetitive social in-
teraction and the emergent routines of human behavior that are organized by time and space, expectations, hopes and fears for the future,
and decisions.*

Most fundamentally, society is—following the logic of the first general
assumption—"a system of purposive actors, in which the outcomes of
events result from motivated action by the actors" (Coleman 1977: 11).
From this perspective, the social order or society is defined as the persistent patterns of repetitive social interaction on the part of such purposive
actors. Additionally, futurists emphasize the emergent aspects of behavior,

the role of time, expectations, hopes and fears regarding the future behaviors of self and others, and the role of choice and decision making.

For futurists, society is a complex web of written and mostly unwritten individual and organizational time schedules that are intricately woven together and synchronized as people pursue their projects in collaboration with others. Society is continually being acted out, as people move with time toward the anticipated achievement of their goals and expected completion of their projects in the future.

As we have seen, the calendar and the clock are used to organize and coordinate the paths of individuals through the inevitable dimensions of time and space and their intersections with the paths of other people. Think of society as intricate webs of expectations that human actors cast into the future, from which they receive feedback and in reaction to which they constantly alter their behavior and their personal timetables. Think of society, too, not as an iron cage, but as a complex, interrelated set of individual and collective opportunities, choices and decisions. Think of society, finally, as a dance of time, the coordination of human action.

Repetitions and routines can be useful. Among other things, they permit an increase in the total amount of behavior, because routines, once established, require little time for additional thought. They free time for us to invent and act out new behavioral repertoires that we add to our kit bags of social routines. They also help account for the orderliness that permits the prediction of other people's behavior, allowing us to guide our own behavior into smooth coordination and synchronization with that of others.

Victor Ferkiss (1977: 15), confirming this, reminds us that "All living creatures operate on the premise, conscious or unconscious, that the future is to a large extent predictable.... If there were no order in nature, if things did not follow patterns, life would be impossible." We must add that if there were no order to human behavior, social life would be impossible also. J. David Lewis and Andrew J. Weigart (1990: 93) refer to the ordering of actions and expectations for the achievement of future goals as "rationality." They say that "rationality is essentially a public reality by which a number of individuals make the same sense of the future.... The synchronicity of an entire society continually recreated by the billions of multifarious actions of millions of citizens makes plausible the rationality of the social order."

Finally, human beings must face the facts of social change. Society is not static. It's patterns of repetitive behavior, no matter how persistent, undergo change. Sometimes the change is slow and sometimes it is electrifyingly fast. This was the message of Toffler's *Future Shock*. Social change, Toffler said, is increasingly rapid. Changes are rolling over us, outstripping our images of the future and our abilities to adapt to them, much less to plan for them and to control them to our individual and collective benefit. We are shocked by surprises, unwanted consequences of our behavior, unanticipated results, formerly unrecognized effects. We are victims of an untamed future that forces us to run harder and faster as we sense we are getting farther and farther behind, battered by crisis after crisis.

Social order makes our predictions of social events possible. Social change can make such predictions more difficult and can produce so much social turmoil and confusion as to greatly reduce their accuracy. When enough of our predictions in everyday life fail, our learned recipes of appropriate behavior become ineffective. Too much uncertainty about the future produces anarchy not freedom.

General Assumption 3: The Existence
and Knowledge of External Reality

Finally, a third underlying assumption is essential, although it largely has been a hidden premise among futurists. It is the burden of chapter 5 of this book to show that any enterprise, including futures studies, that aims to create, preserve, disseminate or apply knowledge must assume that: *An external past reality did exist and a present reality does exist, apart from the human knowing of them, and in principle they can be objectively known by humans more or less accurately. Additionally, futurists assume that a future reality will exist, apart from the human consciousness of it, and that in principle assertions can be made about it that can be objectively warranted more or less accurately.*

I will not discuss this two-part assumption here, since I explain it in detail later. It is fully supported by the theory of knowledge known as "critical realism" that I describe and that I propose as fundamental to the conduct of futures studies.

Conclusion

In this chapter I have given nine key assumptions of futures studies and three general assumptions that futurists share with many other scholars. It is impossible, of course, to state all the assumptions that futurists—or other scholars—make, because there are simply too many things that nearly everyone takes for granted. Thus, I have been selective in setting down the above assumptions and stated the ones that I believe are either essential, consequential or distinctive—or all three—for the proper conduct of futures studies.

Other assumptions might have been given. Dror (1975), for instance, also gives nine assumptions of futures studies, only some of which overlap with the ones that I have given here. But some of Dror's assumptions fall into the category of auxiliary assumptions that are not necessary to state explicitly, for example, his assumption that the human mind is capable of recognizing stability and basing predictions on such recognition. Yes, of course. But his list is useful because it reminds us that there are other futurist assumptions in addition to the ones I have selected here.

Understanding the measurement and meaning of time is absolutely basic to futures thinking. Following Gell (1992: 315), I argued that there is no contradiction between studying differences in conceptions of time in diverse cultures and simultaneously maintaining that time everywhere is always one and the same. In every human society, there exists some form of time measurement, temporal scheduling of human events, and time coordination of individual activities underlying social cooperation. Everywhere, regardless of language and culture and regardless of the cyclical behavior of many natural and social events, people have adapted to the fact that time goes on and on, unidirectionally and irreversibly. No "one can escape from time—everything people are and do is bound up with it—but they are not continuously aware that this is so" (Young and Schuller 1988: 3). We are all time travelers, holding a one-way ticket to the future.

The key assumptions discussed in this chapter are summarized below.

1. Time is continuous, linear, unidirectional and irreversible. Events occur in time before or after other events and the continuum of time defines the past, present and future.

2. Not everything that will exist has existed or does exist. Thus, the future may contain things—physical, biological or social—that never existed before.

3. Futures thinking is essential for human action. For the consequences of action always lie in the future. But futures thinking, both by ordinary people and high-level decision makers, is done only more or less well. The power and utility of futures thinking can be improved.

4. In making our way in the world, both individually and collectively, the most useful knowledge is "knowledge of the future." That is, humans move with time, constantly moving toward the future. In making plans, exploring alternatives, choosing goals, and deciding how they ought to act, humans have a need to know the future and how past and present causes will produce future effects.

5. The future is nonevidential and cannot be observed; therefore there are no facts about the future. Nonetheless, as we shall see in chapter five, it is possible to have "conjectural knowledge."

6. The future is not totally predetermined. It is more or less open. Because it hasn't happened yet, the future is uncertain for humans as cognitive beings. For humans as living, breathing people making their way in the world, the future represents liberty, power and hope, a time when their dreams might come true.

7. To a greater or lesser degree future outcomes can be influenced by individual and collective action. The future is at least partly available to be shaped by human will, either through human control or anticipatory adaptation.

8. The interdependence in the world invites a holistic perspective and a transdisciplinary approach, both in the organization of knowledge for decision making and in social action. In order to know the world, scientists take a granular approach, reductionistic and delimited. But in order to act effectively in the world, humans need a holistic approach that incorporates the attempt to estimate the consequences of a given action on many human goals and values so as to guard against unintended and unanticipated consequences that are unwanted.

9. Some futures are better than others. In volume 2, we shall discuss the values and processes of evaluation on which judgments concerning preferable futures and conceptions of the good society are founded. Although futurists emphasize that people can and do construct alternative maps of many possible futures and ought to do so even more than they do, we must never forget, as Gell (1992: 255) reminds us, that, in fact, there will be only "*one* future, and we had better make sure that that future is the one we want it to be."

Additionally, I formulated three general assumptions that are necessary for carrying out the futurist enterprise that are shared by many other—but not all—scholars and social scientists. They are:

1. People are creative project pursuers; they are acting, purposeful and goal-directed beings.

2. Society consists of the persistent patterns of repetitive social interaction and the emergent routines of human behavior that are organized by time and space, expectations, hopes and fears for the future, and decisions.

3. An external past reality did exist and a present reality does exist, apart from the human knowing of them, and in principle they can be objectively known by humans more or less accurately. Additionally, futurists assume that a future reality will exist, apart from the human consciousness of it, and that in principle assertions can be made about it that can be objectively warranted.

In the next chapter, we move to an extended consideration of general assumption three as we begin a discussion of the foundations of knowledge for futures studies.

4

Is Futures Studies an Art or a Science?

Addressing the participants of a conference held by the World Future Society in New York in July 1986, Alvin Toffler said that the futures field, after twenty years of growth and success, had reached a turning point. On the success side, he pointed out that most people now accept the idea that change is accelerating and that the smokestack industries typical of industrial society are in decline. Futurist terms, such as "scenario," "time horizons," and "time frames," have become part of the general vocabulary. He reviewed some of the developments we discussed in chapter 1: the flood of books on the future, several making the lists of bestsellers; the publication of new magazines and professional journals; the invention and spread of methodologies, such as Delphi; the university courses and degree programs; the centers and institutes devoted to futures research, both nonprofit and profitmaking organizations; the professional societies and hundreds of conferences and meetings of futurists throughout the world; and the utilization of futures research by both business and government. Compared with twenty years earlier, the futures field in 1986 appeared to have become an established, vigorous assemblage of overlapping networks of activities, people, and projects.

Yet by 1986, he said, a certain pessimism seemed to hang over futurists. For example, he said that futurists have overlooked the pain of change, the sometimes violent conflict and personal disruptions that occur with profound change. Also, they have paid too little attention, in his view, to "the incredible," that is, to possible future events that have a low probability of occurrence but a high impact on the way people live their lives.

Although Toffler may well have been correct in these criticisms, he missed what for some futurists was the most important cause of their mid-1980s sense of malaise. Futurist Donald N. Michael (1985: 95) gets it right, I think, in his article, "With Both Feet Planted Firmly in Mid-Air."

165

There comes a time in the growth of any intellectual endeavor, he says, when further progress depends on knowing the philosophical foundations on which it stands. Overcoming its "footless status" is among the deepest problems facing futures research. "At root the problem is epistemological."

If we are confronted with two divergent images of the future, for example, how do we evaluate them? Is one as valid and persuasive as the other? If not, why not? On what basis do we accept or reject a particular image of the future? Are some views of the trends of change and the coming future incorrect? How do we know? How do we convince others that we are right in our judgment? When futurists contradict each other about their forecasts of the most probable future, how can we determine which is most accurate?

We need to know the reasoning behind assertions about the future, the rational processes of thought and the objective evidence on which they rest. Additionally, that reasoning needs to be subjected to critical examination. If the cognitive foundations of futurist assertions remain hidden, we have no basis for choosing among conflicting assertions. Thus, if we wish to assess the validity of an assertion, we must critically examine the assumptions on which it is founded, the empirical, logical, and theoretical grounds that have been put forward in its support, and, most important, the nature of the tests that have failed to refute it (Jouvenel 1967: viii–ix).

Futurists not only need to communicate with each other about what they are doing, but also they must speak plainly and clearly to a variety of nonfuturist groups, such as clients, funding agencies, students, government officials, other professionals, scientists and scholars as well as members of the general public. Futurists are expected to explain themselves and their field. There have been recent gains in understanding of futures studies by nonfuturists, as Coates (1987: 136) asserts when he says that "the day has happily passed when people, on learning I am a futurist, automatically come back with a trite jocularity such as `Where do you buy your crystal ball?'" Although that day may have passed, there certainly continue to exist, perhaps especially among people of practical affairs and members of the established academic disciplines, misunderstanding and skepticism about the nature, bases, and soundness of futures studies.

Despite a few efforts, such as those of Jouvenel (1967), Helmer and Rescher (1960), Mitroff and Turoff (1975), Ogilvy (1992), Scheele

(1975), Masini (1981, 1982, 1993) and a few other futurists, epistemology remains one of the least developed aspects of futures studies. There is irony in this fact, since methodology—specific research tools and techniques—to the contrary, is one of the most developed aspects of the field, and there are, of course, implicit epistemological assumptions and commitments that necessarily underlie every method.

Thus, the purpose of this and the next chapter is to discuss the philosophical foundations of knowledge for futures studies. I have divided the discussion into two parts. In the first, given here in chapter 4, I deal with background considerations, such as whether futures studies is a science or an art, it's nature as an action science, and some shared characteristics of the "transdisciplinary matrix" of futures studies.

In chapter 5, I discuss how the waves of intellectual conflicts involved in the newest revolt against positivism in the 1960s and the rise of postpositivism in the 1970s and 1980s have affected futures studies. Finally, I propose an epistemology for futures studies, a theory of knowledge based on a critical realist view. My aim is to sketch appropriate foundations of knowledge for futures studies.

In chapter 6, I move from the abstract and philosophical discussions of this and the next chapter to the concrete and empirical, that is, to a description and evaluation of specific methods and exemplars of futures research. Since methods and exemplars constitute an area of strength of futures studies, I will describe them in enough detail to capture both their diversity and the ingenuity that sometimes has gone into their development.

What is Art? What is Science?

Futurists Appear to Disagree

Is futures studies a science? Can't a better case be made that it is, rather, an art? Futurists appear to disagree on the answers. As Masini (1993: 23) says, scientificity "is the most debated of the characteristics of Futures Studies and indeed, according to many scholars, is not to be considered a quality of Futures Studies at all." Many pioneers in the field such as D. Bell, Lasswell, and Jouvenel say that the futures field by its very nature cannot be a science. Jouvenel, for example, pointedly titled his book, *The Art of Conjecture*. Moreover, many working futurists today, perhaps a majority, would agree that futures studies is an art.

Some futurists, such as Coates (1987: 133) and Walter A. Hahn (1985: 54) seem to take a somewhat intermediate position in that, although they refer to futures studies as an "art form," they recognize that it draws also on science. Amara (1986), too, apparently sees the futures field as both an art and a science.

Other futurists (Beckwith 1984; Encel et al. 1975; Malaska 1995) see it primarily as a science. As now practiced by many futurists, they say, it encompasses scientific activity. That is, nearly all futurists make some knowledge claims and try to give some objective reasons for them, which in the broadest sense of the term is all that "science" is. Of course, there is a division of labor among futurists, just as there is among any professional group, and some futurists spend much more time doing research— therefore doing the sorts of things that scientists do—than others.

Since the difference of opinion may be more apparent than real, perhaps it is silly to quibble over terms. Let people call the enterprise what they want—as they no doubt will in any event—and get on with the job. Yet it is of some consequence to the future development of the futures field that the issues involved in the terminological difference be made explicit, debated and, if possible, resolved. Everyone involved in the futurist enterprise—and, for that matter, nonfuturist consumers of the results of futures work—has a stake in the discussion. For the question concerns nothing less than the foundations on which futures research stands, the standards of quality and exemplars of good work by which it is to be judged, the reliability and validity that can be placed on its results, the nature of its professional commitments, and the type of curriculum that prospective futurists ought to study.

Figurative Meaning of the Terms

In a figurative sense, of course, we can refer to nearly anything that involves skill and ability as an "art." Thus, we could refer to the *art* of horsemanship, of flying an airplane, of fishing, of cooking, or of war— and some people have done so in each case. We could even speak of the "art of science," such as in the case of Michael Lynch's *Art and Artifact in Laboratory Science* (1985) or we could go further still and say that "Science is an art" (Dyson 1995: 33), but to do so totally muddles the issue at hand. Also, it certainly does not help either that some writers have spoken of the "science" of painting and of other forms of art, with the implication that art is a science.

In such loose usage, although condoned by the meaning of the original Greek and Latin terms, we will find little illumination. If this is all that some people mean when they refer to futures studies as an art, then there is nothing to debate. If we want to discuss the issue seriously, then we must insist on a more restrictive definition, limiting art to its more precise meaning dealing with aesthetics, embracing sculpture, painting, music, poetry, drama, dance and, even literature and some aspects of architecture.

Moreover, using a loose definition, such as the "art" involved in selling futures research projects or in presenting the results of research to clients does not warrant our serious consideration of the proposition that futures studies is an art rather than a science. We might as well have said the "science of selling" or the "science of presentation of results," as some teachers of salesmanship are willing to do. The use of "art" in this sense apparently is intended to convey the idea that selling and presenting are not precisely codified activities, are difficult to teach to others, or are linked to idiosyncratic abilities. Although there may be some truth to these contentions, these skills, nonetheless, are partly codified, can be taught, and are learned passably well by nearly any competent person who has a good teacher and is motivated enough to work at it.

Yet we cannot say that the loose use of the term "art" is wrong. Rather, it simply doesn't address the issue. Thus, let's turn to whether or not futures studies is an art in a narrow and precise sense of the term.

Features Shared by Both Art and Science

Anything may be similar to and also different from anything else. Your dining room table may have four legs and so does an elephant. That doesn't make them the same thing. It depends on the comparative criteria selected as crucial.

Because art and futures studies share some of the same features, some futurists have been led to think of futures studies as an art rather than a science. Both art and the futures field, they say, distinctively involve intuition, creativity, imagination, insight, and spiritual understanding. Both involve a certain amount of subjectivity and invite originality, innovation, and invention. Both are synthetic and humanizing. Art and futures studies require rare talent and exceptional ability. They involve imaging and communication. They are personal and intimate, in the extreme even ineffable, while aiming at a "higher truth."

As a corollary to such views, science is portrayed as having contrasting features: mechanical, producible by any number of interchangeable persons, highly technical and rigorous, rational and dehumanizing, codified, organizational, standardized, pedestrianly empirical, overly abstract, too fragmented and analytic in perspective to represent reality in all its fullness, and limited by linear and outmoded positivist thinking.

Not any one futurist, of course, may hold all of these views. I've gathered them from many different writers. Taken together, however, they do define, though not without contradictions, a perspective differentiating art and science, demeaning science and linking futures studies to art. *The problem is that the distinctions are false.* Both art and science are misrepresented in these characterizations.

First of all, in the same way that the art of the Middle Ages was influenced by religion, modern art has been influenced by science (Vitz and Glimcher 1984). This influence can be seen at every level of artistic activity. For example, in music and poetry the mathematical influences regarding structure, discipline, and form are well known. In painting the very tools and materials used have been altered by science, from the varying chemical compositions of surfaces on which to paint and of the pigments themselves to altered perspectives involving the perception of depth or the effects of light or of mixing or juxtaposing various colors.

Moreover, for nearly five hundred years, European artists accepted the imitation of nature as a central goal, many of them finding inspiration and guidance in two branches of optics: the geometrical science of perspective and the physical science of color (Kemp 1988). Even the pursuit of abstraction and analytic reductionism in art was encouraged by developments in modern science, as has been the subsequent turning away from them which was influenced by the disarray in the philosophy of science, especially the challenge of postmodern beliefs, that began in the 1960s. In recent years, of course, even the computer has joined the ranks of artistic tools.

Artists, like scientists, must learn to use their tools and to apply a set of principles in solving the technical problems they face (Kuhn 1977b). They must be concerned with design principles, such as unity, conflict, dominance, repetition, alternation, gradation, balance, and harmony. They must be concerned—*objectively and rationally* concerned—with line, value, color, shape, size, texture, and direction. They must exercise rigor and control.

Artists, also like scientists, teach other people to do what they do. Nearly anyone can learn to create a satisfactory painting, just as nearly anyone can learn to carry out a satisfactory scientific experiment. To suggest that either art or science is ineffable is to mock the years of training, conscious thought, and work that go into both endeavors. Yet great art, just like the masterpieces of science, may require an extra something of genius, perserverance, or luck. Art, then, no matter how much intuition may be involved, also has its technical and rigorous, mechanical, codified, standardized aspects, just as does science.

Second, the other side of the coin, of course, is that science, like art, has its intuitive, creative, imaginative, and insightful, aspects too. Both the formulation and verification of theories, for example, may involve such capacities. They may be involved also in discovering hypotheses, in designing crucial experiments or in the interpretation of field studies. Ingenuity and subjectivity have a role to play, too. Even beauty—a hallmark of art—may be used by scientists to choose one theory over another, other things being equal (and sometimes even if they are not equal).

Thus, to conclude that futures studies is an art because it may require creativity, intuition, etc. and is not codified, nor mechanical, nor rigorous, etc. won't wash. In fact, futures studies may involve all of these things, as both art and science do (Denton 1986).

The Obligation to Seek the Truth

There is at least one distinction between art and science, however, that remains and that bears on the nature of futures studies. Art can be—and often is—an illusion, a deliberate distortion of reality, perhaps even a negation of it. Art includes expressiveness, especially of inner mental states of the artist that may be fused with his or her perceptions of the outer world. This is not to say that there is no reality in art, even when it is basically deceptive or deliberately ambiguous. For example, as Kott (1984: 212) reminds us, in "theater there is always reality in the illusion and illusion in the reality." The actor may cry real tears on stage even when the emotion is feigned. Nor is it to say that the artist never tries to tell the truth. Sometimes he or she may. Moreover, some artists have suffered for it when their depictions of reality are at odds with official governmental policy. The key point is this: artists, although they *may* create representations of reality or reveal a "higher truth" which give meaning and direction to life, are *not obligated* by their commitment to art to tell the truth.

To the contrary, scientists are committed to tell the truth. Truth is the very heart of the scientific enterprise. No matter how conditional or provisional that truth may be, as we shall see, scientists by their commitment to do science are obligated to seek and to speak the truth.

Of course, because they have erred—or even lied, scientists, in fact, may not always tell the truth. Yet they are expected to try. The ethos of science, as Merton (1973: 259) says is "characterized by such terms as intellectual honesty, integrity, organized skepticism, disinterestedness, impersonality." Or as Churchman (1971: 219) says, "Above all, the scientist is expected to be honest in the sense that he reports what he has observed as truthfully as possible."

This is not to say that deceit, fraud, and misrepresentation of data do not exist in science. They all too frequently do, as Broad and Wade (1982) insist. When such deviant scientists are discovered and their misdeeds made public, however, the guilty scientist's career is damaged, if not ruined, and the resulting scandal reaffirms the norm: the discovery and confirmation of truth are the central principles of the scientific profession.

Michel Foucault has taught us that scientific discourse may serve powerful interests and may constitute, even when "animated by the will to truth," a form of dissemblance. Yet this is not so always or even mostly, as Lemert and Gillan (1982: 62–63) point out. Furthermore, not all truth is in the service of power in the sense of serving the interests of powerful persons and classes in society. Rather, truth serves power in the sense that it has the potential for increasing effective practice for anyone who may so use it (pp. 64–65). Thus, as we saw in chapter 2, futurists are not only concerned with creating, evaluating, and using information about possible, probable, and preferable futures, they are also concerned with how it is distributed and who participates in its creation and use throughout various stratified groups both within and among the world's societies.

Futurists are interested in making social action more intelligent, informed, effective, and responsible. Although they may spin dreams of the future to help orientate that action, they are obligated to seek the truth. For intelligent, informed, effective, and responsible action requires reliable and valid descriptions of present realities and reliable and valid knowledge of causes and effects. It requires, too, assertions about the future that are warranted. Futurists, therefore, function as scientists as they attempt to fulfill the knowledge needs of futures studies.

In doing so, like any scientist, they sometimes may rely on conditional and counterfactual assumptions. But their conjectures and speculations about the future, whatever their imaginative and intuitive sources, can be critically examined. Are they, we can ask, founded on reliable and valid past facts? Are they logically coherent? Are there empirical or theoretical reasons for believing that the assumptions on which they are based are correct? Can the postulated present possibilities and probabilities for the future on which the conjectures or speculations about the future are based be confirmed by others? Does the evidence support the conclusions?

This is not to say, however, that art never enters into futures studies. Art as metaphor probably enters somewhat into every science as a way of helping to communicate otherwise impenetrable ideas. Certainly in physics that is so, and, in new topics such as chaotic dynamical systems and aperiodic structures, artistic images and science are deeply intertwined (Schattschneider 1992). Also, science fiction, while excluded from my definition of futures studies, is often a good source of imaginative ideas, including descriptions of possible futures some of which may be worth exploring using futures research methodologies. Moreover, utopian writings are important progenitors of modern futures studies, and some of them have contributed images of the the good society and of the future that remain powerful in today's world, as we shall see in volume 2, chapter 1.

Finally, it is not all that unusual that many fields once thought of as an art eventually became sciences. We ought not to forget that virtually all the natural, biological, and social sciences had humanistic origins (e.g., in religion, moral philosophy, ethics, etc.) and gradually moved away from them.

But the Future Is Not Factual

In sum, the claims considered above that futures studies is an art and not a science are not well founded. They rest either on such a loose definition of art that art cannot be distinguished from science or on false ideas about the defining characteristics of both art and science.

There is, however, a third argument against futures studies being or becoming a science that many futurists find convincing, and it is an argument made by such writers as D. Bell and Lasswell. It is that the future, as we saw in chapter 3, is not yet existent and, therefore, is not factual. Although this assertion is true, it is wrong to conclude that, therefore, futures studies is not or cannot be a science.

Although, since it doesn't yet exist, we cannot study the future directly, we can study it indirectly by studying factual things that have a bearing on the future. We have already seen in chapter 2, for example, that present possibilities for the future are factual. They can be studied, like other aspects of reality, using the methods of science. There are many other factual phenomena that bear on the coming future, as well. They can be investigated and used as part of the reasoning and marshaling of evidence underlying the construction and analysis of images of the future. These past and present things were—or are—existential. They really happened or are happening. Thus, they can be observed and studied just as any other phenomena that are subjects of scientific investigation. They may not tell us what the future will be, but they help us in testing our assertions about the possible, the probable and even, as we shall see in volume 2, the preferable future. Furthermore, some of them may be instrumental in bringing a given future into being.

The Bearing of Reality on the Future

Existential phenomena that aid in delineating alternative descriptions and assessments of the future that can be scientifically studied include:

1. *Present images of the future and expectations for the future that people hold, that is, their conceptions of the possible.* Mau's (1968) study of beliefs of Jamaican leaders discussed in chapter 2 illustrates a study of this sort.

2. *People's beliefs about the most likely future, that is, their subjective probabilities concerning the chances of particular futures occurring.* Any number of Delphi studies, to be discussed in chapter 6, illustrate the study of such subjective probabilities, usually using experts as judges of the chances of given future events occurring by some specified future date.

3. *The goals, values, and attitudes people hold; the preferences they use to evaluate alternative images of the future, that is, people's hopes and fears.* Delphi studies also illustrate studies of preferences to the extent that the experts are asked to consider particular future events not only according to their chances of occurring at particular times, but also

according to how desirable or undesirable they judge them to be. Another example, to be discussed later, is the comparative study of peoples' hopes and fears using survey research that was pioneered by Hadley Cantril (1965) and a study of preferred futures for telecommunications in six Pacific island societies done by Dator and others (1986).

It is important to some futures studies to know people's relevant preferences, because in some situations, obviously, what is possible may become ever more probable if people want it to happen and less probable if they don't. Also, studies of preferences (see volume 2, chapter 3) are sometimes used as part of the reasoning to justify preferable futures.

4. *Present intentions of people to act.* If asked, people—unlike atoms or quarks—can often tell us what they intend to do. Thus, political polls report on how people intend to vote. Market research predicts what products consumers intend to buy. Studies of decision makers of major organziations describe actions such organizations intend to take. Because many of these intentions are written in five- or ten-year plans of business or government or in the statements of purposes and aims of various organizations, such plans and statements can be examined for intended actions and their possible and probable consequences can be explored.

An example can be found in the well-known case of family planning in the People's Republic of China. As their population reached the billion mark, the Chinese government resolved to limit population size because officials believed that it was an obstacle to economic development. Beyond that, they assessed their resources—from agricultural production to fresh water supplies—and decided that in the future the country could support no more than about 700 million people at their preferred level of living. The government established a one-child family policy, eventually to go to a two-child policy for about one-half of China's families. "Their explicit goal is to end growth as rapidly as possible and then...begin a decline" (Ehrlich 1985: 22). Obviously, both population forecasters and planners of other aspects of the economy and society, even without assuming that the government's aim will be achieved fully (e.g., there has already been some backing away from the policy which has sometimes been brutally enforced), would be foolish not to take these plans and policies into account. They may affect everything from the future demand for school buildings and various consumer products to housing and unemployment.

5. Obligations and commitments that people have to others. Although people may not always fulfill their obligations and commitments to others, often they do, especially when they are not subject to cross-pressures that push or pull them toward different ways of acting. Such obligations and commitments constitute an important part of the persistent patterns of behavior that define society and are congealed into the social roles people occupy and the relationships they have with other people. They define expectations of future behavior. Fathers and daughters, employers and employees, flight attendants and passengers, clerks and customers, husbands and wives, lawyers and clients, doctors and patients, for example, have responsibilities toward one another that are activated in specific social situations and role relations. Knowing those responsibilities can help futurists anticipate future behaviors of the people involved.

In the modern world, some obligations and commitments also result from agreements and contracts among people. They, too, give insight into future behavior. For example, a thirty-year home mortgage commits a home buyer to the future behavior of making a series of payments over a period of thirty years. Of course, there is deviant behavior. But society is not possible unless most people most of the time fulfill most of their commitments reasonably well.

6. Knowledge of the past. There are at least five ways in which knowledge of the past can be of help in constructing alternative futures.

• *Tradition.* The most general way in which knowledge of the past enters into prediction and control of the future is through tradition. Memories of the past, legends and customs of the society, and highly valued patterns of behavior are contained in people's minds, in charter or founding societal documents and myths, and in other repositories of the history, purposes, and meaning of a people. Accepted tradition is "as much a part of the present as any very recent innovation" and it functions to influence present behavior. We learn from others in the present, but much of what we learn is handed down from past to present. The fact of the ignorant infant and the adult with skills, knowledge, and beliefs ensures this, even though the growing child, of course, also learns from personal experience and contrives things by his or her own reason (Shils 1981: 9, 13, 24).

Traditions, of course, contribute to the production of the future. No society has ever had a generation that "created all that it used, contemplated, enjoyed, and suffered.... It would literally be a society without a past to draw on to guide its actions in the present" (Shils 1981: 34). Although futurists believe that the creative imagination must be used to modify or transcend traditions in particular instances either to adapt to new conditions or to find new and better solutions to old conditions that continue to exist, they recognize that at any given time presently accepted traditions are part of the raw data they need in order to understand present behavior and organizational arrangements and to explore alternative possibilities for the future and the trends of change.

• *The use of trend analysis.* When time series data can be assembled on any variable of interest to futurists, then past trends can be described. On the basis of a variety of different assumptions, they can be projected into the future and, thus, they yield alternative depictions of the future. This must be done with caution and, if possible, with the use of other types of data in order to cross-check one's deductions about continuations of trends into the future, but sometimes the probable as well as a possible future can be precisely delineated. See chapter 6 for a discussion and several examples of the analysis of trends.

• *The restatement of scientific explanations into a predictive form.* As we saw in chapter 2, any scientific explanation can be restated as a prediction, although the prediction is limited to a contingent or conditional statement.

• *Analogy.* The use of historical analogy is an available and sometimes useful tool of futures research (Neustadt and May 1986), even though it must be carried out with utmost skepticism and critical assessment of the conditions of applicability. Futurists ask, are the historical situations to be used as analogies sufficiently similar to the future situations or events in question to permit any reasonable deductions from past knowledge to future probabilities? If not, then the past may be a faulty guide to the future. But if so, then knowledge of the past may enlighten the future.

• *Past images of the future.* Any period of history can serve as the focus of an inquiry into the past images of possible, probable, and preferable futures, the conditions that led to their enunciation, and the consequences for social stability or change. Any past was the future of some earlier time. Thus, the actual future of times past and the dynamics of past change can be compared with the earlier images of the future that led to them. An example of such a work can be found in Polak (1961) discussed in chapter 2.

• *Knowledge of the present*. Although their use helps futurists think about the future rigorously, estimate probabilities and possibilities, and increase the plausibility of their assertions about alternative futures, past trends and analogies do not prove anything about the future with certainty. Contrary events and processes can always happen. Thus, using past trends and analogies are relatively weak justifications for beliefs about the chances of different realities of the future. Relatively weak, that is, compared with understanding currently existing phenomena that may cause or somehow prefigure the future. Even scientific explanation is not fully exempt from this particular weakening consideration even though it (e.g. a law of physics) may apply to the present as well as to the past and often, we are usually willing to bet, to the future too.

We can view the present as containing at least two ways of exploring the future in addition to what has already been mentioned. The first is (a) *the design perspective*. Some things that now exist or are developing can be expected to continue into the future and to have implications for shaping the future. Some projects now underway, for example, can be expected to affect the future: a gigantic dam, a bridge, a planned community, a major highway, the production and adoption of home computers, a massive land development. Some of the consequences of such present actions already may be nearly inevitable, unless some totally unanticipated catastrophic event occurs, because the continuing will and purpose of human beings are driving the action.

Such projects, obviously, may overlap with intentions of people and, thus, their study from the perspective of design implications simply adds another dimension of analysis. The design perspective, however, goes beyond looking at only the intended results of behavior, because projects, once begun, take on a life of their own and may have unintended and unanticipated consequences.

The second is (b) *present possibilities for the future*. As discussed in chapter 2, present possibilities for the future are factual. They include present capacities of individuals, groups, and society as a whole for change and development. The potential for future development and growth exists in the present and, thus, can be studied scientifically.

None of the above phenomena, of course, yields easily to measurement and interpretation. Some of them, for example, beliefs about the possible; beliefs about the probable; people's preferences, goals, values, and attitudes; their intentions to act in a given way; or their conceptions of their commitments and obligations—cannot be measured without serious concern for reliability and validity. They are, after all, subjective phenomena often hidden in the minds of individuals. Yet they have some objective manifestations that can be observed and from which they can be inferred. Many objective methods now exist for the measurement of subjective phenomena in the social sciences and they are constantly being improved (e.g., Turner and Martin 1984). Even if reliably and validly measured though, such phenomena are not infallible guides to the future. But, then, there is very little that is.

Science Is Conjecture

To say that futures studies is based on conjecture, as some futurists do, also fails as a convincing argument to support the contention that it is an art rather than a science. For science *is* based on conjecture—it is "fallible, corrigible, presumptive, and contingent" (Campbell 1986: 115). Although science is committed to seeking the truth, its methods and logical structures encompass conditionals, counterfactuals, dispositionals, theoretical speculations, creative formulations of hypotheses, and predictions (i.e., assertions about the nonexistent future). As such, some scientific statements are no different epistemologically from many assertions about the future made by futurists.

Science includes many propositions that go beyond the simple, concrete descriptions of empirical details of uninterpreted reality. Sometimes long chains of inference are involved. Philosopher Karl R. Popper (1965), for example, argues that scientific theories themselves are conjectures, that is, highly informed guesses about the world which cannot be confirmed but, if they are false, can be falsified and discarded. Jouvenel (1967) means little else than "highly informed guesses" by the term "conjecture" in the title of his influential book, although he would doubt that they could be falsified if they dealt with the future as long as that future was still to come. But in this last belief he was mistaken, as we shall see in chapter 5.

With respect to contrary-to-fact conditions, their use is well known in historical research (Michalos 1979: 473). An obvious case is a well-known study of the economic growth of the United States *as if* the railroads had not been built. Nobel prize winner R. W. Fogel (1964) begins with the historical economic data as they were before the railroads, about 1830, and then estimates (i.e., constructs by inference) what the American gross national product would have been in 1890 without the railroads. Thus, he begins with factual economic data and then projects them into the future on the basis of counterfactual assumptions. He inserts a no-railroad economy into the real past, waterways being the most likely alternative to railroads. What his counterfactual analysis suggests is that raliroads may not have been indispensable to America's economic development. Water travel might have done the job just as well.

Fanciful? Perhaps, but ask yourself how we know what effect the railroads had on economic growth without a counterfactual study such as Fogel's. Yes, there really is a measurable amount of economic growth and, yes, the railroads were really built and contributed to it. But how much of that economic growth would have occurred even without the railroads? What did the railroads really contribute to it *over and above what would have resulted from alternative means of transportation*? We do not know until some kind of counterfactual and hypothetical analysis is done as a basis of comparison, an analysis based upon the speculation and conjecture of a "what if?" premise.

Assertions about the future, though, may exceed even the broad limits on conjectures set by scientific standards. For example, since they are asserted and acted on before the time when they can be tested, they are "prefactual" as operating principles, even though the present possibilities on which they are based may come to fruition at some future date and then be observed. Yet their grounds, which include inferences made from facts about the past and present, can be examined scientifically and our beliefs in them can be justified.

In sum, although it shares many characteristics typical of an art, just as does science for that matter, futures studies is basically a science. Unlike artists, futurists are constrained to seek the truth just as scientists are. Their assertions are in part based upon their scientific study of past and present realities. In part, they are based on additional conjectures involving predictions. Futurists can base their assertions on objective and rational reasoning that is similar—if not identical—to scientific thinking. The method of science is, after all, no more than "the rationale on which it bases its acceptance or rejection of hypotheses or theories" (Rudner 1966: 5).

A Transdisciplinary Action and Social Science

An Action Science

Futurists share many intellectual commitments, including a belief in the importance of conscious human decision and action as means to control the human future. Futurists aim to inform such decision and action by futures thinking. Thus, futures studies can be considered an action science.

Argyris, Putnam, and Smith (1985: 4, 80) define a basic tenet of action science as "a conception of human beings as designers of action." This is similar in import—if not identical—to the futurist general assumption concerning people and their projects stated in chapter 3. Thus, a theory of human action is at the heart of action science. Three important features of action science are (1) propositions that are empirically disconfirmable and organized into a theory; (2) knowledge that is usable by human beings; and (3) "alternatives to the status quo that both illuminate what exists and inform fundamental change, in light of values freely chosen by social actors."

Futures studies includes an orientation to action and can be considered an action science in the fullest sense of the term.

A Transdisciplinary Social Science

Since everything might have a future, the subject matter of futures studies, as we have seen, is difficult to pin down. The potential topics of study are as numerous as the world is varied. Thus, the futures field is necessarily multidisciplinary or transdisciplinary, incorporating both diverse subject matter and experts trained in many different fields.

Their orientation to action, of course, encourages futurists, as we saw in chapter 3, to take the complex interdependencies of the world into account and to be holistic in assessing the results of action. Thus, futurists bring into focus whatever knowledge is necessary from whatever fields of inquiry it may come from that may bear on the phenomenon being studied.

This explains why much futures research involves a team effort, where the team is composed of people representing a variety of different disciplines with expertise in different subject matters. It also explains why futurists tend to seek a reorganization of knowledge and constantly cross

the disciplinary barriers erected by the traditional departmental structures of universities which have become serious and frustrating obstacles
to rational decision making and action. An example is the case of twenty
medical specialists each skilled in the care of some particular part of the
body—for example, the eye or skin or bones or heart—no one of whom
is concerned about the well-being of the patient as a whole.

Since he or she is forced to look into many different topics and types
of phenomena and to master many different fields of knowledge, the futurist tends to become a polymath, a generalist, and a universalist. Knowing a lot about a lot of things is a tall order and most of us fail to do it
very well. Yet in a world of specialists and specialized knowledges, there
is an important—and at present neglected—role to be played by the person who sees the big picture, who sees how different things interrelate,
and who sees the whole and not only some of the parts.

Despite its transdisciplinary nature, futures studies has some features that are necessarily aspects of a social science—a transdisciplinary,
unifying social science. Although some futures research projects deal
with topics that are not obviously social at first glance, for example,
technological changes, they are almost always concerned at some point
with the social implications of such changes. Almost always of interest
are social psychological, political, economic, social, or cultural implications. For example, even though there are unavoidable technical,
natural, and physical scientific aspects to the challenges posed by such
topics of futures research as space industrialization, resource depletion, medical advances, electronic computing, robotics, proper storage
of hazardous wastes, and water or air pollution, they all have—and are
evaluated by—their social consequences. They invite both forecasts of
the social impacts of coming physical, material, or technological changes
and recommendations for beneficial changes in relevant human organizations or behaviors, including human interventions in the physical,
material, or technological trends themselves.

Also, their orientation toward action places futurists firmly in the social arena. Social action, by definition, takes place in a social setting and
is affected by the reactions of others. It is a social phenomenon and involves policymaking and policy implementing. Decision making itself is
a social process both because it often directly involves other people than
the actor and, even when it doesn't, because others are almost always
considered in the mind's eye of an actor (i.e., actors' taking into account

the possible future reactions of others in their imagination as they plan their own action).

There seems relatively little disagreement on this point among futurists. Most agree that, whatever futures studies may be, it is largely "social." Jouvenal (1967: viii), for example, explains that the founders of his research organization, *Futuribles*, aimed to influence the social sciences to look more toward the future. Ferrarotti (1986: 20) claims that "the central discipline for futures research is sociology, in that it is the science that studies the systemic interrelations that regulate the (static) functioning and (dynamic) development of society."

Ferkiss (1977) and even operations researcher Helmer (1983) clearly view futures studies largely as a social science or, perhaps more accurately, as a *reformation* of the social sciences once the established disciplines pay more attention to the future.

In the late 1960s to the early 1970s, there was some evidence of a shift in the major field of formal training among futurists toward the social and behavioral sciences. By 1971–72, for example, social and behavioral scientists accounted for 28 percent of futurists responding to a survey—more than in an earlier survey—and were ranked in first place. Ranking in second place were physical scientists who accounted for 13.4 percent of the respondents, followed by engineers who ranked in third place and who accounted for 12.7 percent (McHale 1971–72: 9–10). By 1975, Miles (1975: 3) claimed that recent "trends in both conventional social science" and futures research "suggest that interpenetration of the two is underway." This conclusion, however, may be overly optimistic because the established social sciences have been slow to reform. At least, they have been impervious to a thoroughgoing futures orientation.

Even though futures studies has inevitable social aspects and as a science is largely social, it is, also, necessarily transdisciplinary and unifying. Social actors' informational needs invite the reconstitution of disparate knowledges—including biological and physical—into specially designed synthetic, holistic packages applied *in toto* to given historical situations for planning purposes. Thus, futurists, while necessarily specializing to some extent as individuals, encompass as a group a wide range of knowledge and information. They select from all the sciences and fields of learning in intricate and varied ways as necessary to carry out their specific purposes. They integrate and synthesize many different knowledges.

From "Paradigm" to "Transdisciplinary Matrix"

"Paradigm" has become a popular term and it has an important and useful function as used according to several of its meanings. There is, for example, a "paradigm for futures studies"—or, as I prefer to call it, "a transdisciplinary matrix for futures studies"—that I will describe shortly. But the term has some drawbacks, too, that make me unwilling to use it here. Let me briefly explain why.

Although the term, "paradigm," according to the Oxford English Dictionary, was used as early as 1483 (to mean a "pattern," "exemplar," or "example"), it did not become widely used until after 1962 when Thomas S. Kuhn published *The Structure of Scientific Revolutions*. Even as it spread among university professors and students, journalists, and the literate public in the late 1960s and early 1970s, it was being questioned and respecified at its most recent source, the history and philosophy of science. Margaret Masterman (1978) in a colloquium in London as early as 1965 counted twenty-one different senses of the use of "paradigm" by Kuhn himself in his 1962 work alone, some of which Kuhn (1978: 271) later acknowledged. As early as 1969, Kuhn (1977a) suggested alternative terminology for some of the different meanings he gave to the term.

Masterman's list, plus my own additions after perusing the literature, leads me to the following different meanings that one writer or another has attributed to "paradigm":

A theory; a model; a perspective, a new way of seeing; a world view; a frame of reference; an approach; a research program or tradition; a hypothesis; an explanation; a set of concepts; a myth; a classification; a procedure, a methodology, a technique, a tool, an instrumentation; a formal (logical, symbolic) system; a map; a textbook; something that determines a large area of reality; a set of assumptions; an algorithm; a theme; a set of standards; a pattern; an example; a grammar; a set of rules; a prototype; a set of precedents; a concrete scientific achievement; a universally recognized scientific achievement; something similar to a set of political beliefs or an accepted judicial decision; a characteristic set of beliefs and preconceptions, including instrumental, theoretical, and metaphysical commitments; a successful metaphysical speculation; a set of scientific habits; a genuinely insightful puzzle-solving trick or device; a concrete problem solution; an analogy; a metaphor; a picture; an exemplary object of ostension; and a gestalt-figure.

Admittedly, many of these meanings overlap, as Masterman pointed out in the case of her twenty-one, so the situation is not as confused as such a large number of different meanings might indicate. Moreover, she was able to cluster the meanings into three sets which she labeled: (1) Metaphysical paradigms (including a set of beliefs, a myth, a successful metaphysical speculation, a standard, a new way of seeing, a map, etc.). (2) Sociological paradigms (including both a concrete and universally recognized scientific achievement, a similarity to a set of political beliefs or an accepted judicial decision, and, most centrally, a set of scientific habits). (3) Artefact or construct paradigms (including textbooks, tools, instrumentation, a grammar, an analogy, and a gestalt-figure).

But the situation seems hopeless of coming to any single, agreed-upon definition in the near future, given the diverse, sloppy, and faddish usages that now exist. So some alternative term or terms, more precisely defined, would be helpful. Fortunately, we can look to Kuhn himself for concept clarification. In brief, Kuhn proposes the term *disciplinary matrix* to refer to most of the objects of "group commitment" described in his 1962 book as "paradigms" (Kuhn 1977a, 1978). The major components of a disciplinary matrix are:

• *shared symbolic generalizations* (which include logical expressions used by the scientific community, e.g. "$f = ma$");

• *shared models* (which are preferred analogies and an ontology, either heuristics, like the hydrodynamic model of the electric current, or metaphysical commitments, like atomism);

• *shared values* (which include the accuracy of prediction), and other elements, particularly

• *shared exemplars* (which include concrete problem solutions which scientists learn and accept as distinctive examples of proper work in their discipline).

The motto, "One word, one meaning," such as that recommended by language maven William Safire, may be too constraining. Yet, when they use the term "paradigm," futurists and other scholars can do better than they have at saying what they mean, even to each other, not to mention to such publics as students, the larger academic and professional communities, and policymakers The term "paradigm" carries too many possible meanings for rigorous thinking. It has become quasi-mystical as it has become faddish. In my judgment it is time to discard it in favor of the more precise terms that Kuhn has recommened in its place. They convey to others more accurately what we mean to say. Thus, for the sake of clarity, let's all stop using the term "paradigm."

A Transdisciplinary Matrix for Futures Studies

Even though futurists are diverse in their disciplinary training and background, many of them share some features that define what a futurist and futures studies are. Adapting the term Kuhn has suggested for one of his key meanings of "paradigm," I summarize these features by describing the "transdisciplinary matrix" of futures studies. Although not every futurist may adopt all these defining characteristics of futurists, futures studies as a transdiscipline is composed of some identifiable shared "objects of group commitment" commonly possessed by its practitioners.

I have drawn on a number of sources, including the typical characteristics of a futurist as revised by Didsbury (1991: 2) and the attributes of a futurist as drawn up by Dator (1994b: 34–35), to compile the following items that constitute a transdisciplinary matrix for futures studies:

1. A perspective involving past change and possibilities for the future different from the present.

2. A belief that futures thinking can increase the effectiveness of human action.

3. A faith in the use of knowledge in policy formulation and implementation.

4. A self-identity as a futurist.

5. A shared set of assumptions (e.g., the directionality and irreversibility of time, the singularity of the future, futures thinking is necessary for action, "knowing" the future is the only really useful knowledge, there are no future facts but there are justified conjectures about the future, the future is open, humans make themselves by their actions, the interrelatedness of things in the world, some futures are better than others, people pursue projects, society includes expectation and decision, and an external reality exists and some of it can be known; see chapter 3).

6. Common purposes (e.g., making the world a better place where all human beings will have an equal and good chance of living long and satisfying lives, a commitment to the well-being of future generations; toward that end: studying possible, probable, and preferable futures; exploring images of the future; investigating the knowledge and ethical foundations of futures studies; interpreting the past and orientating the

present; integrating knowledge and values in designing social action; increasing participation in imaging and designing the future; and communicating and advocating a particular image of the future; see chapter 2).

7. Shared methods and exemplars (e.g., pragmatic prediction of one variable by another, extrapolation of time series, cohort-component methods, survey research, the Delphi method, simulation and computer modeling, gaming, monitoring, content analysis, participatory futures praxis, social experimention, ethnographic futures research, writing scenarios, and others; see chapter 6).

8. Key concepts that are shared (e.g., image of the future, future shock, tempocentrism, time frames, time horizons, alternative futures, possible futures, probable futures, preferable futures, post-industrial society, sustainable development, self-altering prophecy, issues management, ethnotronics, scenarios, trends, life-sustaining capacities of the Earth, human values, among others).

9. Similar emergent theories of human behavior and social change (especially those involving human agency and socio-cybernetic processes).

10. An orientation toward conscious decision making and social action aimed at adapting to or controlling the future.

11. A wide-ranging use of the knowledge of many disciplines as needed by the phenomena under consideration in any study or project.

12. A holistic perspective as necessitated by the information needs of social action.

13. A concern with the social implications of scientific and technological changes and, more generally, a concern with the consequences of all human behavior, both intended and unintended.

14. A dedication to understanding the general processes of change, be they social psychological, political, economic, social, or cultural.

15. Shared values (e.g., the welfare and freedom of human beings, a concern for all living things, and a concern for the life-sustaining capacities of the Earth, both now and for the indefinite future; see chapters 4, 5, and 6 in volume 2).

Conclusion

In this chapter, I began a discussion of the foundations of knowledge for futures studies and I ask, How do futurists know what they think they know about the future? How do they justify or ground their assertions? What is an appropriate epistemology for the futures field? My answers are partially given in this chapter and will be continued in chapters 5 and 6.

Here, I have shown that the disagreements among futurists about whether futures studies is an art or a science rest largely on an indeterminate figurative definition of the terms or on inaccurate beliefs about the differences between art and science. I pointed out that both art and science share many similar characteristics. Art, like science, has its technical and rigorous, mechanical, codified, standardized, and objective aspects. Science, like art, has its intuitive, creative, imaginative, insightful, ingenious, subjective, fortuitous and beautiful aspects. Thus, to argue either that futures studies is an art and not a science or that it is a science and not an art on the basis of these false distinctions is useless and says more about the erroneous beliefs of the arguers than about the nature of futures studies.

There is at least one distinction between art and science, however, that is sound. It is that scientists by the very nature of science are obligated to tell the truth. Artists, indeed, may also tell the truth, but they are not obligated to do so by their role as artist. Art is often an illusion expressing the inner mental state of the artist more than perceptions of the external world. In this regard, futurists clearly are akin to scientists, not artists. Although futurists may spin dreams of the future to help guide and motivate human action, they are obligated to seek the truth. In doing so, like any scientist, they sometimes may rely on conditional and counterfactual statements. But their conjectures and speculations about the future, whatever their imaginative and intuitive sources, can be objectively examined and evaluated as truth claims by themselves and others. Futurists function as scientists as they attempt to fulfill the knowledge needs of futures studies.

The fact that the future is nonevidential is also no obstacle to futures studies being a scientific activity. This will be discussed further in chapter 5, but here I have shown that many evidential and existing phenomena that bear on the future can be studied with the standard methods of science. They include:

1. Present images of the future and expectations for the future that people hold, that is, their conceptions of the possible.

2. People's beliefs about the most likely future, that is, their subjective probabilities concerning the chances of particular futures occurring.

3. The goals, values, and attitudes people hold; the preferences they use to evaluate alternative images of the future, that is, people's hopes and fears for the future.

4. Present intentions of people to act.

5. Obligations and commitments that people have to others.

6. Knowledge of the past, including the use of tradition, trend analysis, the restatement of scientific explanations into a predictive form, analogy, and past images of the future.

7. Knowledge of the present, including recognizing human projects with design implications for the future and identifying present *real* possibilities for the future no matter how latent or hidden they may be.

Much of futures studies is based on conjecture and so too is science. Although science is committed to seeking the truth, its methods and logical structures encompass conditionals, counterfactuals, dispositionals, theoretical speculations, creative formulations of hypotheses, and, as we saw in chapter 2, predictions. Thus, many scientific statements are no different epistemologically from many assertions about the future made by futurists. Scientific assertions are conjectures.

Futures studies is a transdisciplinary action and social science. It is transdisciplinary, because it draws on many different disciplines in reaching and justifying its findings and assertions. It is an action science, because it aims to inform human action designed to shape the future. Although it involves many disciplines including natural science, it is necessarily a social science as well, because, whatever the substantive issues involved in choosing appropriate human action, decision making and action are social processes and they take place within social contexts. I listed fifteen shared group commitments of futurists that constitute a preliminary attempt to define the transdisciplinary matrix for futures studies.

Showing that futures studies is an action science, however, is only the first step in establishing its epistemological roots, because the philosophy of science itself has been in turmoil during the last thirty years or so. Positivism has been under attack. Postpositivists, incorporating many postmodern views, have attempted to demolish many of its basic assumptions. More recently, post-postpositivists, going beyond postmodernism, have in turn devastatingly criticized the views of postpositivists.

In chapter 5, I review some of these developments and propose a critical realist theory of knowledge for futures studies.

5

An Epistemology for Futures Studies: From Positivism to Critical Realism

The Greeks gave us the term "epistemology," referring to absolutely certain knowledge as *episteme*, while calling mere opinion *doxa*. In this chapter, I ask, what are the philosophical foundations on which futurists base their knowledge of the past, present, and future? After examining currently dominant views, I propose some principles for an epistemology for futures studies, that is, for a theory of knowledge for the truth claims about the past and present and, distinctively, for the truthlike statements about the future that futurists make.

In chapter 4, we learned that futures studies can be considered a transdisciplinary action science. In this chapter, we shall focus on the "science" part of that label and briefly consider the nature of scientific knowledge. In the standard account, what constitutes scientific reasoning, evidence, and proof? Have recent controversies in the philosophy of science that have raised serious questions about that standard account—or "the received view" as we shall label it—any bearing on futures studies? What, if any, modifications must be made in current philosophies of science or in basic scientific assumptions because futurists deal not only with past and present realities, but also with assertions about the nonexistent and nonevidential future?

Science, of course, like any other human activity does not exist in a social vacuum. It is a part of the larger society within which it takes place and the members of its various communities, both social and natural scientists, are also members of families, neighborhoods, towns and cities, and nations and states. Thus, science is partly shaped by the larger currents of belief and change in its social environment. Futures studies, as we have seen, experienced many formative developments during the post-World War II years, especially during the 1960s and 1970s. They are the key years in shaping the nature of futures studies today. By look-ing at some of the dominant intellectual currents and important events of those times we can further understand the social forces that have helped make futures studies what it is.

In this chapter, I begin by examining the social context of the 1960s and 1970s and how it affected futures studies. After that, I turn to some intellectual conflicts within the scientific and scholarly communities them-selves that, for better or for worse, have influenced the thinking of today's futurists. Specifically, I compare three theories of knowledge: positiv-ism, postpositivism, and critical realism. Nearly all futurists—and most literate people throughout the world—adhere to the major tenets of one or another of these philosophical perspectives, whether they realize it or not, or they hold some, often contradictory, mixture of them.

Futures studies came of age during the most recent attack on positiv-ism, and futurists have been greatly influenced by the postmodern and antipositivist philosophies of the attackers. These postmodern influences have had some beneficial effects for the development of futures studies. Nonetheless, postmodern theories of knowledge, in the opinion of some writers, are basically flawed and do not provide adequate philosophical bases for the development of knowledge in any scientific or scholarly field, including futures studies. Among other things, in the extreme, Rosenau (1992) claims, they breed nihilism.

Thus, I describe an alternative philosophy, known as critical realism, that does provide an adequate theory of knowledge. Critical realism, which is a post-postpositivist theory of knowledge, seems well suited for futur-ist inquiry, because, without giving up the idea of justified beliefs in the truth of propositions, it postulates that all knowledge is conjectural. Thus, it can incorporate within the same epistemology truth assertions both about the evidential present and past and about the nonexistent and nonevidential future.

The Climate of the Times: The 1960s and 1970s

As Gellner (1985: 125) says, "science is consensual, and the philoso-phy of science is *not*." The dissensus in the philosophy of science reached gigantic proportions in the late 1960s and early 1970s. "Turmoil" may be a better word even than "dissensus" to describe the acrimonious intel-lectual strife. An outsider reading the then-current literature on the philosophy of science is hard put to find a "standard account" that has not been under attack from some quarter; even the attackers were attacking each other. Not only futurists, but also any seekers of some sound theory of knowledge during this period found their feet, if not "firmly planted in mid-air," at least unable to rest on any ground that was not being undermined, because the world's intellectual communities were shaking and splitting in a series of seismic shocks and countershocks.

The turmoil in the philosophy of science, of course, was a part of larger social phenomena. This was a period of social discontent, revolution, and reaction, especially in the Western world, as we saw in chapter 1. Social protest spread to the university campuses, the city streets, and the capitals of the world. Self-styled revolutionaries placed establishments under siege and threatened destructive anarchy. The social sciences, themselves in crisis, entered their heyday of student interest and large enrollments, ironically, at the same time that they often failed to understand, much less predict, what was going on in the society around them. Everywhere long-standing theoretical assumptions were challenged, pluralism grew more marked, and polarization occurred (Bell 1971; Giddens 1973; Szomptka 1979), things that have continued in some disciplines until the present day (Horowitz 1993).

The rapid growth of the Welfare-Warfare state, the culture of the young, the New Left, the psychedelic counterculture, the civil rights movement, the discovery of the young humanistic Marx, Marxist-motivated social change in the third world, existentialism, Black Power, the bitter domestic conflict about the Vietnam War, civil disobedience, or simply dropping out—all of these things helped create a sense of crisis, sometimes combined with hope for a better world and more often accompanied by bitter disillusionment and despair (Friedrichs 1970; Gouldner 1970).

"Straight thinking" was disparaged and "stoned thinking," whether achieved through drugs or through spiritual techniques, was elevated to a place of honor among followers of the so-called "psychedelic movement." Straight thinking, that is, rational, scientific understanding of external reality, was depicted as pedestrian and pessimistic, characteristic of epistemological dinosaurs. In contrast, stoned thinking, the intuitive understanding and the experience of inner reality, led to optimism, joy, and a cosmic oneness. "When the psychedelic movement crashed shortly afterward, many of its leaders renounced drugs in favor of religious or philosophical concerns" (Weil 1972: 412). And more than a few started drifting into meetings and conferences of various futurist groups.

Looking back at the late 1960s and early 1970s, most of us can agree that there were some constructive changes, such as the developments in civil and political rights in the United States, although some people still disagree even about them and certainly about others, such as America's involvement in the war in Vietnam. What is in retrospect most striking is how vulnerable the world of academia was. In field after field, revolutions of the larger society were mimicked not only in the social structures of the campuses, but also within the intellectual disciplines themselves. Established concepts and theories were attacked. New concepts and theories were put forward to replace them. All established intellectual traditions became suspect as devices used by the powerful to control the powerless.

Is it, thus, any wonder that one of the most cited books in the history and philosophy of science became Thomas S. Kuhn's *The Structure of Scientific Revolutions* (1962)? Kuhn who made so much of "the climate of the times" as an influential factor in his interpretation of the history of revolutionary shifts in scientific "paradigms" was himself perhaps his best example. He was the quintessential product of his day.

And is it any wonder that the intellectual establishment of the day would have a clenched philosophical fist defiantly shoved in its face with the publication of Paul Feyerabend's provocatively entitled *Against Method: Outline of an Anarchistic Theory of Knowledge* (1975)? People weren't talking the same language out there in the political and social world—or so it appeared from all the conflict—and scientists using different theories, Feyerabend emphasized, weren't talking the same language either. They couldn't talk to each other because their theories were *incommensurable*. Israel Scheffler (1967: v) recognized the threat as early as 1966. Critics, he wrote, were calling "into question the very conception of scientific thought as a responsible enterprise of reasonable men [and women]."

The upheaval in the philosophy of science was mirrored in the humanities and social sciences. The empiricist and positivist citadels of disciplines ranging from English and history to political science and sociology were threatened, just as English-speaking social philosophy was undermined, "by successive waves of hermeneuticists, structuralists, post-empiricists, deconstructionists and other invading hordes" (Crews 1986: 36).

Futures studies was growing up during this period of threats and underminings and in the midst of the agitation and bewilderment they engendered. It was both part of the period of social protest and change and a reaction to it. Toffler (1970), reacting to the sense of crisis of the times, devoted his *Future Shock*, as I said earlier, to a warning about accelerating social change and the need for futures thinking both to understand what was going on and to cope with a coming future that was going to be dramatically different from the past. But futures studies as an emerging science was itself affected by the social upheavals of this historical period, none perhaps being more consequential than the attacks on science in general and positivism in particular.

Some futurists, for example, Gordon and Helmer, developed their innovative methods and justifications of their studies of the future largely following the canons of science as defined by empirical or logical positivism. Other futurists such as Willis W. Harman and James Ogilvy did so more by following the contrary views of writers launching attacks on positivist science—or by unwittingly adopting prepositivist views of romanticism or rationalism. Realism vs. idealism, explicit rationality vs. implicit intuition, and scientism vs. humanism were polarities built into the emerging futures field.

Futurists, in other words, were struggling to discover their proper epistemological roots just at the time when some philosophers and historians of science believed that the philosophy of science itself had been—or ought to be—uprooted. Some futurist were searching for the knowledge foundations of futures studies just at a time when foundationism itself was being demolished. At the extreme, science itself as a superior—or even adequate—way of knowing was called into question. For some writers at the extremes, there was *no* acceptable theory of knowledge at all that justified scientific statements (Weimer 1979).

What is Positivism?

Distorted Definitions

Listening to the debate between positivists and antipositivists, we find each side unfairly caricaturing and polarizing the other in extreme ways. The result is, if you believe the critics of each view, that there is little of worth in either view. The antipositivists argue that positivists are doomed to being superficial and naive because they look only at the observable and isolated manifestations of phenomena, only the surface of things. They ignore the deeper truth and endorse the political status quo: "*Chiens de Garde*, Jackals of the Established Order." They are one-dimensional, cannot ask deep questions nor consider radical alternatives. They have shackled and blinkered themselves (Gellner 1985: 4–5),

The view of the antipositivists of whatever stripe from the perspective of the positivists, however, is no more complimentary. Antipositivists are viewed as talking and writing nonsense in high-sounding language; hiding trivialities in a maze of obscurities; speaking unclearly, writing unclearly, and thinking unclearly. They are "an Army of verbiage-intoxicated, pseudo-rebellious windbags" (Gellner 1985: 6).

One problem with this debate is that it is not usually clear exactly what the antipositivists take positivism to be, that is, other than a dirty ten-letter word. Peter Halfpenny (1982: 114–15) sheds some light on the topic by pointing out that there may be as many as twelve different versions of positivism, some more—and some less—different from the others. For example, he shows how positivism has been defined as several different theories of knowledge, different theories of scientific method, a theory of history, a theory of meaning, a unity-of-science thesis, and a secular religion.

The debate is difficult to summarize in detail because both positivists and antipositivists alike in a given instance have taken some aspect, distortion, or genuine version of positivism to defend or attack while ignoring others.

Features of Positivism: "The Received View"

Let us try to define positivism both accurately and fairly as it was when the most recent revolt against it began. Fortunately, some writers, including Gellner (1985: 12), have already done so and an authoritative account does exist. Suppe (1977: 16) calls it "the received view."

Positivism is a theory of knowledge following or building on the work of the "Vienna Circle" of the 1920s and 1930s associated with such people as Moritz Schlick, Rudolf Carnap, and Otto Neurath. It includes the "Berlin Association" and the work of Hans Reichenbach and Carl G. Hempel. It prominently features the work of Karl R. Popper, perhaps its most central figure in recent decades, despite his denials that he was a positivist. Many other writers, of course, contributed.

Nine key defining features of "the received view" of positivism as a theory of knowledge about mid-1950 are as follows:

1. A focus on science as a product, a linguistic or numerical set of statements;
2. A concern with axiomatization, that is, with demonstrating the logical structure and coherence of these statements;
3. An insistence on at least some of these statements being testable, that is, amenable to being verified, confirmed, or falsified by the empirical observation of reality;
4. The belief that science is markedly cumulative;
5. The belief that science is predominently transcultural;
6. The belief that science rests on specific results that are dissociated from the personality and social position of the investigator;
7. The belief that science contains theories or research traditions that are largely commensurable;
8. The belief that science sometimes incorporates new ideas that are discontinuous from old ones;
9. The belief that science involves the idea of the unity of science, that there is, underlying the various scientific disciplines, basically one science about one real world (Hacking 1981: 2).

This is by no means a comprehensive list of the features of the received view of positivism. Even adding more features would not be fully adequate, because the authors who have suggested them (Gellner 1985; Hacking 1981; Suppe 1977) are summarizing many different and sometimes conflicting views. Nonetheless, the above characterization conveys a rough conception of the positivist view of the nature of science as it was by the mid-1950s.

What is Postpositivism?

"The Scientific Revolution" and Scientific Revolutions

If we take the long view, then it is possible to talk about "the" scientific revolution, rather than many different scientific revolutions. That is, we can view the interrelated events and developments in science, spanning perhaps as many as five hundred years from 1300 to 1800, as a single set of phenomena. The key century, most scholars agree, was the seventeenth, because it was then that the most significant shifts in perspective and the greatest transitions of ideas took place. "The central point is that there is exactly one scientific revolution, at least when the term 'revolution' became a technical term for historians of science" (Hacking 1986: 23). Moreover, it may have been the most important series of events in the history of modern Europe.

Throughout the course of the development of modern science, there have been vociferous debates and swings of opinion both among various proponents of different theories of scientific knowledge and between them on the one hand and a variety of antiscientific writers on the other hand. In the 1890s, for example, there was an early revolt against positivism within the philosophy of the day "which swept even Britain, the bastion of empiricism" (Halfpenny 1982: 24). The presumed "mechanistic and materialistic thinking" of positivist philosophy was attacked by those who believed that intuition was superior. They thought that the conscious reason of the positivists should be replaced by some "combination of rational and affective processes too minute to be identified" that existed on the fringe of consciousness (Hughes 1958: 30).

The modern revolt against positivism contains some of these same elements, a leaning toward idealism and romanticism and away from realism and empiricism, toward subjectivism and relativism and away from objectivism and certainty. Its beginning can be conveniently (if somewhat arbitrarily) marked by the publication in 1962 of Kuhn's aforementioned *The Structure of Scientific Revolutions*. Kuhn has come to be prominently associated with many of the major ideas of the new revolt against positivism. Partitioning the centuries of scientific developments by shorter time periods and particular scientific subject matters, he viewed particular sciences as progressing through alternating periods of "normal science" and periods of change so rapid and profound that it is reasonable to refer to them as "revolutions." Using his original—but now outdated—terminology, we can summarize his stages of scientific development as follows:

Paradigmatic pluralism is a prescientific stage in that there is no dominant paradigm defining a field, its practitioners, its problems, nor its direction of work.

Emergence of a unified paradigm is a stage during which a single paradigm begins to dominate a field. The field itself, its boundaries, its adherents, and its problems are defined more clearly.

The stage of *normal science* occurs when a single paradigm achieves complete dominance in a field. The reigning paradigm defines the field's problems and its solutions to them. Great progress occurs, but it is largely a mopping-up operation.

The stage of *beginning of doubt* occurs as inconsistencies and uncertainties are no longer easily resolved within the dominant paradigm and become recognized as persistent anomalies. Some members of the scientific community begin to ask questions that have no apparent answer within the dominant paradigm, to depict other scientists as dopes, and to treat some aspects of the paradigm as questionable.

A *scientific revolution* is the next stage and takes place when a new and competing paradigm is proposed as an acceptable alternative to the old paradigm. This is a period of struggle between the adherents of the old and the new paradigms over disciples and control of the professional journals, textbooks, research grants, and the programs of meetings of professional organizations. In Kuhn's conception of process and change there are many specific scientific revolutions.

The recurrence of normal science is the final stage which comes about when the new paradigm becomes dominant in the field. The core of the field, its boundaries, and its membership are again clearly defined. Once again, efforts are focused on the problems generated by the science, and puzzles are defined and solved within a reigning paradigm, but now the new one. Great progress again occurs with a mopping-up operation within the new paradigm. The unconverted scientists have died, retired, become college deans or presidents, returned to undergraduate teaching, or simply defined themselves—or have been defined by others—out of the field. Then, of course, doubt may begin again and the whole process repeated (Kuhn 1962; Friedrichs 1970; Blum 1970).

Kuhn views scientists as working from a given point of view or *Weltanschauung* that shapes their interests, how they view phenomena, and their conceptions of what a good theory is. Moreover, this *Weltanschauung* is dynamic and is constantly subject to revision. Occasionally, the revisions become so large and significant that the *Weltanschauung* itself is undermined and is eventually overturned and replaced by another one. Thus, Kuhn concludes "that scientific change fundamentally is revolutionary" (Suppe 1977: 135–36).

Of course, there were recent precursors to Kuhn's notions. Among them are W.V. Quine who published "Two Dogmas of Empiricism" in 1951, Stephen Toulmin who anticipated some of Kuhn's ideas as early as 1953, and Norwood R. Hanson whose thoroughgoing criticism of positivism published in 1958 included, among other similarities to Kuhn's views, the contention that all descriptive terms are theory laden. Even before that, as early as the mid-1930s, the critical theorists of the Frankfurt School in Germany had been attacking positivism and empiricism, a tradition that has been continued, although with considerable revision, by Jürgen Habermas. Also, Feyerabend must be mentioned again because he takes some of the ideas of Toulmin, Hanson, and Kuhn and pushes them to the extreme (Suppe 1977: 166).

Again, many other writers had helped prepare the ground. For example, Kuhn appears to have superimposed his account of epistemological crises in natural science on some earlier writings of Michael Polanyi. For example, he took from Polanyi the idea that all justification takes place within a social tradition and that such a tradition functions within a scientific community to enforce agreement about what contrary evidence or difficult questions will be put aside (MacIntyre 1977: 465). Also, Kuhn was influenced by Alexandre Koyre in his emphasis on changes in perspective as a key to intellectual transformations (Hacking 1986: 23).

Features of Postpositivism

By the end of the 1960s, many writers on epistemology were saying that positivism was no longer a plausible theory of knowledge. Reporting on a 1969 symposium on the structure of scientific theories, Suppe (1977: 115), for example, said that in the face of all the criticisms the Received View was hard to defend and that most philosophers of science then agreed that it was inadequate. But there was no general agreement about what the source of its inadequacy was. Thus, although many voices joined in a "swan song for positivism," they were not all singing the same tune.

One voice or another, though, was raised against one or more of the above nine tenets of the received view of positivism until all of them, except perhaps for the eighth dealing with discontinuity, had been called into question. In order to highlight contrasts with positivism, I describe postpositivism with the same nine categories used above. Postpositivism, then, is a theory of knowledge characterized by the following nine characteristic features:

1. Instead of focusing only or mainly on science as a product, critics of positivism shifted to a concentration on science either as an activity or a developmental history. Sociologists, particularly, said that research should center on what scientists actually do. Thus, they studied scientists' activities in the laboratory, their networks of influence and commu-

nication, their social organization and subcultures, their norms and their violations of them. Examples are the UNESCO six-country study of the factors contributing to research performance (Andrews 1979), Latour and Woolgar's (1979) study of a research laboratory at the Salk Institute, Collins' (1985) participant-observational study of three cases of scientific replication, and Gaston's (1973) study of originality and competition among British high-energy physicists. Gaston, for example, contrasted the scientific norm of sharing information with the contradictory secret behavior of the scientists. Such secrecy was engendered by the competition among scientists for recognition of being first with a discovery.

Kuhn himself is a prime example of the historical—as opposed to the philosophical—approach. He focused on the processes of change and development in a science and what explains them, for example, change in paradigms or in what, as we saw in chapter four, he now calls "disciplinary matrices."

2. Instead of a concern with axiomatization and the beliefs that theories "have a *deductive structure*" and that "tests of theories proceed by deducing observation-reports from theoretical postulates," postpositivists generally believe that theories do not have "tidy deductive structures" (Hacking 1981: 1,4).

3. The positivist insistence on testability of theories and the belief on which it rested that there is an external world to test them against gave way to a number of contrary postpositivist beliefs. One is that facts alone don't overthrow a theory; some alternative theory must be available. Another is that scientists construct their own versions of realities by their theories, methods, analogies and metaphors. Facts, in other words, are theory laden. Still another, more extreme view, as depicted by Hacking (1981: 4), is that there really *are* different realities, that "in deploying successive paradigms we rather literally come to inhabit different worlds."

4. Postpositivists often argue that science is not highly cumulative, because science proceeds by revolutionary jumps.

5. Postpositivists believe that science contains cultural biases. Scientific knowledge is not culture-free.

6. Postpositivists also believe that scientific results are influenced by the personality and social position of the investigator. Thus, according to this view, a totally unbiased perspective is impossible and so, too, is a socially and psychologically free science. Thus, one's personal life history and psychology and one's positions in society and

various socioeconomic structures, as well as one's cultural setting, are viewed as affecting a researcher's or a theorist's characteristic ways of conceiving of things and set limits on the dimensions of possible thought. Also, the political clout of different scientists and, ultimately, a consensus of a community of scientists shape the definition of "reality" accepted by a given science. Knowledge, more broadly, inevitably serves special interests, usually those of established and powerful groups in society, and is especially insidious when it is masked behind the false claims of disinterest and value-free statements.

7. The positivist belief that theories or research traditions are largely commensurable was attacked on the grounds that, even if two theories used the same term, the meaning of the term—and the concept to which it refers—are not necessarily the same. Feyerabend (1975), for example, argues that concepts have no meaning in isolation; their meanings come from the theoretical context in which they are used. Thus, if any statements from one theory appear to contradict empirically any statements from another theory, it may simply show that the concepts that appear to be the same in the two theories really have different meanings. Thus, theories taken from different research traditions are not commensurable and evidence cannot be used decisively to choose among them. In terms of early Kuhn, each "paradigm" defines for itself what counts as relevant evidence with which it ought to be tested (Hacking 1981: 4).

8. Both positivists and postpositivists agree to some extent with the notion that science incorporates new ideas that are discontinuous from old ones, although positivists say "sometimes" while postpositivists say "almost always."

9. In general, postpositivists believe that science does not constitute a unity. They disagree that less profound sciences can be reduced to more profound ones, that, for example, sociology is reducible to psychology, psychology to biology, biology to chemistry, and chemistry to physics. Rather, science is viewed as being composed of many different "knowledges" each relative to its particular topic and community of scientists (Hacking 1981: 2,4).

The Spread of Postpositivism

The fallout from the recent revolt against positivism appears to have a long afterlife. It drifted into many fields of learning where, in some, it became a hot topic. It spread not only into literature, art, and the social sciences, but also into architecture, mathematics, and linguistics. In philosophy itself, some writers took it to absurd extremes. Walter B. Weimer (1979), for example, pushed his postpositivist critique of science close to the point of total and unalterable skepticism. He says (pp. ix–x) that "traditional conceptions of science and its methodology have been examined, found wanting, and are in large part being abandoned." Labeling the heretofore dominant metatheory of science "justificationism," Weimer leaves us little hope because he manages to tar with the same brush not only logical positivism and logical empiricism but also some of the seemingly contradictory positions that have been offered as substitutes such as existentialism and hermeneutics. In the end, Weimer's nihilistic view is that no genuine knowledge is rationally possible. All we have left is a multitude of competing fictions.

This pessimistic view became widely shared between about the mid-1960s to the mid-1980s among the world's intellectuals, many of whom had embraced the ideology of "postmodernism." Inspired by Martin Heidegger and Friedrich Nietzsche and, more recently, by such writers as Paul de Man, Jacques Lacan, Jacques Derrida, and Michel Foucault, postmodernists added to the antipositivist movement. In literary criticism, postmodern scholars, often known as "deconstructionists," claimed that language has no relationship to reality and that textual meaning was inherently indeterminate. In the social and natural sciences, extreme postmodernists claimed that there is no causality, no determinism, no objectivity, no rationality, no responsibility and no truth. For them, "no grounds exist for defensible external validation or substantiation." Their goal is "to register the impossibility of establishing any...underpining for knowledge" (Rosenau 1992: xii, xiv).

Some extreme postmodernists "talk of terror, suicide, and violence as the only truly authentic political gestures that remain open." They believe that evidence is a meaningless concept, that knowledge claims have no foundations, and that we might as well give up trying to know or understand anything. They reject not only the idea of truth, but also the idea of moral agency and teach that no value system is better than any other. They believe that everything—both truth and goodness—is arbitrary human construction (Rosenau 1992: 23–24, 114).

Ironically, in the extreme, the postmodernist even rejects Kuhn's postpositivist model of science "as a series of successive paradigms and announces the end of all paradigms. Only an absence of knowledge claims, an affirmation of multiple realities, and an acceptance of divergent interpretations remain" (Rosenau 1992: 137).

In the social sciences where the humanist-oriented social scientists, social activists, and political philosophers after the Second World War generally lost out in defining the core problems to empirical researchers and abstract theorists working in a positivist framework, the publication of Kuhn's book fanned smouldering resentments and discontent into flaming revolt. Scientific methods came under attack. Attackers offered in their place a variety of alternatives, including sophisticated versions of positivism (but with labels other than "positivism"), some genuine revisions and improvements, some vague or unintelligible prescriptions, and some perspectives that were extremely subjectivist and relativist.

Shulamit Reinharz (1979), for example, proposes the use of "subjective feelings" of the researcher as a valid source of information about the social world as a part of what she calls "experiential analysis." Since experiential analysis involves data gathering of a sort even though limited, it is not necessarily entirely antipositivist. Yet the author's stated aim is to provide an epistemology that is fundamentally antiobjective. It's what's inside the researcher that counts as much as—or more than— what's outside in the social world itself. Knowledge about *my* data become more important than knowledge about my *data* (Gellner 1985: 20).

Another example is "naturalistic inquiry" proposed by Lincoln and Guba (1985). It is designed to be postpositivist and assumes multiple realities, the inseparability of knower and known, and, among other things, the impossibility of distinguishing causes from effects. Phillips' (1973) *Abandoning Method* stems from a disillusionment with standard scientific procedures dominant in the social sciences and contains some convincing critiques of them, but he throws the baby out with the bath by endorsing beliefs that make it difficult to falsify any truth claims no matter how improbable. Mitroff and Kilmann (1978: 107) announce that "the entire field of science is in need of revolution and revision—not just the philosophy and sociology of science."

Kuhn and others, such as Feyerabend, had opened the gates, and a flood of writing poured through that tended, as Gellner (1985: 20) says, to reduce the whole scientific enterprise to a total subjectivism. This subjectivism fails to recognize the drive behind scientific inquiry, "which is to know something true independently of ourselves."

The Decline of Postpositivism

Postpositivist views, after a period of grace and dominance, began to fall as they came under increasing criticism (Newton-Smith 1981). They began losing out as serious contenders for a viable philosophy of science with some philosophers confidently announcing the coming of a post-postpositivist phase. By the late 1970s, even while some writers were still discovering and exploring postpositivism, Suppe (1977) was able to say that contemporary philosophy of science, although strongly influenced by postpositivism, had gone beyond it and was heading in new directions. He concluded that postpositivism was *passé*, although, of course, it was then still waxing in the social sciences and some other disciplines.

Although Suppe may have exaggerated the degree of newly achieved consensus and failed to alert us to the possibility that the "new directions" may mean scatteration to the four winds—an era of "creative chaos," to put the best possible label on it—he is certainly correct in saying that the leading thinkers have gone beyond postpositivst views. By the end of the 1980s and the beginning of the 1990s, although postpositivist views remained widely held, we had entered the beginning of a post-postpositivist or post-Kuhnian period. As early as 1977, in the preface to *The Essential Tension*, Kuhn himself seemed to be saying, "*Je ne suis pas Kuhniste*" (Stehr 1978: 180).

By 1992, Rosenau had effectively questioned postmodern beliefs as a viable theory of knowledge. She demonstrates that postmodern beliefs involve obscurities, contradictions, destructiveness, confusions, intellectual anarchy, pessimism and the rejection of truth even as a goal. Thus, any futurists who remain seduced by antipositivist ideas, as well as all serious thinkers, ought to be motivated to move beyond postpositivism to a theory of knowledge that has sounder foundations and more useful consequences. Moreover, futurists in particular have still another reason to go beyond postmodern ideas, because even moderate postmodernists tend to reject policy-oriented futures thinking and extreme postmodernists tend to be totally present-oriented (Rosenau 1992: 169, 171).

But that sounder theory of knowledge must also correct the errors of positivism. For postpositivism was an understandable reaction to the hubris and smugness of scientists and the uncritical confidence and acceptance of the results of modern science (Rosenau 1992: 9). It was a reaction, too, to the undue and often unwarranted influence of scientists over policy decisions, many of which, such as nuclear testing and bombing and "scientific" management by American decision makers in Vietnam, had some disastrous results.

Postpositivism has been beneficial in leading to more humility and caution in asserting scientific facts, more understanding of the provisional and partial nature of scientific knowledge, greater appreciation of the ways in which unconcious cultural assumptions can sometimes bias scientific results, more insight into how the personality and social position of the investigator can shape his or her research, more concern about how the community subculture of particular scientists shapes and directs scientific inquiry itself, and more openness to the view that technical knowledge may not contain all—or even most—of the answers needed when making policy decisions.

These beneficial contributions ought to be incorporated into any proposed alternative theory of knowledge, as, indeed, they are in the critical realist view discussed below.

Physics and Mysticism

We cannot end this discussion of the spread and decline of postpositivism without mentioning the role played by modern physics. Fritjof Capra, for example, captured the public's imagination with his explorations of the parallels between modern physics and Eastern mysticism in his bestseller, *The Tao of Physics*. Wayne I. Boucher (1986) has said that this mystical trip has resulted in some physicists talking as wildly as the mystics of the Middle Ages did about the nature of reality and the universe. He may be right. For example, one physicist or another has suggested that subatomic particles are conscious in their own way, that the universe is intelligent, or that poltergeists are real (Talbot 1986).

A more restrained claim is that some of the efforts of mainstream physicists to interpret quantum mechanics paved the way for the attacks on positivism and a revival of the extreme subjectivism and relativism that came with the rise of postpositivist philosophies. As early as the 1927 Copenhagen interpretation, which was accepted as the official understanding of quantum theory, the notions of uncertainty, a noncausal reality, indeterminism, and the impossibility of "an unequivocal description of nature" were widely accepted.

Yet a number of physicists—including Einstein, Planck, Schrodinger, and de Broglie—have found the Copenhagen interpretation unacceptable (Suppe 1977: 182–83). Heinz R. Pagels (1982: 154) discusses alternatives and claims that "quantum weirdness" simply does not exist for the macroworld. David Bohm attacked the Copenhagen interpretation in the 1950s and constructed an alternative philosophical view. Bohm totally rejects the renunciations of causality, continuity, and the objective nature of reality (Suppe 1977: 184).

It is beyond the scope of this book to pursue this particular debate here, except to say that many formerly suppressed or defeated antipositivists in any number of fields from political science and sociology to history and literary criticism were emboldened as the news spread that hardheaded, "real" scientists such as physicists had come to question the presumed certainties of science. With publication of Kuhn's book in 1962, the most recent revolt against positivism began in earnest. But today that, too, is passing. We have moved beyond it to a post-Kuhnian theory of knowledge, critical realism.

What is Critical Realism?

Features of Critical Realism

Using the same nine characteristics that I used earlier to describe some of the key features of positivism and postpositivism, I give below a description of critical realism. It is a post-postpositivist (post-Kuhnian) theory of knowledge according to which:

1. Science is a body of linguistic or numerical statements about the nature of reality, and, also, includes an interest in the activities of scientists and the history of science and its institutions.

2. Science includes a concern about the logical structure and coherence of these statements and, also, a special concern about their utility in manipulating the world to achieve human goals, as, for example, in the conscious process of using known cause-and-effect relationships to create desired ends.

3. Science rests on the assumption that, although many aspects of reality may always remain beyond human ability to observe and understand, how the world really is plays a decisive role in the achievements of science; truth can be known within the limits of human senses and intel-

lect, even though it is not absolute and is fallible, conjectural, conditional, corrigible, tentative, qualitatively judgmental, and presuppositional; warranted assertability is possible; some assertions are true and some are false, and we can frequently justify our beliefs, empirically or logically, about which is which; scientific "wars end only when some compelling new line of argument eventually emerges to persuade everybody of the truth of one side" (Papineau 1993: 15).

4. Science is cumulative to an important degree, partly because even revolutionary jumps usually incorporate many current scientific findings and beliefs (as Kuhn acknowledged).

5. Science faces a threat to its validity because of possible cultural biases that can distort the truth.

6. Science also faces threats to its validity because of possible personal and social biases; however, each of these biases—cultural, personal and social—can be self-consciously guarded against more or less effectively; intersubjective evaluation of research findings permits correcting them; objectivity, thus, can often be achieved. The abilities to learn and carry out procedures of objective observation and of making accurate and warranted assertions about the nature of reality are not limited to the members of any particular nationality, race, religion, social class, gender, or age group. "Science is the most reliable path to truth that we know…precisely because it transcends personal beliefs" (Davies 1993: 12).

7. Science contains theories or research traditions that generally overlap enough so that some contradictions among them can be noted and critically tested.

8. Science incorporates new ideas both by small continuous additions and by bringing in discontinuous ideas; preparadigm, normal, and revolutionary science don't alternate in a tidy cyclical sequence; rather, they may all be going on at the same time.

9. Science may involve a basic unity of science, whereby the less profound sciences can be reduced to more profound ones, or it may not; it is an empirical question that remains to be answered.

Thus, unlike postpositivist views, a critical realist view assumes both that there is a reality that exists and that we can test many of our ideas about it to see whether they are most likely true or false. Unlike positivist views, though, it emphasizes the conjectural aspects of knowledge, the many threats to validity, and the limitations to knowing with certitude.

Critical realism owes much to various positivist views. Alvin W. Gouldner (1985: 258–59) reminds us that in its day positivism was an important intellectual advance. It opposed metaphysical invisibles and many conventional and established religious beliefs. Even though it was mistaken to argue that it was suppositionless itself, positivism was—and is—correct in its emphasis on the resolution of public contention by appeal to facts. That emphasis is beneficial because it makes problematic any society's understanding of itself and, thus, brings into question dominate definitions of social reality. In this, of course, positivism is similar to the culture of critical discourse which is discussed below.

But critical realism also owes much to the postpositivist critiques of positivism and, more generally, to humanistic critiques of scientism. Critical realism recognizes that all science is to some extent presumptive and qualitatively judgmental in nature; that presuppositions are inevitable; that the historical context affects science (even physics); that science is a social process and that scientists are human beings (who are just as greedily ambitious, competitive, unscrupulous, and self-interested as other human beings); that the manipulation of events should have precedence over mere correspondence in time and space as a criterion of powerful knowledge; that social causality can and should be linked to people's intentions and purposes as well as to passively observed concomitancies; that plausibility is sometimes the best result that we can obtain in science, that it can be achieved in different ways, and that it is enhanced by using multiple methods, measures, and approaches; that science invites creativity, imagination, intuition, and insight, as we saw in chapter 4; and that it recognizes uncertainties (Campbell 1984b).

Critical Realism and the Traditional Theory of Knowledge

We can also place critical realism in the larger philosophical conceptions of knowledge and into the century-old debates among empiricism, rationalism, and skepticism. In this larger debate, other names for critical realism that have been used include "fallibilist realism," "fallibilism," "mitigated skepticism," "critical rationalism," and "critical empiricism." It has also been called "sophisticated indirect realism" to convey the idea that, even though observation statements are fallible, they concern the external world and are not mere subjective experiences (Musgrave 1993: 274). Critical realism, as the reader might guess from the "falliblist" labels, owes much of its general philosophical elaboration to the work of Karl R. Popper.

To be genuine knowledge rather than mere belief or opinion, according to traditional philosophy, three conditions must be satisfied: (1) a person must believe that some proposition is true; (2) the proposition must be true; and (3) the person is able to justify that the proposition is true. A belief, then, "is reasonable if and only if it is certain or justified" (Musgrave 1993: 3).

As we have seen, critical realists give up the commitment to certainty. They accept the skeptical belief that we cannot have certain knowledge, if we define knowledge as justified true belief. But they are unwilling to say, therefore, that we must give up trying to know and understand. They redefine knowledge as "conjectural knowledge," allowing for the possibility of the fallibility of their conjectures.

To know conjecturally that a proposition is true, critical realists also have three conditions: (1) a person believes that a proposition is true; (2) the proposition is true; and (3) the person is justified in believing that the proposition is true (Musgrave 1993: 298).

The first two conditions, clearly, remain the same in the critical realist view of knowledge as in that of the traditional philosophical definition, but there is a difference in the third one. The difference is between one of certain knowledge versus reasonable beliefs. Critical realists do not demand that the truth of the proposition be justified, but only that a person is justified in believing that the proposition is true. This, of course, allows for the possibility that conjectural knowledge may be false. When that happens, however, critical realists say that what they believed was wrong, not that they were wrong to believe it (Musgrave 1993: 282).

To show that it is reasonable to believe that some proposition is worthy of belief, we could adopt two different approaches. One is to try to show that it is true. Another is to "try to show that it is false, and hence unworthy of belief." The critical realist adopts the second approach, the way of criticism. If we find a reason to think that a proposition is false, then our belief in it is not justified. But, if we fail in our serious efforts to criticize it, that is, find no reason to think the proposition is false, then our belief in it is justified (Musgrave 1993: 281).

This leads to an important distinction between justifying a proposition (P) and justifying a person's belief in it. They are distinct judgments, because a person may be justified in his or her belief in P without having justified the proposition (*P*) itself, either conclusively or inconclusively. Critical realists believe that if *P* withstands serious criticism they are justified in believing *P* even though *P* itself is not justified in the traditional sense.

When critical realists say that the evidence "supports" a proposition, then, they do not mean "proves" or even "makes more probable." They mean that it "fails to refute" (Musgrave 1993: 290).

Why "Critical" and Why "Realist"?

Critical realism is "realist," because it assumes that a reality exists quite apart of human constructions of it. This is known as "ontological realism." It is the belief "that the world of which we have knowledge exists independently of our knowledge of it" (Skagestad 1981a: 77–78). As Cook and Campbell (1979: 29) say, critical realism "assumes that causal relationships exist outside of the human mind."

It is also realist because it assumes that the senses are a source, not of certain knowledge, but of "reasonable beliefs." Moreover, the senses are given a privileged place. What the senses tell us, that is, our perceptions of reality, is accepted as reasonable unless they fail to withstand criticism, while "other beliefs are reasonable because they succeed in withstanding criticism" (Musgrave 1993: 284). The reason for this special treatment of sense data is an evolutionary one: human senses have evolved over the millennia as basic mechanisms by which humans have been able to survive and flourish by adapting to or controlling the environment. Thus, we ought to override them only with good reason. Yet critical realists do not agree with classical empiricists that all beliefs are derived from sense data. "Beliefs typically transcend experience, though they are invented to account for it" (Musgrave 1993: 286).

Critical realism is "critical," as we have seen, because it emphasizes the strategy of falsification. Critical realists criticize propositions. They attempt to falsify them. Only propositions that are not refuted are considered worthy of belief. Scientific statements aim to be open and clear about procedures, methods, assumptions, theories, and the nature of the data on which they are based. Thus, others can examine and criticize them as well as attempt to get the same results using them. Thus, the effort to confirm or, especially, to falsify is to an important degree a community activity, as Kuhn said, and the critical examination of other scientists' work is part of the community norm.

It is also critical, because it assumes that human sensory and intellectual capacities are imperfect and, therefore, often fail to perceive valid causal relationships accurately (Cook and Campbell 1979: 29). Not only is some unknown part of reality beyond our senses, it is also beyond our present means to detect with all of our technological devices. We cannot detect radio waves without a radio nor television signals without a television set, for example, and we don't know what else out there remains unmeasured by any natural sense organ or technological instrument or indicator humans now have. But these limits do not "affect our pursuit of truth" (Larmore 1987: 21).

Most important, critical realism as a scientific mode shares the "culture of critical discourse." Gouldner (1985: 30) defines it as the proposition "that any assertion—about anything, by anyone—is open to criticism." That is, no assertion can be defended simply by invoking someone's authority, position in society, or personal character. All claims to truth must be considered equally on their own merits and equally subjected to tests of evidence, logical coherence, and comprehensibility.

Habermas' concept of an ideal speech community (1970a, 1970b) and Campbell's "disputatious communities of 'truth' seekers" are similar to the culture of critical discourse. Campbell (1986: 119), for example, says that it is part of the ideology of science that the "community of scientists is to stay together in focused disputation, attending to each others' arguments and illustrations, mutually monitoring and keeping each other honest until some working consensus emerges (but conformity of belief per se is rejected as an acceptable goal)." As he says further, "The ideology of science was and is explicitly antiauthoritarian, antitraditional, antirevelational, and individualistic.... Old beliefs are to be doubted until they have been reconfirmed by the methods of the new science. Persuasion is to be limited to egalitarian means, potentially accessible to all, that is, to visual and logical demonstrations."

In Kuhn's (1978) view, critical discourse is a dissensus that occurs in science only when the bases of a scientific field are in jeopardy and he insists that normal science focuses instead on puzzle solving. To the contrary, critical discourse does not necessarily imply dissensus. It can also be a source of consensus, when, for example, the use of logic and evidence is so strong as to convince even the most adamant doubters. Nor does critical discourse preclude puzzle solving. Rather, creating solutions to puzzles and showing how they withstand criticism is part of critical discourse. Moreover, critical discourse implies civility, because it is civility in discourse that allows disagreement on matters of great moment to be transformed into warranted agreement through the use of logic and evidence. Civility does not mean imposing conformity in thought, but allows both orderly disagreements on the one hand and persuasion based upon the merits of a given case on the other.

Critical realism is not based on a narrow definition of science as limited to quantification, white coats and laboratories, experimentation, or only to the new sciences and technologies. Although the search for truth may include these things, it also includes qualitative modes of investigation and the older humanistic disciplines to the extent that they search for truth openly and marshal evidence, qualitative or quantitative, to test their ideas, especially if they have put their ideas at genuine risk to be falsified by the data if they are wrong. The simultaneous pursuit of intelligibility and evidence, of course, is not always a harmonious activity in any discipline because the discovery of hitherto unsuspected facts "is just what may disrupt an hitherto intelligible account" (MacIntyre 1977: 455).

An Evolutionary View of Knowledge

In one sense, critical realism is evolutionary, because it assumes that the survival of the human species is linked to formulating and using knowledge about causes and their effects, particularly about causes that allow human manipulation of the world. That is, critical realists recognize that the human predisposition to infer causal relations is adaptive (Cook and Campbell 1979: 28–29).

In another sense, critical realism is evolutionary, too, because it views the development of knowledge itself as evolutionary, as a winnowing process involving three major evolutionary principles of variation, selection, and retention (Campbell 1986: 120). From this perspective, the creation of knowledge is a "fit-increasing process" during which change in the collective belief system of scientists tends to enhance the validity of statements about the world, physical or social. A diversity of ideas about the nature of reality are proposed; after they are examined, some are selected and others rejected, and, then, some of those selected are retained. Although they may never reach it fully, scientists, by such a process, tend to get closer and closer to the truth.

The variation of beliefs through time, of course, is at the expense of retention, so there is often disagreement within any scientific community over whether some beliefs should be retained, modified, or replaced by other beliefs. That is exactly the point. Such disagreement within the constraints of civil discourse, when arbitrated by observation of empirical data and logical analysis, contributes to the evolution of scientific knowledge.

Ontological realism, too, is related to an evolutionary view. Humans must have a world external to them or they would die. The law of entropy leads us to believe that without the life-support systems from which the human body draws energy, the human organism would move toward uniform inertness, that is, in this case, toward death. Being able to select accurately true statements about cause and effect is not simply an intellectual exercise, it also can contribute to the survival and evolution of individuals and groups. Moreover, appeals to observations of reality, whether qualitative or quantitative, are part of intellectual dispute-resolution procedures that make social life and collective action, and their mutual benefits for individual and group well-being and development, possible.

Reality and Text-as-Reality: A Difference

A hermeneuticist might remind us that depictions of reality, whether pictorial, verbal, or mathematical are themselves realities. Novels, photo albums, diaries, maps, theories, and so on, may represent some other reality, but like all texts, they are in and of themselves realities too. They are real novels, real albums, real diaries, real maps, real theories, and so on. We can hold them, look at them, or read them. To generalize, any set of statements or depictions purporting to represent or to model some reality other than itself is itself a reality.

Some scholar-hermeneuticists take on the task of analyzing the realities of models of reality themselves. So doing, they create still another reality: an analysis of maps, for example, which becomes a model of maps as models. Any text, though a description of something else, can itself become the object of interpretation. Thus, constructions of models of reality are themselves part of a new reality. Moreover, it is perfectly legitimate to be interested in studying these realities too.

Critical realists argue that it is important not to confuse these two senses of reality. There is that reality out there that the map, description, or model is trying to depict, even if "out there" from the outset involves observers' active creation in trying to know it. What observers then claim they know is both some aspect of the reality out there they are trying to learn about and a new reality of the model they have created in its own right which becomes subject to scrutiny.

An example from Samuel Beckett's play, *Krapp's Last Tape*, illustrates the point. Many of us have tried to keep a diary. Krapp, the character in Beckett's play, is an up-to-date type of person who uses a tape recorder to chart his daily life's events. In the beginning, he bounds into his apartment from outdoors exuding enthusiasm and joy, breathlessly recording glowing accounts of his many adventures and encounters with other people in a variety of places.

The tapes begin to pile up. He starts spending part of each day listening to old tapes, that is, the taped reports he had earlier made of his adventures outside the apartment. Then, he makes a new tape, describing the tapes that he listened to that day.

As time passes, tapes continue to pile up, and more and more of his time is spent in the apartment listening to his tapes of his listening to tapes, until, in the end, he never leaves the apartment at all. His whole life consists of listening to his tapes of tapes of tapes and lifelessly making new tapes about what he heard on the tapes that day. For example, at sixty-nine, sitting in a room piled high with tapes, he may be listening to himself at thirty-nine commenting on his youthful self.

Krapp's life, of course, is just as real when he spends it listening to tapes in the apartment as when he spends it outside of the apartment doing other things. The tapes and what they say are in some sense just as real as Krapp's interaction with other people in a variety of locations outside of his apartment. Yet, surely, it is important for scientists—and all of us ordinary people choosing the way we live our lives—to recognize the difference between Krapp's reality out of the apartment interacting with people and having adventures and Krapp's reality in the apartment listening to the texts of his recounting of his past adventures or of his past listening to still earlier tapes. One reality is only a pale and distorted copy of social life while the other *is* social life.

Certainly, we construct new realities with our statements, pictures, and models that aim to represent some other reality. But each is a *different* reality than the one we aim to represent. Of course, these statements, pictures, and models invite study and analysis in their own right just as any reality may. A world of maps is real and a science of a world of maps is possible, but it should not be mistaken for that other world of people, land masses, oceans, rivers, mountains, valleys, roads, towns, cities, or whatever which maps were constructed to represent.

Distortions of the Truth

Truth matters not only for the survival and, hence, evolution of humanity, but also for the success or failure of particular individual and group projects. Although sometimes things turn out well for us despite our erroneous beliefs about the nature of reality, over the long haul, people and groups with more accurate maps of reality, both physical and social, will be more successful in achieving their goals than those with less accurate maps.

To many people the above assertion is obvious, yet since we are faced with the ontological nihilism of extreme postmodern views as we prepare to enter the twenty-first century, it seems necessary to affirm it once again. Thus, let's look at a few examples. There are many to pick from, including beliefs that the human landings on the moon never happened (filming supposedly having been done in Nevada), that the Japanese did not really launch a sneak attack against Pearl Harbor in 1941 (a diplomatic message from Japan to the United States declaring war having been badly mishandled and unread by the appropriate Americans), and that Elvis Presley didn't really die in August of 1977 (having been seen in various places throughout the world since then). I've selected three others:

Three Examples

Did Nazi extermination camps exist? Mariette Paschoud, a teacher of history in Switzerland, has asserted that "there is no evidence" that the Nazi exterminations of Jews and others ever happened. She has also defended Henri Roques "who publicly asserts there never were any gas chambers or any innocent victims of the Nazis" (Bettleheim 1986: 1). In Canada, Ernst Zundel published a booklet claiming that the Nazis had no plan to exterminate Jews and did not use gas chambers to kill them. The booklet describes Auschwitz more as a Jewish country club than a Nazi death camp (*New Haven Register*, 1 February 1987: A18). These are only a few of the people who have made the claim that the Holocaust never happened.

The Holocaust is "the planned and systematic extermination by the Germans, during World War II, of approximately six million European Jews" (Reich 1993: 31). Other groups died in the extermination camps, too, including Gypsies, Poles, and Soviet prisoners of war, but the aim of the Holocaust was the mass annihilation of Jews. Some died of disease, starvation, and overwork in forced labor camps. Others were systematically herded into gas chambers and exposed to lethal doses of poisonous gas.

Shall we assume with the extreme postmodern relativists and subjectivists that any such assertion is as good as any other? Shall we agree with some extreme antipositivists that no truth can be justified in any case? Or shall we assume that we can marshal some evidence to prove within some reasonable degree of certainty that the claims are false if indeed they are false, or corroborate them if they are true?

An enormous pile of evidence, mostly irrefutable, has been examined by numerous historians to support the belief that the Holocaust occurred. It includes speeches and orders by the highest Nazi leaders; testimonies or diaries of SS officers, Nazi officials—including the commandant of Auschwitz—and ordinary German soldiers; it includes reports by the chiefs of mobile killing squads that executed over a million Jews; it includes the accounts by Jews who had been forced to drag corpses out of the gas chambers and burn them; it includes paper work that documents the building of gas chambers; and it includes, among many other things, detailed records of Jews transported to Auschwitz, Birkenau, and other death camps (Reich 1993: 31).

Did World War II end in 1986? In 1986 two Japanese soldiers in their mid-sixties left over from World War II were found in Malaysia. They had been deluded for forty years "into thinking that World War II was still on, that the communists were aiding Japan and that Malaysian security forces were Allied troops." They did not know that Japanese forces "in the Malay Peninsula surrendered to the British at Singapore in 1945." Nor did they know that by 1986 Japan not only had survived an American occupation, but also had developed one of the most successful economies in the world (*The New Haven Register* 13 January 1986: 39).

This story, if true, beautifully illustrates the proposition that the *consequences* of beliefs about reality may be real whether or not the beliefs themselves are true. To put it in W.I. Thomas's famous terms, "If men define situations as real, they are real in their consequences."

But it also highlights another proposition: false beliefs about reality can be damaging. For if based upon true beliefs, then behavior is more appropriate to the facts and more likely to be of greater potential value for the actor's survival and for achieving his or her intentions and desires. If the Japanese soldiers had held true beliefs about the war, then they could have returned home to Japan, reunited with their families, and continued their lives forty-one years earlier.

Were "body counts" in Vietnam accurate? James William Gibson in his book on Vietnam, *The Perfect War* (1986) gives many illustrations of self-delusions on the part of the American war managers. One chilling example is the case of "body counts." The production model of war that dominated the official American perspective meant that it was unthinkable to deploy the "massive apparatus of planes, helicopters, artillery, and troops" without a product. The number of enemy killed became the "product," and success meant "high productivity." Hence, there was a constant pressure to get high "body counts." Both the production model of war and the reward structure it fostered, whereby the performance of units and commanders—and the chances of promotion and of other rewards for officers and men—were measured by the body counts they produced, contributed to this pressure.

Under such circumstances it is no wonder that outright lying produced a lot of false numbers turned in as "verified" dead soldiers of the Vietcong and the People's Army of North Vietnam. To take one example, according to Lieutenant William Calley, his battalion commander threatened to take the command of his platoon away from him unless he became more "productive." As a result, Calley had his platoon open fire on the surrounding jungle. Then, he called in artillery fire as if he and his men were repelling a real attack. When the "mock little firefight" was over, Calley said that he reported a bogus body count of three and added a combat loss of some compasses that earlier he had mislaid somewhere (Gibson 1986: 125).

Gibson tells of numerous similar fabrications, the invention of body counts out of thin air, five enemy bodies becoming a reported 131 and 30-odd mushrooming to 312. Generals, colonels, majors, captains, first and second lieutenants all had to meet their production quotas and all faced the temptation to accept tacitly the lies of others as part of their own productivity. Other motives existed too. With one's dead and maimed buddies lying there could there possibly be no dead Vietcong? Were American lives to be lost or ruined for nothing? So an enemy body count was turned in, even when no enemy dead had been found.

Added to simple fabrications of body counts were what Gibson (1986: 126) calls "inferential counting rules," which are rules by which the soldiers inferred "the numbers of enemy dead according to some found object or sign—an enemy weapon, a blood trail, a dismembered body part, or other mark of enemy presence." Some examples of how such rules inflated the body counts include taking a severed limb as an entire body, counting graves (many being counted more than once by different units), counting an enemy gun as five bodies, assuming that so many rounds fired into an area must result in a certain number of enemy troops killed even when no bodies could be actually counted, assuming that the disappearance of blips on a radar screen after firing into the area meant so many enemy deaths, and on and on. As Gibson (1986: 128) says, such extrapolations were made not just once, but often by each different military unit that participated in a particular engagement. Thus, body counts were inflated several times over. Moreover, as Gibson (p. 158) goes on to say, on the grounds that not all enemy dead bodies could actually have been found, higher echelons of war managers further inflated the total body count by multiplying them by 1.5, a 50 percent increase. Please note that quantification is no guarantee of truth.

Gibson reports even more demented distortions of the truth than falsified body counts in the official version of the war. Lewis H. Lapham (1986: 28) in his review of *The Perfect War* summarizes a few of them:

> After the embarrasment of the Tet offensive, the war managers had no choice but to resort to increasingly fanciful literary devices. Their reports began to read like surrealist fiction; their metaphors became more vicious. In more than one village American troops murdered every man, woman and child and called the process "pacification"; the Air Force destroyed forests and crops and towns and called the process "Nation-building"; the military police herded thousands of peasants into concentration camps and called the process "urbanization."

Distortions of the Truth Matter

Some extreme postmodernists claim that any version of the truth is as accurate as any other, that there is no accurate version of the truth at all, or that truth simply does not exist. Such claims, I have tried to show, are false.

Moreover, the distortions of the truth matter. They matter in our human abilities to make our way in the world intelligently and effectively. To base our actions on false conceptions of reality means to function under a handicap in the struggle to survive and thrive. Distortions of the truth, when uncovered, are lessons about the consequences of ignorance or deception that should be heeded.

Distortions of the truth also matter as enduring facts by which we ourselves and our leaders can—and should—be judged. Distortions of the truth are not always—or even often—the outcomes of impersonal forces and institutional arrangements. Although such forces and arrangements help both to understand and explain the production of untruth, they are themselves the product of the decisions and actions of people who have responsiblity in given situations. But accountability means more than distributing blame. It also means designing new social conditions under which error will be reduced and conjectural knowledge made more accurate.

Also, distortions of truth (e.g., lying, dishonesty) matter because they undermine trust. Without trust, social life would be impossible. We must trust in order to learn from others, to work with others, to be cared for by others, to live with others, and to carry out most of the tasks and functions that keep a society going. One pillar of trust is the assumption that others are telling the truth, not just as they see it, but as it is.

The above accounts and many more that could be told illustrate why many futurists and other researchers and scholars are—or ought to be—unwilling to give up the idea of truth, even if we can know it only conjecturally and uncertainly. For, among other things, the idea of truth of a reality external to ourselves plays an indispensable role in reminding us that the "understanding we seek need not coincide with our present beliefs" (Larmore 1987: 21). Moreover, the alternative, ontological nihilism, "as a systematic position would seem to undermine curiosity and the motivation to persuade others of the errors in their beliefs"—if indeed the idea of such errors was even permissible with such a view (Campbell 1986: 115).

Subjective theories of truth (that a statement may be true for me but not for you and vice versa), too, have a similar consequence. They lead to agreeing to differ, although there is no real disagreement between different views if people believe that truth is relative or that there is no "real" truth. Therefore, it is useless to try to have a rational discussion about what is really true and whose view the evidence supports.

Often, this does not matter, but suppose people "have to reach a consensus because they have to act" collectively on whether some statement is true or false. Rational discussion is foreclosed if they believe that there is no truth or that all is relative. If they must act together, then one or the other must force the other to act on his or her belief. "In short, relativism about truth encourages the use of violence to achieve consensus in action (if not in belief)" (Musgrave 1993: 253).

Finally, even postpositivist writers, in some curious and contradictory sense, are themselves implicitly committed to the idea of truth. Masterman (1970: 87–88) refers perceptively to their "protest against the unconscious dishonesty and the swings of bias with which the history of science has been done in scientific textbooks up to now." Clearly, they are trying to tell us how science is *really* done and how scientific theories are *truly* constructed, changed, or discarded.

Critical Realism and the Future

"Knowing" the Future: Reprise

Critical realists conclude that conjectural knowledge is possible. They believe that, even if a proposition cannot be justified as being true, the belief in the truth of a proposition can be justified as being reasonable. From this perspective there is little philosophical difference in justifying beliefs in assertions about past and present realities on the one hand and beliefs in assertions about the future on the other. We can try to falsify our grounds in much the same way for believing in one or the other. Yet, because the ontological status of the future, as we saw in chapter 3, *is* different from that of the past and the present, we must consider further the application of critical realism to assertions about the future.

The problem of knowing the not-yet-existent future within a scientific framework, of course, goes back at least to the eighteenth-century philosopher, David Hume, as van Vught (1987: 187) has reminded us. He showed that the theory of inference then dominant, which was based on Francis Bacon's idea of induction by enumeration, was unjustifiable. An inference based on past and present observation, he insisted, does not necessarily apply to future observations. He undermines an inference such as the following: "All crows so far observed were black, therefore all crows in the world are black." According to Hume's critique, the contrary inductive conclusion may be true. The next crow we see may be white. We do not know before the fact of observing the next crow what its color will be.

Although any scientific explanation may be rephrased as a prediction, the prediction itself is not, according to this view, knowledge. Furthermore, this is so no matter how much past evidence supports the explanation. The prediction remains speculation until after it is tested by new (future) observation. When tested, it may be corroborated and we then *know* (within the limits of our ability to know the past) whether or not it *was* true. Thus, according to this view, predictions appear to be even more conjectural and uncertain than is our knowledge of past and present.

This is elementary, yet some of the continuing dissensus in the philosophy of science stems from the fact that some kernel or other of Hume's critique of empiricism can be leveled against most philosophical justifications of knowledge. Thus, the task of justifying knowlege—about the past *and* the future—is sometimes given up as futile (Weimer 1979). Ludwig Wittgenstein, for example, used the problem of the uncertainty of the future to undermine the whole justification of scientific generalizations (Gellner 1985: 167–68).

Popper's (1957) famous statement of the problem is an attack on what he calls historicism. It is unfounded, he argues, to use past trends or historical stages as bases for prediction. They are not, according to him, universal laws, but are facts unique to a period of time. Ian Miles (1975: 2), repeating Popper's criticism, says that "historicism fails to recognise" the "conditional nature of trends." A "specification of initial conditions and a statement of an operating law" would yield "a conditional prediction." But this is not what historicism does. John H. Goldthorpe (1971) sees what he calls crypto-historicist assumptions in the futures research literature. *The Year 2000* by Kahn and Wiener (1967), for example, relying as it does on a basic, long-term multifold trend for its speculations about the future, rests largely on historicist assumptions.

Reichenbach (1951: 91) says that the older empiricists did not see the difficulties arising from the fact that knowledge of the future is not of the observational type. Because predictions can be verified or falsified at a later time, they failed to see that knowledge of the future was different from knowledge of the past and present. They apparently didn't realize that "we wish to know the truth of predictions before the predicted events occur, and that when knowledge has become observational knowledge it is no longer a knowledge of the future."

Reichenbach (p. 252) tries to solve this problem by substituting "probable" for "certain" or "absolute" knowledge. He applies a frequency definition of probability based on a time series of observed past events. Observations of a time series, he points out, may result in convergence. As sample measurements of past events build up, we get closer and closer to the true figure until we reach the point where there is stability. Projection of the series into the future, he argues, is thereby made more accurate.

Indeed, futurists, as well as forecasters of all kinds, often do follow such a procedure in practice and sometimes assume that increased confidence in a prediction is thereby warranted. But that does not prove that we know the future with any high degree of confidence as a result. Thus, ultimately, Reichenbach fails to justify predictive knowledge logically, because Hume's criticism can be applied to the idea of probable knowledge as well as to the idea of certain knowledge, even though the explicit recognition of uncertainty is an improvement.

Popper has the most successful answer to Hume's anti-induction conclusion that no belief or prediction about the unobserved is any more reasonable than any other. Popper simply denies that we reason inductively. He argues instead that "What we typically do is jump to conclusions and use these conclusions to regulate our future behaviour." He says that having "formed beliefs or expectations in this non-inductive way, we employ them in perfectly valid deductive arguments in order to anticipate the future" (Musgrave 1993: 171).

For example, let's evaluate a hypothesis that bread will nourish us and a competing hypothesis that bread will not nourish us. Is one hypothesis more reasonable than the other? Consider this line of reasoning: If bread did nourish us Monday through Friday, then the hypothesis that bread will nourish us is unrefuted while the hypothesis that bread will not nourish us is refuted. This is a valid deductive argument (Musgrave 1993).

To generalize, we can say that it "is reasonable to believe unrefuted rather than refuted hypotheses." We tentatively accept them, even though they are unjustified in the strict sense, because they have withstood serious efforts to falsify them (Musgrave 1993: 172, 174).

Hence, Hume's argument about the invalidity of inductive reasoning may be considered irrelevant, because, in fact, we are reasoning deductively. "We jump to conclusions and we form hypotheses about the world.

Some of these general beliefs are reasonable beliefs in the fallibilist sense. We use such beliefs to anticipate the future by deducing predictions from them. Predictions deduced from reasonable beliefs are themselves reasonable. Hume is right that every prediction about the unobserved is uncertain; he is wrong that every such prediction is unreasonable" (Musgrave 1993: 286).

This is a sensible and sound argument. If they want to live—and to live well and effectively—people have no choice but to make their worldly ways over time using the best methods they have to form their opinions about the future, opinions on which they make their plans and pursue their projects. They apply past knowledge to anticipated future events, and make speculative leaps. They use their knowledge of cause and effect of past events to adapt to or control the future by their own actions. They act on their past experience and on their best guesses about the future, including the probable consequences of their own acts, and, if they wish to increase their effectiveness, they constantly revise their beliefs as they gain new experience.

Posits

With some modification, Reichenbach's (1951: 240) concept of a "posit" is useful for making assertions about the future within a critical realist theory of knowledge. "A posit is a statement which we treat as true although we do not know whether it is so." It is a statement about the future on which people can act *as if* it were true. Posits include, too, statements about the future on which people might or could act appropriately and effectively *if* certain circumstances were to prevail.

Reichenbach limits the concept by saying that it is the most likely outcome. For futurists, such a limitation is neither necessary nor desirable. True, actors often want to know the most likely outcome, but sometimes improbable events do occur. Also, possible events, no matter how unlikely, that may have potential for far-reaching changes are worthy of consideration, as Toffler says. Being prepared for them, just in case they improbably might occur, is a form of "fail-safe" contingency planning. For example, astronauts spend 95 percent of their time in shuttle-flight simulators "anticipating what might go wrong, battling a steady stream of problems, glitches, snags and errors. These range from minor frustrations, such as faulty computer readings, to life-threatening scenarios involving severe damage to the ship" (Broad 1988: 15). They spend endless hours practicing procedures for emergencies that may never occur. Futurists, too, posit alternative possibilities for the future, both likely and unlikely, and try to assess the probability and the importance of the impact of each.

Posits are treated as true on an "as if" or "what if" conditional basis for planning contingent actions. Thus, contradictory posits can be used to construct alternative plans. For example, the consequences of unlikely possibilities, such as a large-scale nuclear war between China and the United States, may be planned for on a contingent basis, even while efforts are being made to reduce the chances of its happening under a posit about negotiated peace. Posits, both likely and unlikely, can be used in planning a repetoire of alternative actions.

In sum, a posit, for futurists, is a statement about a possibility for the future and an estimate of the probability of its occurrence, no matter what the source of its rationale. Some posits, of course, are better grounded and more plausible than others. Some constitute conjectural knowledge and some do not. All posits may be useful in preparing for the future.

Surrogate Knowledge

Another distinction that is useful for an epistemology of futures studies is that between errors in knowing the present and the past on the one hand versus obstacles to knowing the future on the other. Alan Coddington (1975) refers to the former as "knowledge deficiencies" and to the information substitutes constructed despite the latter as "knowledge surrogates." Knowledge deficiencies refer to inaccuracies in knowing the past and present, to mistakenness, ignorance, error, deception, and delusion. We ought not to minimize these problems. They are largely the well-known and ever-present threats to knowing that positivists have discussed and have worked to overcome. They produce errors in prediction because the initial conditions, if wrong, are erroneous starting points and extrapolation based on them can compound errors.

Knowledge surrogates refer to posits about the future, but only those posits that have survived serious attempts to falsify them and that, therefore, constitute conjectural knowledge. Nonetheless, they remain hypotheses, even though they may be well grounded in past and present knowledge and logical reasoning. As conjectures, they involve speculating, expecting, anticipating, predicting, projecting, forecasting, and prophesying. Such knowledge surrogates are useful because they substitute for the unattainable knowledge of the future that is needed for competent and intelligent action. Such surrogate knowledge stands in for what we cannot yet factually know because it hasn't actually happened yet. Surrogate knowedge is constantly subject to revision in the light of new experience as yesterday's future becomes today's present.

For example, last winter's influenza vaccine is not likely to protect a person from next winter's flu because the flu virus evolves every year. Since it takes six months to produce the vaccine, by the time next winter actually arrives and scientists know what the strains of the virus are exactly, it is too late to produce the vaccine to protect against them. Consequently, by studying virus samples from around the world and looking for recently evolved and spreading strains many months before the flu season begins, scientists predict what strains of the virus will cause the illness during the next flu season. They produce a vaccine based on their predictions, on the basis of surrogate knowledge. Given the mid-1990s technology of making flu vaccines, they have no other choice if they wish to produce a vaccine in time that might work to protect people.

The justification of surrogate knowledge, that is, of our judgment that it is reasonable to believe in a proposition about the future, is found in the reasons that we give for believing it. Such reasons, explicitly stated and critically assessed, permit what Rorty (1979) calls "warranted assertibility." But such warranted "assertibility differs from truth, since a warrantedly assertible sentence (i.e., conjectural knowledge) may by bad luck be false, and a true sentence may not be warrantedly assertible, since we may have no good evidence for it" (Smart 1984: 96).

Within the critical realist theory of knowledge, surrogate knowledge constitutes conjectural knowledge. Since conjectural knowledge refers to justified belief in statements about the past and the present as well as to justified belief in statements about the future, we retain the term "surrogate knowledge" in reference to the last. For, by doing so, we acknowledge the additional threats to validity and the possibly greater uncertainty we face when we make assertions about the future compared to those we make about the past and present. Additionally, using "posits" and "surrogate knowledge" to refer to our statements about the future further reminds us that the future is monumentally different from the past. It has not happened yet.

Presumptively versus Terminally True Predictions

Taking predictions that come true as indicators of the validity of knowledge, as is commonly done in science, is often misleading and in a strict sense irrelevant to the immediate needs of decision making. It is misleading, because predictions are contingent on conditions, and, since conditions may change, a prediction that is perfectly sound when it is made may turn out to be false. Moreover, the existence of the prediction itself may contribute to changes in conditions, as is the case with self-altering prophecies. Thus, a prediction *would* have come true *if* conditions had remained the same. It is irrelevant to the needs of decision making, because decision makers need to know the consequences of ongoing changes and actions *before* they decide, not afterwards. Knowing the outcome of a given prediction may be of some help in making the next decision, but it is of no help in making the last decision.

To deal with these problems, Jeffrey K. Olick and I propose as part of an epistemology for futures studies the concepts of "presumptively true (or false) predictions" as distinct from "terminally true (or false) predictions" (Bell and Olick 1989).

Presumptively true (or false) predictions are assertions about the future that are assessed, that is, subjected to refutation as far as is possible, *before* the time arrives with which the prediction deals. Obviously, we cannot assess them directly since we must do the assessment before the time when the predicted event is supposed to occur. Thus, we make the grounds that justify our belief in a prediction explicit and subject them to verification, especially to attempts to refute them. Presumptively true predictions, thus, are those whose grounds withstand such attempts.

Terminally true (or false) predictions are those that are assessed by observation of the predicted phenomenon *after* the time arrives with which the prediction deals. Terminally true predictions, for example, are those for which the predicted outcome actually occurs as predicted, when the future time of the prediction becomes the present, and terminally false predictions are those for which the predicted outcome fails to occur as predicted.

Consider that there are also predictions that are indeterminate, both presumptively and terminally. In the case of presumptively true or false predictions, they are indeterminate when there is insufficient evidence to reach a clear judgment that a prediction is warranted, for example, when there is little opportunity to subject the prediction (or, more accurately, its assumptions and its grounds) to falsification.

For example, one justification of the invasion of the Caribbean island of Grenada on 25 October 1983 given by the Reagan Administration was the prediction that the American students attending the St. George's University Medical School would be harmed or taken hostage by the Grenadians or Cubans in Grenada, if American military forces did not rescue them. It is impossible to tell whether or not it would have been terminally true or false *if* the invasion had not taken place, because the invasion itself changed the conditions: the American military forces intervened and *did* "rescue" them. Many predictions share the same fate. When predictions are used in decision making, as they often are, conditions may shift too rapidly for a fair test of their final predicted outcomes if conditions had not changed.

In the case of the presumptive truth of the prediction, however, there are grounds to support a reasonable belief that the prediction concerning the American students being harmed or taken hostage was presumptively false. Therefore, it is also reasonable to believe that the invasion should never have been ordered, at least not on the basis of that prediction.

For example, President Reagan specifically compared the Americans in Grenada with "the nightmare of our hostages in Iran." Thus, he asked his audience to view the plight of the Americans in Grenada with the past situation of the Americans in Iran. Was the analogy an appropriate one? It is debatable. For example, the geographical location, size, religions, political situation, language, and culture of Grenada and Iran were totally different.

Also, in Grenada, the Americans were ordinary citizens. In Iran, they were employees of the American government. In Grenada, there had been no threats made to the safety of the Americans. In fact, their safety had been guaranteed repeatedly up to hours before the invasion began by General Hudson Austin, head of the Revolutionary Military Council governing Grenada at the time. Furthermore, Austin had sent messages to the American administration that the students could be evacuated if they wished. By contrast, in Iran, there had been prior threats to the Americans and these had been reinforced by hostile statements made by top Iranian officials. Additionally, there were other differences between the situations of the Americans in Grenada and Iran that raise serious questions about the applicability of the Iranian hostage experience to the

Grenadian case (Bell 1986). Moreover, this is not mere hindsight or Monday morning quarterbacking. These things were known at the time of the decision to invade.

In the above case, the prediction was probably presumptively false, but terminally indeterminate. A somewhat different situation involves another prediction which American leaders held with tragic consequences, one in which the prediction was arguably presumptively false and clearly terminally false as well. It is the so-called "domino theory" as applied to South Vietnam in the 1960s and early 1970s. If South Vietnam were to fall to communism, the theory stated, then the rest of Southeast Asia would go with it.

The presumptive truth of that prediction was hotly debated at the time. We now know that it was terminally false. South Vietnam did fall. Southeast Asia did *not* fall with it. Moreover, the fall of South Vietnam was *not*—as was also falsely predicted at the time—a grave threat to the security of the United States, possibly leading to World War III (McNamara 1995). In fact, we should have known that the threat of communist domination of the world would not in the least be affected by what was to happen in Vietnam but was significantly linked to unrelated developments within the then Soviet Union. How many American and Vietnamese lives might have been saved with more accurate—and more honest—futures thinking?

Even though it is false, a prediction can be useful. At the most mundane level, a prediction may be false, both presumptively and terminally, but used cynically by persons interested in a given course of action simply to persuade others that such a course of action ought to be adopted. That is, the prediction may not be believed by the people who make it even though they expect others to believe it. Such people deliberately deceive others in order to control them by making false prophecies. (E.g., a mother says to her son, Johnny, "Stop crossing your eyes or they will stick that way forever.")

More complexly, a prediction may be presumptively true at the time it is made, but may portray such an undesirable future that it influences people to change their behavior, thus altering the conditions on which it is based. This, of course, is the familiar "self-altering prophecy," that is, a prediction that leads to changes in its own conditions and, hence, in itself (Henshel 1978). Such prophecies can be self-negating or self-fulfilling.

To take a commonplace example of a self-negating prophecy, predicting an oil shortage may lead either to increasing the supply of oil or to lowering the level of consumption of oil so that the shortage never occurs. Thus, the prediction of the oil shortage may have been presumptively true and very useful precisely because it was so. But it turned out to be terminally false because the prediction itself led to a change in the conditions on which it was based. No oil shortage occurred. Other examples are a prediction of prison overcrowding that doesn't come true because prison officials, learning of the prediction, "increase the number of offenders released early on parole," or a prediction of empty classroom space negated because school officials, becoming aware of the prediction, "simply decide to close some schools" (Berk and Cooley 1987: 260).

Many predictions are of this doomsday variety, picturing a dreadful future followed by an admonition to change some behavior to prevent it from coming about. Presumably, the prediction is presumptively true, but it is used to direct and motivate behavior that will make it terminally false.

A self-fulfilling prophecy is found in the fact that how well children do in school is partly a response to the expectations of their teachers. Another is a prediction about coming inflation that results in people buying more to beat the price increases, with the extra spending helping to fuel the feared inflation. Henshel (1987; 1993; and Johnston 1987) gives many examples, both positive and negative, of the "feedback loops" and "bandwagon effects" involved in self-altering prophecies.

Taking into account indeterminate outcomes, we have nine types:

Predictions that are terminally:

		True	Indeterminate	False
Predictions	True	1	2	3
that are	Indeterminate	4	5	6
presumptively:	False	7	8	9

Whether or not a prediction is based on empirical grounds or is presumptively true or false, once made and communicated to others it itself *is* a part of reality. It is similar to a road map which, as I pointed out earlier, is both a model of reality and a reality itself. As such, a prediction becomes a part of the real intellectual environment to which people react and which they take into account in their decision-making processes.

Presumptively false predictions, truly believed and acted on by their makers, might lead to beneficial as well as unfavorable results. That is, since unintended, unanticipated, and unrecognized consequences of action are always possible, favorable outcomes might fortuitously occur no matter how presumptively false the predictions on which the action is based. To take an example from American Revolutionary War, Gordon S. Wood (1984: 10) says that the revolutionaries wrongly believed in "a conspiracy of British officials against them" and that "these false and unreasonable beliefs in conspiracies were crucial in mobilizing people into action." Such action resulted in the revolutionaries winning the war.

But presumptively true, compared to presumptively false, predictions have the obvious advantage that action based on them is both more intelligent and more effective in achieving intended results. When presumptively true predictions are logically and empirically grounded and their grounds have withstood serious criticism, they allow actors to behave knowledgably since the actors then have some reasons for their beliefs about what the consequences of their actions will be.

Since we cannot rely for a test of the presumptive truth of a prediction on it's accuracy in describing an eventual outcome, we turn to the critical realist theory of knowledge. Within this framework, we can do two sets of things. First, we can make the grounds for making the prediction explicit, intelligible, and logically coherent (all of which make formulations open to the critical assessment of others). Second, we can make a serious attempt to refute them by seeing whether or not they are congruent with relevant past and present facts (corrigible though they are), consistent with other predictions concerning the same time frame that have been shown to be presumptively true, and logically correct. If predictions can withstand such critical analysis, then we have a reasonable basis for believing that they are presumptively true.

Obviously, no claim of presumptive truth can be absolutely certain. Persuasive plausibility and intersubjective agreement may be the most that can be expected. Moreover, because of any number of threats to validity—from faulty data and logic to erroneous analogy—a claim of presumptive truth may be false.

One final point must be added. The observation of present reality itself may contain either retrodictions about the past or predictions about the future. For example, "'This is a sedimentary rock' and 'This is a volcanic rock'" are statements about the present, but they "entail different retrodictions about how the rock was formed, in virtue of the meanings of the terms 'sedimentary' and 'volcanic.'" When we make observation reports about the present, we always make some implicit predictions. "Since any of these predictions might turn out to be mistaken, future experiences might...lead us to reject a past observation report."

Thus, while the grounds that warrant predictions include past observations, the truth or falsity of past observations themselves may depend on our subsequent (i.e., future) knowledge of whether or not present predictions turn out to be correct. For example, to say "This is a table" commits us to "This will not play music." But if the "table" does play music at some future time, we conclude that it was not a table after all but a stereo (Musgrave 1993: 56–59).

Prediction Leads to Control and Vice Versa

Some phenomena are what Henshel (1976) calls "unaltered." In them, human design and engineering are minimal or lacking. For the physical world, they include such things as the path and speed of a forest fire, the path of a falling leaf, the shape of rising cigarette smoke, the weather or earthquakes at specific times and places, or the presence of oil deposits at a given location as yet undrilled. For the social world, unaltered phenomena include such things as the number of homicides on a specific date at a given location, the frequencies of specific words occurring in particular conversations, the total population of the United States at a certain date, or the proportion of people elected to the U.S. House of Representatives who are short. Henshel reminds us that natural scientists often have as much difficulty in predicting such unaltered phenomena as do social scientists and his view has been confirmed by others (Land and Schneider 1987).

Other phenemona are not "unaltered," but are deliberately altered or designed by human action. Much of human life is more or less intentionally designed or engineered, although "side effects" in the form of unintended and unanticipated consequences frequently occur. Predictions about designed systems would include, for natural phenomena, such things as the stability of a Gothic cathedral, the flow of smoke in a new chimney, the weight-carrying capacity of a ladder, or the tensile strength of a new rope of a given thickness. Comparable social phenomena include how a Mass will be conducted in your neighborhood Catholic church, the total number of school days in Bethany, Connecticut next year (since it's prescribed by law), the number of players on an American football team or where the home games of the New York Giants will be played next season, and the side of the street on which people drive their automobiles. If the "designs" (i.e., rules, schedules, intentions, plans, or controls) change, then we can change our predictions accordingly.

Henshel (1976: 27) rightly says that people tend to predict by controlling. Predictions are part of the steering and social control mechanisms of society and are connected to outcomes by feedback loops. Prediction leads to control and control leads to prediction. Social scientists can make reasoned predictions partly because so many social phenomena have been consciously or unconsciously ordered and controlled.

Either directly or indirectly, people design social organizations and institutions—which are, after all, human tools of social action—to achieve their goals. Organizations and institutions constantly evolve within the social interactions of everyday life and become legitimated by values and congealed in traditions. Thus, behavior tends to be ordered, patterned, regulated, and constrained not only by the limits of the biological and physical contexts of human life, but also by human goals and the social contexts and conventions within which their achievement is attempted. Yet organizations and institutions are always more or less in flux, as they are redesigned to better achieve their goals.

For example, the year of the next American presidential election, the number of members of Parliament in Britain five years from now, or the fact that next year a woman will not marry her father (or her mother) can be predicted with considerable confidence because they are the results of social control. Since futures research aims to increase human control, it follows that it aims also to increase predictive accuracy.

This is not to say, of course, that human decision and action cannot change constitutions, governing rules, or even the traditions concerning eligible marriage partners. But, obviously, many elements of social systems are quite stable, even though others may be changing rapidly. Moreover, futurists or forecasters can change their predictions to conform to new behavior codes or traditions, as social changes occur.

Thus, human control in the design of systems and behavior leads to the ability to predict with increased accuracy. And the ability to predict, for example, from knowledge of cause-and-effect relationships, can lead to control: "To control the future—to shape future happenings according to a plan—presupposes predictive knowledge of what will happen if certain conditions are realized" (Reichenbach 1951: 246).

Presumptively true predictions are useful precisely because they lead to human control. But the situation can become complicated because such predictions, as we have seen, may be self-altering, leading to actions that negate the predictions themselves. Thus, presumptively true predictions may turn out to be terminally false, even though they can serve to organize effective action, reduce anxiety, give meaning to events, and ensure that the prediction will be more likely to turn out to be either true or false *as the actors wish*. Thus, predictions, even—perhaps especially—self-altering predictions, increase the human ability to control.

What about those human actions that appear to take place without futures thinking and conscious control? Some examples, as Brown (1963: 75–76) suggests, are "unintentional actions, such as offending someone unwittingly, habitual actions like sleeping only on the left side, automatic actions like turning the car wheels in the wrong direction in a skid, expressive actions such as scowling or gesturing in conversation, and goal-directed actions which are not intentional, e.g., guilt projection." Additionally, he includes actions done for their own sake, such as skipping rope in the park.

Although futurists would agree with Brown that usually little conscious thought is given to such actions, they would add that humans can improve the competence and effectiveness of their actions by analyzing their future consequences even in the case of such seemingly unconscious acts. Take each of Brown's examples: Do you want to offend someone unintentionally? If not, posit the consequences of your actions on the sensibilities of other people and act accordingly. If you sleep only on your left side, will it have any deleterious consequences for your digestion or the chances of getting an ulcer? If so, sleep on your right side sometimes. People can—and do—learn not to turn the car wheels in the wrong direction in a skid. Expressive actions such as scowling or gesturing are important means of nonverbal communication and we can learn to use them to produce the meanings that we wish, although it is difficult to do so if they do not express our genuine feelings. Psychiatrists can help us rid ourselves of the delusions of guilt projection and the possible damage they do to our social relationships. You may skip rope in the park simply because you like to do it, but such pleasure itself is a goal (and in this case you may achieve some beneficial consequences for your cardiovascular system).

True, sometimes people do all of these things without much conscious thought, but each behavior is open to analysis of its possible future consequences. Such analysis, if less costly in time or other resources than the undesirable outcomes or the rewards of the possible benefits, helps make rational choice and competent behavior possible. Such futures thinking helps to place control of people's lives into their own hands. It helps them build their "projects of action for the future" (Masini 1982).

Conclusion

Although futurist methodological procedures are reasonably well developed, no generally agreed upon theory of knowledge for futures studies yet exists. Yet all contemporary futurists function with some theory of knowledge, even if implicit. Mostly, they have been influenced by two that are basically contradictory, positivism and, more recently, postpositivism, an aspect of postmodernism. A considerable number of pioneer futurists during and after World War II were trained as scientists or social scientists in the positivist tradition. They emphasized, among other things, the logical structure of relational statements, the language of mathematics, causality, empiricism and testability, objectivity, rationality, transcultural applicability, and the building up of a body of cumulative, verified knowledge.

In the early 1960s, a revolt against positivism, conveniently marked by the publication of Kuhn's book, *The Structure of Scientific Revolutions*, ushered in a postpositivist theory of knowledge. Postpositivists questioned most of the major beliefs of positivists and focused attention both on the developmental history of communities of scientists and on what scientists actually do. Postpositivists attacked the positivist ideas of scientific theories having tidy deductive structures, of causality, of certainty, of facts existing independently of theories, of knowledge being cumulative, of knowledge free of cultural and other biases, and of theories being chosen because of empirical testing.

Although bearing many of the marks of positivism, futures studies, both for better and for worse, has been greatly influenced by postpositivist thought, because its own formative years coincided with the rise of postmodernism of which postpositivism is part.

In the extreme, postpositivism, while being a beneficial corrective to an uncritical acceptance of positivist science, appears to be inadequate as a theory of knowledge. Some writers have concluded that all thought would be brought to an end if the postpositivist program were taken to its logical conclusion. Not only would causality, determinism, necessity, objectivity, and rationality be abolished, but also humanism, liberal democracy, responsibility, truth itself and, we can add, futures thinking. Postpositivism and postmodernism in the extreme appear to be false philosophies that provide little basis for knowledge, much less a basis for a future- and action-oriented field such as futures studies.

I propose critical realism as an appropriate theory of knowledge for futures studies. It is a post-postpositivist and post-Kuhnian epistemology and part of the larger humanistic culture of critical discourse. It synthesizes some aspects of older positivist views with some of those of the newer postpositivist philosophers, including in the latter case the claim that plausibility, not absolute certainty, is the most that can be claimed from scientific labors.

Yet critical realists believe that there is an external reality independent of the human mind and that it can be known objectively. They believe that conjectural knowledge is possible. Furthermore, such knowledge is often beyond a reasonable doubt, if it remains unrefuted after serious efforts to falsify it. Rival hypotheses, they assert, often can be refuted. Moreover, causation, including the purposive behavior of individuals and social groups, is a necessary assumption.

Although futures studies is devoted to the study of the future, it is necessarily concerned with much that is part of the past and the present, as each bears on the future, as we saw in chapter 4. The past and present, of course, have a reality that the future does not share. Unlike the future, they are evidential, they did or do exist. Thus, futurists can study them scientifically as critical realists. Doing so, futurists face the same threats to the validity of their assertions as do other scientists.

But futurists face additional threats to validity, just as do all scientists and scholars who make predictions central to their work. For they attempt to make assertions about the as-yet-nonexistent and nonevidential future. Even so, critical realism is an appropriate epistemology for futures studies. Since it is based on fallibilism, critical realism claims only a knowledge that is conjectural and it incorporates justified beliefs about the future.

Critical realism can be elaborated beyond most of its current formulations in order to highlight the ontological difference between the past and present on the one hand and the future on the other. Toward that end, Olick and I suggest the concepts of "posit" and "surrogate knowledge." Posits are simply statements about the future which we treat as true, although we do not know whether they are true, in order to explore alternative possibilities for the future, including improbable ones. Knowledge surrogates are those posits that we accept as conjectural knowledge, because we have subjected the grounds for believing them to critical analysis and they remain unrefuted.

Posits can be formulated on an "as if" or "what if" basis for constructing images of alternative futures. Such posits can be used in decision making for social action, hypothetically exploring different circumstances, alternative policy recommendations, and different probable outcomes. Surrogate knowledge can be used to provide the substitutes for knowledge of the most likely future that decision makers need.

Also, in order to acknowledge the special ontological status of the future, Olick and I propose another conceptual distinction for a futurist epistemology: "presumptively-true (or false) predictions" and "terminally-true (or false) predictions." The former concern predictions that are assessed before the fact. They are (presumptively) true when the assumptions and grounds on which they stand have withstood serious criticism, that is our belief in them is justified. Or the assumptions and grounds fail such tests and the prediction is (presumptively) false.

Terminally-true (or false) predictions are assessed after the fact. Those that turn out to be true after the predicted events and processes have become present reality are terminally true and those that do not are terminally false.

This distinction allows us to recognize the possibility that our belief in a prediction is justified at the time it is made, but that the prediction may turn out to be terminally false. Circumstances may change. Moreover, the prediction itself may change the circumstances that affect its chances of coming true. Hence, this distinction also directs futurists' attention to the power of self-altering prophesies. Futures studies is, thus, self-consciously reflexive.

Knowledge surrogates can be subjected to test, first, by making the grounds for them explicit, intelligible, and logically coherent, which make formulations open to the critical assessment of others, and, second, by attempting to refute them by examining their consistency with relevant past and present facts and with other presumptively-true predictions and by seeing if they conform to the rules of logic.

Where possible, the futurist's specification of the assumptions and conditions underlying assertions about the future includes the human actions that probably will be taken or that might be taken to achieve what ought to be. The futurist, thus, reflexively takes into account the preferences, choices, images of the future, and probable actions of relevant groups and individual members of society.

Futurists focus on the transformation of hindsight into foresight. On the one hand, they speculate, think laterally, intuit, reason counterfactually as well as factually, cogitate linearly *and* dialectically, entertain outrageous—and even despised—notions, and creatively invent in order to unveil possible and probable futures. On the other hand, they specify past and present data using a multitude of standard and special methods, collecting, analyzing, and interpreting evidence in order to make posits about possible and probable futures and to construct surrogate knowledge as reliably and validly as they can. Their candidates for surrogate knowledge of the future are based on patterns of reasoning and the marshalling of evidence, and they can become justified belief if they remain unrefuted after being subjected to serious efforts to falsify them.

In the next chapter, we move to an examination of some of the specific methods of futures studies, highlighting a few character-defining exemplars that futurists share. Although they have neglected the philosophical basis of knowledge, futurists have devoted a great deal of time and effort to developing specific methods of futures research or to adapting standard methodological techniques to the purposes of futures studies.

6

Methods and Exemplars in Futures Research

In this chapter, I describe some of the different methods used by futurists, highlighting a few exemplars of futures studies. Some have been pathbreaking and character defining in their impact on the field, such as the computer simulations of *The Limits to Growth*. Others show extraordinary promise, such as Robert B. Textor's ethnographic futures research, or are part of popular culture, such as John Naisbitt's *Megatrends*. In earlier chapters, of course, I have already described many exemplars, including H. D. Lasswell's developmental analysis, some of the work of Herman Kahn, and F. L. Polak's sweeping historical investigation of images of the future for entire civilizations. In an effort to minimize repetition, I won't dwell on them further in this chapter.

Here, I aim to give the reader some examples that illustrate the range of methodologies that futurists use. I describe: the pragmatic prediction of one variable by others, the extrapolation of trends using time series analysis, cohort-component methods, survey research techniques, the Delphi method, simulation and computer modeling, gaming, monitoring, content analysis, participatory futures praxis, social experimentation, ethnographic futures research, and, of course, the methodological strategy that futurists most commonly use, the construction of scenarios which is an aspect of each of the above methods.

Although in this chapter I emphasize possible and, especially, probable futures, I do not ignore preferable futures. Many of the methods include the investigation of desirable futures, such as respondents' preferences, and some of them, such as participatory futures praxis and social experiments, include social design and engineering, that is, planning and action aimed at social betterment. Note that even those studies using the most mathematical of methods, such as simulation and gaming for example, often have a normative focus, that is, some built-in image of a desirable future (such as creating a sustainable economy or winning a military battle).

Some of the methodologies, for example, time series analysis, simulation and computer modeling, gaming, and so on, can be highly complex and technical. Some require mathematical or statistical skills well beyond those of the average educated person. Thus, they may require additional study and training before one can use them in a competent way. For example, survey research, when done correctly, requires expert knowledge and practical experience about sampling; construction of questionnaires and interview schedules, from the mechanics of wording questions to the effects of question sequence and general format on the respondents' answers; techniques of telephone or face-to-face interviewing of respondents and recording of responses or of mass administration of questionnaires to be filled out by respondents themselves; coding of the raw data for machine manipulation; statistical analysis using computers, including multivariate methods and causal analysis; construction of tables, charts, graphs, and figures for the presentation of the results of data analysis; the interpretation of data to explain what the empirical results mean which may require theoretical knowledge in the substantive field being studied; and the writing of research reports. There are specialists in each of these phases of the survey research process, and there are threats to validity and sources of bias contained in each of them that, if not guarded against, can make research results misleading if not meaningless (Judd et al. 1991).

Methodological Diversity

Although the methodological tool of the "scenario" provides a unity to futurist methodology, there is considerable methdological diversity in the futures field. Different futurists have different methodological tastes and styles, preferring some methods more than others. Thus, how futurists create the stories they tell and how they try to make them plausible vary greatly, depending on the methods they choose to collect, organize, and present the data on which their stories are based. At one extreme, futurists' stories may be difficult for some readers to grasp, being wrapped in highly technical and mathematical language, while at the other extreme they may be easily understandable, being written in ordinary language and reading more like a popular short story or a novel than a mathematical treatise. The highly mathematical work of some futurists, such as global modelers for example, or the data-rich statistical analyses of survey researchers are seldom used by qualitative field workers, while the methods of the qualitative field worker, such as participant observation, or the intuitive reflections of the armchair futurist may be viewed as "mere" journalism or unfounded speculation by the statistically trained, scientifically oriented futurist.

Despite such methodological diversity and preferences, any fair-minded assessment of futures research methods can only reach one conclusion: *No method has a monopoly on producing good—or bad—work.* Although some methods may be more appropriate to certain research problems than are others and although each method may have its own distinct advantages and disadvantages, no method within a naturalistic framework (i.e., excluding mysticism, astrology, etc.) inherently produces good or bad futures work. That depends on the skills, talent, ingenuity, insight, diligence, and even luck of the futures researcher him- or herself. That is, it depends not on the method per se but on the adeptness of the user of a particular method and appropriateness of its application to a specific research question. A sharp tool in the hands of inept researchers may produce less useful results than a blunt tool in the hands of skilled and sensitive researchers.

Yet some methods are more codified than others and the average person, thus, can be taught to use them to produce at least a minimally adequate result. For example, it is generally easier to teach people to do passably good futures research using a sample survey (which for the most part has highly codified instructions) than to teach them how to create nontrivial scenarios about the future based on insight and wisdom.

Experienced forecasters have learned that both scientific explanation and reliable prediction are best served when several methods are used on the same problem (Land and Schneider 1987: 26). Thus, the best advice is to be ecumenical in one's methodological commitments and to use several different methods, where possible, to study the same domains of alternative futures, basing a final set of forecasts or alternative scenarios on combined multimethod results.

General Considerations

Futurists Use Methods from Many Fields

The scientific, scholarly, and rhetorical methods of any discipline in the humanities, social sciences, and the sciences might be—and sometimes are—used by futurists doing research on some particular topic.

That is, as a holistic and synthesizing field, futures studies draws on whatever methodological techniques may be available in existing disciplines if they are relevant to the futures investigation at hand.

Thus, futurists borrow techniques from other disciplines. In fact, most of the examples of futures research given below use methods commonly in use in other disciplines. They are not distinctive to futures studies. What determines their relevance to the futures field is their substantive content and the purpose of their use (e.g., making assertions about possible, probable, and preferable futures) rather than their methodological characteristics alone.

Other methods described below, however, such as the Delphi technique and ethnographic futures research, have a legitimate claim to being primarily futures research methods, having been invented by futurists for the specific purpose of studying the future.

Even literary techniques of the essayist, short-story writer, or the novelist have been borrowed by futurists. Television and motion picture techniques, especially when dealing with science fiction and fantasy, have also been used. In these media, futures work and fiction may merge. But, as I said earlier, with the exception of some examples of utopian writings to be discussed in chapter 1 of volume 2, purely literary works are mostly beyond the scope of this book.

Classifications of Different Methods

The different methods used by futurists have been classified in a variety of ways. Gordon (1992), for example, distinguishes between quantitative and qualitative methods on the one hand and exploratory and normative methods on the other. The first distinction refers to the degree to which a method relies on explicit measurements and numbers. Quantitative methods are mathematically based, using equations and precise measuring instruments, and are illustrated by methods such as time-series analysis, the cohort/component method, computer simulations, and survey research. They may or may not use real data sets; some modeling and gaming, for example, may be based on contrary-to-fact assumptions or hypothetical data even though they are highly mathematical in presentation.

Qualitative methods do not use numerical measurements to any great degree and seldom are based on statistical analyses. They are illustrated by most examples of participatory futures praxis, ethnographic futures research, and writing scenarios out of one's own imagination. Qualitative methods, too, may be empirically based or not, allowing for, on the one hand, detailed empirical facts of the past and present situation and, on the other hand, the inclusion of the intuitive, the speculative, and the hypothetical when probing for respondents' images of the future.

The quantitative-qualitative distinction is better conceived as a continuum than a dichotomy, most methods allowing for some degree of quantification, however limited.

Obviously, methods that are objective, systematic, and explicit are more easily assessed for their reliability and validity than methods that are subjective, unsystematic, and implicit. This is not to say that the results based on them are necessarily more true. Rather, it is to say that they are more open to the scrutiny of others. We are, thus, more able to decide how much reliance to place on images of the future that are produced by them, because we can examine critically the procedures and data ourselves for their cogency.

Another distinction that Gordon makes is between forecasts of futures that seem plausible (i.e., exploratory) and forecasts of futures that seem desirable (i.e., normative). As Gordon says, both exploratory and normative forecasts can be produced by either quantitative or qualitative methods.

This distinction is somewhat misleading, because it is less an inherent feature of a method than a description of the intention, purpose, or arbitrary constraints of the futures researcher. It is quite true that some researchers aim to describe plausible futures without regard to desirability, while others deliberately concentrate on depicting desirable or undesirable futures that may be more or less contrary to fact. Yet every image of a possible or probable future, no matter what method is used to construct it and no matter what the intention of its author, can be evaluated as to its desirability.

Such evaluation is the heart of the ethical foundations of futures studies that I discuss in volume 2. There are philosophical problems and solutions in making value judgments about preferable futures that go beyond those involved in making assertions about possible and probable futures and they deserve separate treatment.

Some Preliminary Cautions

Large numbers of cases versus small numbers of cases. When they are empirical and not mere exercises in logical analysis, quantitative studies tend to be based on a relatively large number of cases (N) while qualitative studies tend to be based on a relatively small number of cases. But we ought not to leap to the conclusion that, therefore, quantitative studies are necessarily more valid than qualitative studies.

In fact, the total amount of information in a study may be much greater with small N's than with large ones. That is, with small N's researchers often have the opportunity, in participant-observation for example, to make repeated observations on the same individual cases over long periods of time and, thus, capture the complex texture of human experience. By comparison, the user of census or large-scale survey data typically has only the tiniest bits of superficial information about his or her cases and usually they are collected only at one brief point in time. Look at the mountain of information Robert E. Lane (1962) collected on only fifteen American men in his study of political ideology. His interview guide was fifty pages long and the recordings of his extended conversations with his respondents resulted in fifteen books of data ranging from 154 to 322 pages *each*.

The point is, as William Foote Whyte (1989) says, to get the facts straight. Thinking accurately about the future often rests on reasonably accurate pictures of the past and present without which extrapolation into the future gets off to a false start. A small bit of information about each of a large number of cases is not always—or even usually—the best way of obtaining the most data.

The small-N researcher, especially in social research, may have better as well as more information, usually personally experiencing the people and social situations being studied with considerable intimacy and understanding, while the large-N researcher is often distanced from his or her respondents or other data by the technologies of large-scale data collection and analysis. The same may be true of physical scientists, such as geologists, for example, working out in the desert or in the mountains personally holding individual rocks in their hands and examining the rocks themselves on site compared to geologists in a statistical laboratory manipulating numbers based on satellite photographs of the Earth and mathematical models with a computer.

Although small numbers of cases are not necessarily superior to large data sets, the number of cases alone does not allow us to judge either the total quantity nor the quality of the data of a study, in the futures or any other field.

Other cautions. Gordon (1992: 26) also lists some warnings that ought to be considered when evaluating the methods of forecasting. They include:

"Forecasts can be very precise but quite inaccurate." Very specific forecasts can be made, even to the point of giving particular numbers, for example, the price of a barrel of oil or the number of automobiles sold in the year 2174. Because a forecast is precise in its prediction does not alone mean that it has any special claim to coming true at the time specified—if ever.

"Extrapolation is bound to be wrong eventually." Gordon may have somewhat overstated this, because whether it is true or not may depend on the nature of the phenomena being forecast, for examle, the half life of nuclear radiation versus the performance of the stock market. But he is certainly correct to point out that simply extending historical trends into the future ignores the possibility of something new deflecting the trends, especially if the forecaster has no explanatory theory of change about the phenomenon being extrapolated. As we saw in chapter 4, the forces that shaped the past may not be the ones that will shape the future.

Forecasts are almost always incomplete. Since we cannot know all the discoveries not yet made that may be relevant to our assertions about some future phenomena, our present forecasts cannot take them into account. Therefore, our forecasts tend to be incomplete.

Forecasts, planning, and human action interact and affect each other. Thus, each must be repetitively and endlessly altered to maintain accuracy and effectiveness of decision making and action. Since our plans are necessarily based on beliefs about the future that contain some error and are incomplete, they should be constantly reviewed, and revised as necessary, as new information becomes available. Moreover, as we have seen in earlier chapters, forecasts can be self-altering. Thus, forecasts themselves can change plans and actions and hence can also change the chances of different possible futures becoming reality.

Finally, Gordon says, *forecasting "is not value free,"* pointing out two different and opposing tendencies. The first is the tendency for some futurists to choose the doomsday strategy emphasizing undesirable futures to describe their findings in an attempt to stir people to action to prevent them from occurring. The second is the opposite tendency of people, when assessing chances of different outcomes, to give higher probabilities to outcomes that they judge to be desirable.

Although Gordon is certainly right about forecasting not being value free, the question of the role of values in futures research is both fundamental and complex. Making value judgments is an inescapable aspect of any scholarly or research process and, beyond that, has a special role to play in futures research. But that does not imply that the reliability and validity of the results of research, including futures research, are thereby necessarily lessened.

Pragmatic Prediction of One Variable by Another

Let's now turn to a description of some common methods of futures research, starting with the pragmatic prediction of one variable by another or others. Sometimes, something we can measure in the present correlates with something that we cannot measure because it hasn't occurred yet. Take a hypothethical example of an historical record showing a correlation between the price of gasoline in Arizona and the price of rice in China a year later. There seems to be no sensible connection between the two variables. In such a case, a researcher simply may have stumbled onto an empirical relationship between them, and, if it is stable, that is, holds up over time, it can be useful, because we can use the price of gasoline now to predict the price of rice in China next year.

We may not understand why or how the two variables are connected, that is, we may not have a theory or hypothesis about a causal connection between them. Yet for practical purposes we sometimes can use what we can measure to forecast what we cannot. Thus, even though there is no apparent causal link between the price of gasoline in Arizona and the price of rice in China a year later, such a pragmatic prediction of one price from the other is possible if a correlation continues to exist between them.

Some writers (Martino 1993) classify the "leading-indicators" approach to forecasting as pragmatic prediction in contrast with a cause-and-effect approach where the influence of one or more independent variables on a dependent variable is theoretically understood. An example can be found in the Index of Leading Indicators. As early as 1977, the U. S. National Bureau of Economic Research was using indicators that either lead, lag, or are coincident with the aggregate American economy (Granger 1980). By the mid-1990s eleven key variables had been selected to compose an Index of Leading Economic Indicators. They included average hours employees worked per week in manufacturing, weekly claims on unemployment insurance, orders for manufacturing consumer goods and materials, percentage of firms getting delayed deliveries, contracts and orders for new plants and equipment, monthly total building permits issued, change in manufacturers' unfilled orders for durable goods, change in sensitive materials prices, prices of 500 common stocks, the monthly average of the real money supply, and consumer expectations of the economy's health. Average movements in the time series of such variables are used to forecast the subsequent movements in the growth rate of the American economy, based upon past empirical relationships rather than explanatory theory.

The leading-indicators approach, however, is not quite pure pragmatic prediction, since there are clearly possible causal and structural relationships between such variables as manufacturing workers' hours of work, unemployment, orders for new consumer goods, contracts for new plants and equipment, building permits, and so on, on the one hand and growth in the gross domestic product on the other.

One dramatic case of the pragmatic prediction of one variable by another without any apparent or understood causal connection is the relationship between the new year's mood of the West German population and the real growth of the West German gross national product in the *following* year (Noelle-Neumann 1989, 1994). Every year in December since 1949, the Allensbach Institute has interviewed a representative sample of 2,000 people aged sixteen years and older in the Federal Republic of Germany and in West Berlin. Included in the interview schedule was the following question: "Is it with hopes or with fears that you enter the coming year?"

It turns out that there is a striking correlation between "the percentage of hopes at the *end* of a year and the development of the real gross national product the *following* year" (Noelle-Neuman 1989: 137). From the time-series graph in figure 6.1, note the similarity in the two time-series data: the optimism at the end of the previous year (the percentage responding "with hopes" from 1957 through 1994) and the next year's growth rates of the real gross national product (from 1958 through 1995). A regression analysis shows that from 1963 to 1983, the coefficient of determination is .57, a one percent increase in optimism resulting in a .1925 percent increase in the GNP growth rate. This forecast record was more accurate than the joint expert opinion of economists from five major economic institutes in West Germany.

What is surprising about this correlation is that changes in hopes *precede* changes in the economic growth rate. If it had been the other way around, that is, if the increase or decrease in people's optimism had oc-

FIGURE 6.1
Optimism at the End of the Previous Year and Growth Rates of the Real Gross National Product in Germany for Thirty-Seven Years

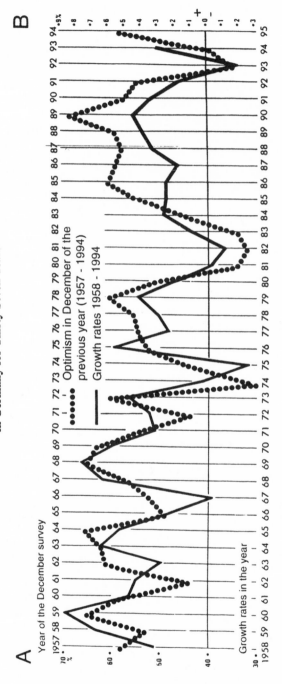

A: Optimism: The percent that responded to the question posed at the end of the previous year: "Is it with hopes or with fears you enter the coming years?" by answering "with hopes" (Allenbach Archives).

B: Growth rates of the real gross national product.

Source: Noelle-Neumann (1994)

curred *after* an economic upturn or downturn, we would have had a ready explanation for the correlation, that is, the percentage of hopes at the end of the year simply reflect the economic experience of the past year. But that is definitely not the case. In fact, after a year of outstanding growth, people's optimism is nearly as likely to decline as to increase. "Instead of looking back," as Noelle-Neumann (1989: 145) says, "the population responds by looking forward."

One explanation for the correlation might be that the hopes indicator translates into consumer buying, capital investments, and entrepreneurial decisions and, thus, affects the next year's economic growth. But this explanation fails because exports strongly affect the economic growth rates of (then) West Germany and exports vary with factors other than the mood of the German people. Another explanation that fails is that people have some concrete economic knowledge on which they base their optimism or pessimism. But no, from other data collected by Allensbach Institute we know that the respondents generally had little knowledge of economic facts. They knew little, for example, of such things as the national debt, inflation rates, the unemployment rate, and economic growth rates. Also puzzling is the fact that the question about hopes and fears that was asked says nothing whatsoever about the economy. Rather, it simply asks a general question.

Although no adequate explanation has yet been found explaining the relationship between this year's hopes and next year's economic growth, the relationship exists and appears stable over more than three decades. Moreover, it has practical merit and has been used to improve economic forecasts (Noelle-Neumann 1989: 150).

Whenever some empirical correlation exists between two variables, it is possible to use one to predict the other. But be cautious. The prediction will be accurate only so long as the relationship continues. When we do not understand the nature of a relationship, we are on shaky ground. An observed empirical correlation might be unstable or conditional. It might change tomorrow. Thus, using a correlation based simply on our knowledge of its empirical existence for pragmatic prediction is risky business.

Yet sometimes an empirical correlation that seems farfetched not only remains stable into the future, but also is subsequently supported by understanding and theorizing. One such empirical pattern is between the amount of rainfall in the American Midwest and the eruption of volcanoes in tropical latitudes, from as far distant as the Philippines and Indonesia, a relationship that was recognized before it was understood. The explanation linking them turns out to involve the sulphur dioxide that is thrown into the atmosphere by the volcanoes. It changes to sulfuric acid in the stratosphere and then reacts with water to form droplets. These droplets, or so-called aerosol fog, last for several years and block some sunlight. The resulting cooling of some regions alters the Pacific jet stream, stalling it for long periods and dropping rain in huge amounts in concentrated areas (*New Haven Register*, 31 July 1993: 38).

Extrapolation of Time Series

The most simple method of extrapolation is very little different from the pragmatic estimation of one variable by another and involves no necessary theoretical understanding or causal explanation of changes in the variable. The only difference is that the future value of a variable is estimated not by the value of another variable but by itself, that is, by its own value or values at some earlier time or times. The most naive assumption is that the value of a variable at some future time is equal to its value in the present or at some recent past time, under the assumption that there will be no change: That the predicted value of variable Y for period t will be the same as its value in period $t - 1$ (Holden et al. 1990: 10) or:

$$Y_t = Y_{t-1}$$

Usually, extrapolation is somewhat more complicated, being based upon a series of repeated observations on the same variable over some period of time. The researcher looks for some pattern in the past and, under the assumption that that pattern will repeat itself into the future, projects the time series into the future by estimating values of the variable or a range of values for specific times in the future. That is, we go beyond the most simple assumption of no change to a somewhat less simple assumption of a continuation of past change, that is, unchanging change. In other words, we assume that the direction and rate of change in the recent past will continue, perhaps at a constant or the same changing rate of change depending on what the recent past data reveal. This can be done by finding an equation that describes the known data and then projecting it into the future. A linear regression trend can be estimated from

$$Y_t = a + bt$$

where t is time, a is the intercept, and b is the slope of the trend. Usually, an error or random term is added. Nonlinear trends can also be

FIGURE 6.2
Past Trend and Projection into the Future for Hypothetical Variable Y

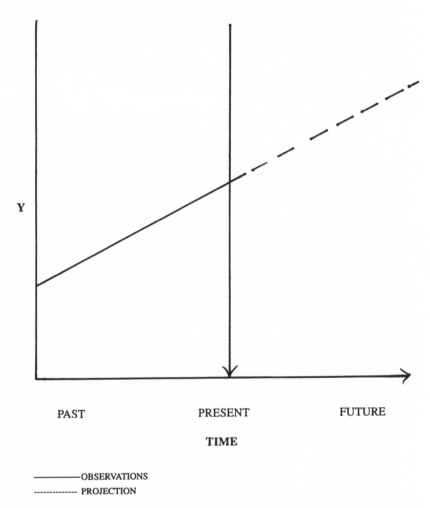

OBSERVATIONS
--------------- PROJECTION

estimated using other models, such as exponential, parabolic, modified exponential, the Gompertz, the logistic, or a variety of other curves (Granger 1980).

Figure 6.2 gives some hypothetical values for variable Y for the past up to the present (shown by a solid line) and a projection giving some possible future values if the past pattern continues (broken line).

FIGURE 6.3
A Growth Curve Showing the Percentage of Consumers Possessing Some Product (with a Penetration Level of 90 Percent)

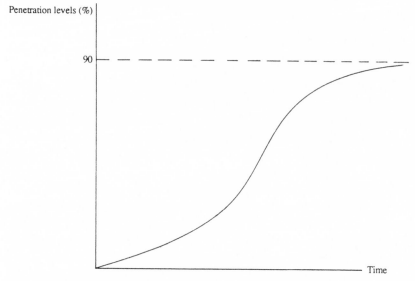

Source: Holden et al. (1990: 121)

Often a trend is fitted to only part of the past data available, say the first two-thirds, then the resulting forecast is tested and revised by using the last one-third of the data. Thus, some estimate of the adequacy of a given forecast can be made by comparing it to the known data. Sometimes a simple moving average is used. For example, given "Y_1, Y_2,...., Y_n a fifth order moving average forecast for period $n + 1$ is obtained from

$$Y_{n+1} = (Y_n + Y_{n-1} + Y_{n-2} + Y_{n-3} + Y_{n-4} / 5)"$$ (Holden et al. 1990:12)

Some variables exhibit a growth curve that approaches or tends to approach some maximum possible level or saturation point. Durable consumer goods are an example. When a new product is introduced for the home, say an interactive computer-television, there is a marked tendency for its spread to begin slowly and then to increase at an increasing rate of increase. As more and more homes obtain the product, the percentage of homes left without it dwindles, and, although the product continues to spread into new homes, it does so at a decreasing rate of increase. Finally, as the spread of the product approaches saturation and nearly all homes have it, the growth curve levels off. The result is a stretched S-shaped curve.

This is illustrated in figure 6.3 where the percentage of consumers possessing some product is shown over time, with a 90 percent penetration level indicated. Note that the curve moves upward slowly at the beginning with a middle-level slope, then moves upward more rapidly with a greater and greater slope, then reaches a maximum rate of increase, and then continues to increase but with a decreasing rate of increase (the curve still moves upward but with a shallower and shallower slope), finally the curve levels off as the 90 percent penetration or saturation level is approached.

Time-series data are sometimes decomposed into characteristic types of patterns, such as trends and cycles. Trends refer to the relatively smooth non-repetitive movement of the time series while cycles refer to the repetitive movement of the series. What is left over is considered residual or random movement, sometimes called "noise" or "hash," and is assumed, often rightly, to cancel itself out. The most simple decomposition of time-series data is

$$Y = \text{Trend} + \text{Cycle} + \text{Residual}$$

where Y represents the value of variable Y for various times. Sometimes, a seasonal cycle is shown separately, since for many variables— from employment rates to agricultural harvests—such seasonal cycles may exist quite apart from the existence of some other cyclical pattern.

From any set of time-series data, the trend can be removed mathematically. After that, the cycle or cycles can be removed mathematically, leaving some residual or random fluctuation. Note figure 6.4 where a set of observed data for hypothetical variable Y are shown along with a trend line and the cycles by which they can be represented. Note also that there remains some small residual movement of the time series in addition to the trend and the cycles. Both the trend and the cycles can be projected into the future.

When forecasting using time-series data, it is not always appropriate to use all available data. For example, since food consumption patterns have varied, for example, in response to access to refrigerators in the home and supermarkets in the neighborhood, we would not use all the

FIGURE 6.4
Observed Data, Trend, and Cycles for Hypothetical Variable Y

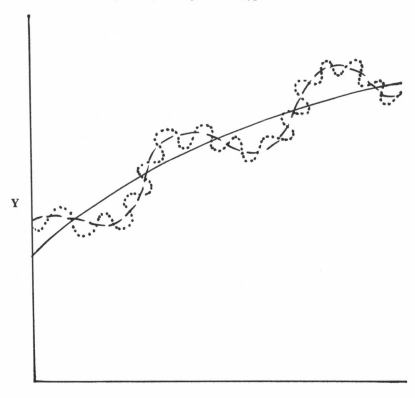

TIME

OBSERVED DATA   ··················
TREND   —————
CYCLES   --------------

available data on food consumption going back maybe eighty years or so. Rather, we would want only the most recent data, perhaps only the last twenty years (Holden et al. 1990: 184). In this case, the data more than twenty years into the past may be irrelevant to understanding today's and tomorrow's behavior.

In 1970, major advances in time-series analysis were made with the more sophisticated and complex models introduced by Box and Jenkins

(1970). Instead of merely extracting simple monotonic changes, their autoregressive-moving-average models allow identifying past cycles and turning points to project future movements in a time series in an intricate way. Although the Box-Jenkins technique gives good forecasts, it is rarely used, first, perhaps, because it requires many observations (say more than forty) for one to have confidence in the results, and, second, because of the heavy demands it makes on forecasters' time and skills (Holden et al. 1990: 7, 184). Also, if the results are to be used for decision making, then the cost of making the wrong decision may be less than the cost of doing all the work necessary to carry out a Box-Jenkins type of analysis.

Even when based on complex and sophisticated methods, a projection or forecast of a future value of a variable remains based only on known past values of the same variable. In the end, after what could be seemingly endless hours of compiling and analyzing data, there is no guarantee that the future behavior of a variable in fact will be like its past behavior.

Time-series analysis works best when accurate, quantitative data have been recorded over a reasonably long period of time. One such set of data concerns the size of human populations, since many governments have been counting people for a very long time in national censuses. As a consequence, techniques for projecting from time-series data have been highly developed by demographers, especially for short-term projections. The predominant method used by demographers, however, is not time-series analysis but the cohort-component method.

Cohort-Component Methods: An Example From Demography

In demography, time-series analysis has often been combined with other methods, especially with what have been called the "accounting/cohort-component projection methods." The underlying model, usually aimed at forecasting populations as far as fifty or seventy-five years—sometimes farther—into the future involves double-entry definitions of population inflows and outflows. The basic population model for the entire Earth is

$$P_2 = P_1 + (B - D)$$

where P_2 is the population at time 2, some later time, and P_1 is the population at time 1, some earlier time, B is the number of live births between times 1 and 2 (the inflow of people), and D is the number of deaths between times 1 and 2 (the outflow of people). Obviously, the population at time 2 will be larger or smaller than the population of time 1 by the amount that births exceed deaths or deaths exceed births.

For any given geographical region of the Earth, of course, the geographical movement of people from some places to other places must also be taken into account, since such migration is another way people get added to or removed from a given population. Thus,

$$P_2 = P_1 + (B - D) + (IM - OM)$$

where the new term IM means in-migration (inflow) and OM means out-migration (outflow) of people into and out of a given geographical area between times 1 and 2.

Thus, demographers disaggregate population growth into its component parts—fertility, mortality, and migration. Moreover, they further divide the data by various subgroups of the population, such as age groups, sex, race, and often marital status, geographical sub-region, and other socioeconomic categories. Each subgroup is given probability rates for changes in fertility, mortality, and migration and a Markov procedure is used to project, starting with the disaggregated base year population (Land 1986; Long and McMillen 1987).

Different assumptions about future levels of fertility, mortality, and migration can be made, so that a range of projections will result from the same base year population data. The United States Census Bureau, which has used cohort-component methods for nearly four decades, usually makes low, middle, and high projections. Recently, time-series analysis has been incorporated into this method as a way of projecting changes in the components of population change (Long and McMillen 1987).

Although some writers would not be willing to call the accounting/cohort-component method a causal model, it does rely on a model in which there are mathematically known relationships, for example, among age distributions on the one hand and birth rates and death rates on the other in a closed population. Hence, the future implications of changes in one can be precisely calculated in terms of changes in the others and vice versa. Thus, demographers have a relational model with which they work and can go beyond mere pragmatic prediction or extrapolation of time series. Moreover, as demographers learn more about the effects of exogenous variables on the components of population change, for example, how economic conditions or changes in human values and lifestyles causally affect birth, death, and migration rates, they can incorporate explanatory models into their forecasts.

Survey Research

By "survey research" I refer to the various techniques of systematically collecting data directly from individuals by asking them questions, usually with the individual him- or herself as the unit of analysis. It can be a census of an entire population in which every individual in some defined category—perhaps every person in a whole country as in the case of national censuses—are enumerated, but usually survey research is based on a sample. Thus, opinion polls, for example, may use a sample of a few thousand people to describe the beliefs, attitudes, and plans of an entire population of millions or even hundreds of millions of people in a country. How accurate the results are, of course, depends on many factors, including how the sample was drawn, how questions were asked, and how responses were classified.

Survey research is an ancient technique. Censuses, for example, were mentioned in the Old Testament and were regularly conducted in ancient Egypt. A Roman census may have accounted for Jesus being born away from home, because Joseph and Mary were on a trip to Joseph's ancestral home to be counted in the census (Babbie 1986: 203). Although it is not well known, both great social thinkers, Karl Marx and Max Weber, carried out empirical social research using surveys. Since the end of World War II, survey research has become a highly codified and reliable set of techniques by which to collect data on individuals and groups.

Survey researchers sometimes interview the individuals being surveyed in face-to-face situations in their own homes, in their workplaces, or in other settings such as shopping malls or outside voting places. Or they may interview them by telephone, an increasingly common practice. Sometimes researchers "administer" questionnaires by asking respondents to fill them out in a supervised group situation similar to giving a classroom test; at other times, researchers deliver questionnaires to individuals, for example using the mail and including stamped envelopes for the return of the filled-out questionnaires.

The modern world, as we all are aware, is awash in surveys and polls of various kinds and their results are often widely disseminated. They are a major source of the collection of original data in economics, political science, psychology, and sociology. But they are also used by newspapers and television networks, by political candidates, and by business firms. They deal with both theoretical or practical questions, with the most important issues of the day or the most trivial. Survey research topics range from studies of unemployment, crime, and health to studies of the taste of colas, the smell of perfumes, or the preferred shape of a liquid soap container.

Any survey on any topic may have some bearing on the future if the information produced by such surveys is used by decision makers in formulating or implementing policy, since their decisions can affect the future. High and increasing unemployment rates, crime rates, and health complaints as enumerated by surveys, for example, may lead to new governmental or corporate policies designed to create new jobs, reduce crime, and prevent the spread of a particular disease just as the public's preferences for cola tastes, perfume smells, and the shape of soap containers may lead to new products or packaging.

Some survey researchers deliberately aim to "study the future," as we have seen, by studying the intentions of people to act in particular ways—for example, voters' intentions to vote for particular candidates, consumers' intentions to purchase some class of product such as a dishwasher or an automobile or some particular product such as a Whirlpool dishwasher or a Mercury Cougar within the next few months, or business executives' intentions to make capital expenditures of various amounts and kinds. Surveys of consumer confidence, attitudes and preferences and of businessmen's decisions are periodically carried out and have become a standard part of the economic, political, and social forecasting industry (Land and Schneider 1987a).

More generally, some social scientists and pollsters, especially futurists, study people's broad images of the future, their aspirations, their expectations, their hopes, and their fears. One of the first large-scale comparative surveys of images of the future was carried out in fourteen countries by an international team of researchers during 1967–68 (Ornauer et al. 1976). These researchers asked more than 11,000 people nearly 200 questions dealing with their attitudes, expectations, hopes, and anxieties concerning the future.

Another early exemplar of research on images of the future using sample surveys can be found in the work begun by Hadley Cantril (1963, 1965; and Free 1962). In a series of studies in the late 1950s and early

FIGURE 6.5
Ladder Used as Part of Cantril's Study of Hopes and Fears

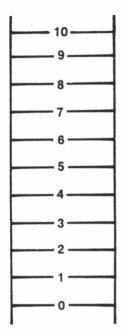

1960s, Cantril (1965) investigated the hopes and fears of people in many countries, including Brazil, Cuba, the Dominican Republic, Egypt, India, Israel, Nigeria, Panama, the Philippines, Poland, the United States, West Germany, and Yugoslavia. Using a then innovative "Self-Anchoring Striving Scale," Cantril's interviewers asked respondents to describe in detail the best possible life for themselves that they could imagine. Then, they asked them to describe the worst possible life for themselves.

The respondents were then handed a sheet with a scale or ladder printed on it going from 0 at the bottom to 10 at the top (see figure 6.5). They were asked to equate their best possible life, which the respondents had just described in considerable detail, to the top rung, giving it a 10, and to equate their worst possible life, as they had just described it, to the bottom rung, giving it a score of zero. The respondents' own images of alternative futures—the best and the worst that they could imagine—thus defined the end or polar points of the scale.

Within these "self-anchoring" limits, the interviewers then asked each person:

1. where he or she presently stands on the ladder,
2. where he or she stood five years ago, and
3. where he or she expected to stand five years in the future.

The respondent, having the ladder in front of him or her, could simply respond with a number from zero to ten representing a rung of the ladder. The whole procedure was then repeated, asking the respondent not about his or her personal hopes and fears but about those of his or her country.

Cantril found that the vast majority of people's personal hopes and fears were not linked to grand political ideologies or religious dogmas and that there were some similarities from one country to another. For example, in every country individuals expressed hopes that tended to deal with their personal well-being as this was rather simply defined. Most people hoped for good health, a job, a decent level of living, decent housing, a happy home life, better educational facilities, and opportunities for their children. True, a few respondents had more idealistic or sophisticated hopes that included an improved sense of social and political responsibility or aspirations for self-development, but often little more than 5 percent of respondents so answered. Even fewer people spoke in terms of abstract ideas such as capitalism, communism, democracy, or socialism.

There were differences among countries, however, in how close people thought they now were to their best possible futures, in how much progress they had made during the last five years, and in how much progress they expected to make during the next five years. And there were differences in emphasis, with five types of personal hopes being identifiable, roughly correlated with level of economic development:

1. Acquiescence to circumstances (characteristic of many Indians and Brazilians).

2. An awakening consciousness of the possibilities of increasing the range or quality of satisfactions (characteristic of many Yugoslav peasants and of Filipinos of lower socieconomic status).

3. An awareness that the new possibilities can be realized (characteristic of Nigerians).

4. The assurance and self-reliance derived from achieving desired goals (characteristic of Israel).

5. The satisfaction some feel with a way of life that promises continuing development (characteristic of Americans).

Remember that these data were collected thirty or so years ago, so the results in particular countries might be quite different today. Nonetheless, the five levels of personal hopes that Cantril identified may remain a valid classification today and probably reflect underlying universal human hopes, ranging from survival to comfort to self-fulfillment.

The Self-Anchoring Striving Scale is an effective and appropriate tool for the study of people's hopes and fears. Using it, futurists can describe and compare the images of the future that people hold and understand them within the respondents' own frames of reference (Cantril and Free 1962: 8).

In sum, survey research in all of its versions offers futures researchers a variety of ways of studying people's intentions, plans, anticipations, hopes, fears, and images of alternative futures, both general and specific. Survey research is a way of studying any population or group, societal members at large or members of any specified subpopulation from students to leaders. It is especially adapted to studying large collectivities of people.

The Delphi Method

General Description and Purpose

In 1953, RAND researchers invented the Delphi method specifically to assess the future and elaborated it in a forecasting study during 1963–64 (Helmer 1983; Gordon and Helmer 1964). They did so in an effort to provide additional foresight for the practical process of planning. Also, they did so, apparently, quite independently of the advances in survey methodology then becoming standard procedure in psychology, sociology, and economics. Fundamentally, the Delphi method is a version of survey analysis, particularly that form of survey research that involves repetitive questioning of respondents, sometimes referred to as "the panel method" in the social sciences (Glock 1955). It is similar, for example, to what Judd et al. (1991: 112–18) refer to as "one-group pretest-posttest design" and "interrupted time-series design."

But the purpose is quite different. Panel studies in the social sciences, especially the pretest-posttest design, aim to investigate some change in the subjects being studied, usually after the subjects have been exposed to some stimulus or treatment. The simplest design includes (1) observing a group of individuals; (2) exposing the individuals to some treatment; and, finally, (3) reobserving the same individuals to measure what change in the group occurred as a result of the treatment (often comparing the group with a "control group" of people who did not receive the treatment).

To the contrary, Delphi researchers, although allowing for and measuring change among the subjects being studied, aim to predict and explore alternative future possibilities, their probabilities of occurrence, and their desirability by tapping the expertise of respondents. Moreover, nearly all Delphi studies are action-oriented with the results aimed at affecting the actions or thoughts of decision makers (Scheele 1975).

The Delphi method, named after the oracle at Delphi in Greece—or perhaps "misnamed" since the method certainly does not intend to produce the ambiguous answers for which the oracle was noted—was invented in 1953 by Helmer (1983: 134) and Norman Dalkey in a RAND study of probable effects of a massive atomic bombing attack on the United States (Linstone and Turoff 1975). There followed a decade of relative neglect of the method until Gordon and Helmer (1964; 1966; Helmer 1983: 220–59) collaborated on a technological forecasting study in 1963 and 1964 which was also done at RAND.

Following the Gordon and Helmer report, which has become the classic exemplar of the method, Delphi methodology quickly spread not only throughout the United States but also abroad. As early as 1969, according to Linstone and Turoff (1975), Delphi studies numbered in the hundreds. By the mid-1970s, they may have been in the thousands (Stewart 1987). Today, they certainly are in the multiples of thousands, many of which are not in the public domain or the professional literature because they have been done by private corporations for their own decision making and are considered their property (Linstone and Turoff 1975).

Of course, as other researchers used the method, modifications, adaptations, and improvements were made, so that today there are many versions of the Delphi method, for example, focus groups and the nominal group technique to mention only two (Millett and Honton 1991: 53–60; Martino 1983 and Preble 1983). Generally, the Delphi method involves a research and communication process that includes at least eight steps:

1. The specification of some topic or subject whose possible, probable, and preferable futures are to be investigated.

2. The construction of a questionnaire as an instrument of data collection.
3. The selection of some individuals (respondents) whose opinions are to be studied, usually experts (to serve as "oracles") on the topic being investigated.
4. The initial measurement of the opinions of the respondents by means of a questionnaire.
5. The preliminary organization and summary of the data resulting from the initial measurement.
6. The communication of the results of the initial measurement of opinions as feedback to all the respondents (reminiscent of the stimulus or "treatment" in standard pretest-posttest designs).
7. A re-measurement of the opinions of the respondents as they have been informed and may have been changed by their knowledge of earlier results including of other respondents' supporting comments for their opinions.
8. An analysis, interpretation, and presentation of the data and the writing of a final report.

Often there are more steps than this because the process of observation, preliminary organization of results, communication of results, and reobservation may be repeated several times. In the original study by Gordon and Helmer, for example, there were four rounds of data collection. Subsequent experience with the method has convinced some futurists that three rounds are sufficient to produce stability of responses; also, more rounds than three may result in respondents being subjected to boring repetition (Linstone and Turoff 1975: 229).

Other features of the conventional Delphi (paper-and-pencil data collection) are that the respondents are usually separated by geographical space from one another, do not know which specific other respondents gave which answers on earlier rounds of data collection, and work separately to answer the questionnaires. Among the modifications of the conventional Delphi are a face-to-face discussion among the experts of the initial results in which the group dynamics of the interaction can affect the outcomes (Holden et al. 1990) and computer Delphi studies in which the initial processing and analysis of data are done electronically and fed back to the respondents almost instantaneously, which makes the entire procedure somewhat similar to computer teleconferencing (Linstone and Turoff 1975).

That pesky term, "objective," turns up again in considering the Delphi method. For if it is carried out correctly, it can be completely objective and scientific as far as the various research steps are concerned, from specifying a sample and writing questions to statistically analyzing data and reporting the results. But the data themselves, of course, are basically the subjective beliefs and judgments of the expert respondents, even though such beliefs and judgments may be based on the respondents' individual objective knowledge of their fields. The objective measurement of such subjective beliefs and judgments are commonplace in the social sciences.

The Gordon and Helmer Forecasting Study: An Exemplar

Gordon and Helmer (1964, 1966) in their classic study in 1963–64 aimed to establish long-range trends—ten to fifty years into the future— with respect to six topics: (1) scientific breakthroughs; (2) population control; (3) automation; (4) space progress; (5) war prevention; and (6) weapon systems.

They selected a group of experts for each of the six topics, inviting a total of 150 experts to participate from which eighty-two responded to one or more of their four rounds of questionnaires which were spaced about two months apart.

Typically, as a Delphi study proceeds through several rounds, the feedback of results from earlier rounds tends to produce more similarity of opinion. Although such convergence is a typical phenomenon in Delphi studies, there is another pattern that sometimes occurs, a bipolar one when expert opinions coalesce into two different views. Also, on some questions, neither the convergent nor the bipolar pattern may occur, and opinions may simply remain spread over a wide range. This happened, for example, on the question concerning when a universal language from automated communication would evolve.

Delphi researchers generally ask questions in addition to when certain events might occur or what events might occur. It has become standard procedure in Delphi studies, for example, for researchers to query the experts on their judgments as to the desirability of certain events occurring as well as their timing.

Also, researchers may ask about alternative policy responses to specific problems and which policies the experts believe would be effective. Gordon and Helmer (1966) asked the automation panel what they thought about the unemployment that might result from automation and some experts gave suggestions for countermeasures that could prevent or reduce it. Then, Gordon and Helmer incorporated the countermeasures proposed by the panel into later rounds of the study and asked panel experts to appraise them, thus bringing in a design or engineering perspective into the study. Table 6.1 gives each of the panel's proposed measures along with their judgment as to its effectiveness, desirability, and probability of occurrence.

In addition to reporting the detailed responses of each of the six panels, Gordon and Helmer (1966), using the most significant forecasts, construct overall pictures of the world for the years 1984 (twenty years into the future from when the study was done), 2000, and 2100. The authors continue with an interpretative commentary on their experts' judgments, specifying the responses that they found surprising (e.g., from the judgment that "the water-covered portions of the earth may become important enough to warrant national territorial claims" to the "fact that control of gravity was not rejected outright") (Gordon and Helmer 1966: 80) and identifying four policy areas (that remain important today) where future human effort ought to be concentrated to avoid future disaster: (1) War prevention; (2) equitable distribution of resources; (3) social reorganization to incorporate automation without social disruption; and (4) properly managing coming genetic engineering (pp. 81–82). They conclude with an evaluation of their then-new Delphi methodology.

Cross-Impact Analysis

Although it has achieved the status of a separate method in the eyes of some writers, cross-impact analysis is most often used as a part of an expert-opinion study and can be appropriately thought of as an extension of the Delphi method (Gordon 1968; Gordon and Hayward 1968). The results of a standard Delphi study may conclude with a set of forecasted events or trends that will occur with some estimated probability at some forecasted future times. But there are all kinds of contingencies and dependencies that may affect such forecasts. That is, the occurrence of particular events may depend on the occurrence of other events.

Cross-impact analysis is designed to deal with this fact by constructing a matrix showing the interdependencies of different events. A cross-impact matrix lists a set of events or trends that may occur along the rows and the events or trends that possibly would be affected by the row

TABLE 6.1

Measures Proposed to Counteract Unemployment Caused by Automation and Their Effectiveness, Desirability, and Probability of Occurrence

Proposed measure	Average effectiveness	Average desirability	Average probability
Creation of new types of employment	moderate/high	high	moderate/high
Retraining of persons unemployed through automation	moderate	moderate/high	high
All-out vocational training program	minor/moderate	moderate/high	moderate/high
Education for better leisure-time enjoyment	minor/moderate	moderate/high	moderate/high
Massive aid to underdeveloped regions (including parts of the United States)	moderate	moderate	moderate
Two years of compulsory post-high school education	moderate	moderate	moderate
Legislation shortening the work week by 20 per cent	minor/moderate	neutral/moderate	moderate/high
Massive WPA-type of programs	minor/moderate	neutral	moderate
Legislation lowering the retirement age by five years	minor/moderate	neutral	moderate
Legislation protecting household and service jobs from automation	nil/minor	negative	minor

Source: Gordon and Helmer (1964: 23).

events along the columns. Usually, both sets of events are the same. Respondents, then, are asked to give their judgments as to how the occurrence of each row event, if it were to occur, would affect the probability of occurrence of a column event. The researcher can then average the responses to create a summary presented in what has become known as a cross-impact matrix (Smith 1987: 50).

In such a matrix, except for the diagonal cells (which represent the same event paired with itself), the cell entries tell us by how much the chances of a column event occurring would be increased or decreased if that row event should occur (Smith 1987: 50). Such a summary matrix allows for a range of different scenarios of the future of particular events, conditioned by the probabilities of occurrence of all other events in the matrix.

For example, an abbreviated cross-impact matrix is shown in figure 6.6 that resulted from a graduate seminar in the early 1980s and that dealt with some aspects of possible futures for the university. All of the items in the matrix had been created from discussions of the possible consequences of faculty tenure being eliminated. The next task was to determine how these consequences might affect each other.

For each cell in the matrix, respondents were asked, "Given the condition that faculty tenure is eliminated, then 'If the trend on the left were to occur, what impact—negative or positive—would it have on the trends listed across the top?'" (Wagschall 1983: 47). The cell entries have the following meanings:

+ + = very positive effect
+ = positive effect
0 = no effect (and I assume that "---" does too)
- = negative effect
- - = very negative effect

To illustrate how to read figure 6.6, note that the first row tells us that if the average faculty salary declines, then there would be (according to the average response of the respondents) a very negative effect on innovative ideas, no effect on the compatibility of student-faculty ideals, a very negative effect on the amount of research funds, no effect on average student performance on the Graduate Record Examination (GRE), and no effect on the last four events listed: the use of educational technologies, the frequency of the faculty's off-campus consulting, the faculty-student ratio, and the amount of tuition.

FIGURE 6.6
A Cross-Impact Matrix: Consequences of Tenure Elimination

	Average faculty salary declines	Innovative ideas increase	Study-faculty ideals more compatible	Research funds increase	Average GREs increase	Use of educational technologies increases	Faculty off-campus consulting increases	Faculty-student ratio improves	Tuition increases less rapidly	Sum of impacts
Average faculty salary declines	X	– –	0	– –	0	—	0	0	0	4–
Innovative ideas increase	– –	X	0	+ +	+	+ +	+	+	0	5+
Student-faculty ideals more compatible	0	+	X	0	+	0	0	+	0	3+
Research funds increase	—	+ +	0	X	0	+	+ +	0	0	5+
Average GREs increase	0	+	+	0	X	+	0	+	0	4+
Use of educational technology increases	0	+ +	0	+ +	+	X	+	0	—	6+
Faculty off-campus consulting increases	+ +	+	—	+	0	+	X	0	0	5+
Faculty-student ratio improves	0	0	0	—	0	0	—	X	—	0
Tuition increases less rapidly	+	0	0	0	0	—	0	0	X	1+
Sum of impacts	1+	5+	1+	3+	3+	5+	4+	3+	0	

Source: Wagschall (1983: 48)

The data, of course, can be aggregated in several ways. In this case, the row entries were added algebraically to measure the overall effect of other variables (Wagschall 1983: 47). Thus, we can see from comparing the numbers shown at the end of each row that declining average faculty salaries would have the largest overall negative effect (scoring 4 minuses), while increasing the use of educational technology would have the greatest overall positive effect (scoring 6 pluses). The algebraic sums at the bottom of each of the columns shows which of the events listed across the top of the matrix are affected, positively or negatively, by the other events. Thus, a tuition increase is least affected by the other changes (scoring 0), while the amount of innovative ideas and the use of educational technologies are most affected (scoring 5 pluses each).

Evaluating the Delphi Method

Just as is the case with the other methods discussed, we can find good and bad examples of the use of the Delphi method. At every stage of a Delphi study something might be done incompetently, thereby threatening the reliability and validity of the results. For example, the "experts" selected may not be truly expert in the subjects that they are asked about or they may not be a random or representative sample of all experts in a given field. The questions asked may be ambiguous, biased, trivial, or irrelevant to the topic. Worse, perhaps, they might not include the right questions, that is, the most revealing and pertinent questions on the topic may never be asked (Simmonds 1977: 24). The questionnaire may be constructed poorly so that one set of questions contaminates another or it may be too long or too boring to keep the interest of the respondents. The preliminary summary of data might be inaccurately done and communicated in an ambiguous or confusing way to the experts before the second and subsequent rounds, thereby invalidating their reactive responses.

Also, experts may drop out of the study and refuse to participate in subsequent rounds of data collection, introducing bias into the results since such dropouts are seldom a random or representative sample of initial participants. When the data collection is complete, the coding and analysis of the data might be inaccurately or ineffectively carried out. Finally, the data might be inadequately summarized and presented to the reader, and the report might be badly written with the true story that the data tell garbled or obscured. Such errors as these flow from sloppy execution and are not inherent in the Delphi technique itself. Many of them are avoidable and occur because Delphi researchers generally have not been trained in the methods of social research (Sackman 1975; Hill and Fowles 1975).

Moreover, in their reports Delphi researchers, just as some other empirical researchers, may gloss over many of the errors and failures of their work, rescuing what they can by putting the best possible face on their results, even while disingenuously pointing out *some* of the study's limitations. It is usually a small miracle when a piece of research of any kind is actually carried out properly from beginning to end without compromises that threaten the validity of the results. The fact that a considerable amount of Delphi research is proprietary, and, therefore, is often privately used by clients and kept secret from competitors and the general public, means that peer review and professional criticism are often lacking. Thus, there are few safeguards against incompetent work and few guarantees of quality. A "Delphi study may be the product of the creativity and ingenuity of a skilled practitioner or of the misconceptions and stumbling of an ill-informed novice, but there is no easy way to tell the difference" (Stewart 1987: 105).

Although it may not be inherent in the method itself, Delphi researchers can also be criticized for failing to probe for the causal connections among events and trends and for ignoring theoretical explanations that would provide an understanding of the social processes involved in the events and trends being forecast. So far, Delphi researchers use, create, test, or know precious little—if any—social theory.

On the plus side, the Delphi method is a good way to structure group communication. Beneficially, it can bring in opinions of groups that would be too large and too diverse to function effectively in face-to-face interaction. The Delphi researcher can organize such diverse opinions into a cohesive statement. Carrying out a Delphi study is often less expensive than having face-to-face group meetings and it avoids some of the biases that can occur in group meetings when they come to be dominated by a few loud and overbearing voices. Moreover, the repetitive rounds and feedback to respondents allow the respondents to refine and further inform their opinions and encourages them to think more deeply about the subject of research and to give reasons to justify their opinions.

When a researcher uses the Delphi method, however, there are cautions that must be considered in addition to sloppy execution. First, it is not entirely clear how much of the convergence of expert opinion in any particular Delphi study is the result of informed opinion change and how much is superficial conformity to the majority opinion (Rohrbaugh 1979). Second, some critics argue that rather than promoting serious and deep thinking about a topic, the Delphi method tends to promote "shallow, narrow, conventional thinking" (Stewart 1987: 100). Third, experts may be so caught up in their own field that they are unaware of developments in related fields that may bear on the topic under study. For example, Linstone (1975) cites the 1930s forecast by experts on reciprocating engines that propeller-driven aircraft would be standard until 1980; they erred because they disregarded the coming technological changes that were to lead to jet engines well before then.

As a forecasting technique, the Delphi method contributes additional information to supplement data from other sources, such as trend analysis of objective data, simulation, or gaming. One recent assessment, for example, found the Delphi technique to have been accurate in about half of the forecasted events that could be evaluated (Ono and Wedemeyer 1994). Experts who are fully immersed in their expertise, working at the leading edge of a field, and dealing daily with new developments, have a distinctive perspective on the future of their field. They often know what is now being researched and developed and they often have special insight about what technologies, including social technolgies, may be on the verge of a breakthrough. Moreover, expert opinion is often an accessible and inexpensive method of measuring some things for which historical objective data are difficult or impossible to obtain. Thus, probing expert opinion constitutes another window through which futurists can view the future. Like other forecasting methods, the accuracy of a Delphi forecast tends to decrease the farther into the future it reaches.

As Millett and Honton (1991: 60–61) say, expert "judgment can be very educational, and the information may be more important to the forecasting process than the accuracy of the forecast." For example, the brainstorming aspects of a Delphi study facilitate people's thinking about a problem, the relevant factors that enter into it, possible solutions to it, and the alternative courses of action that may be available—all useful contributions to successful policy formulating and decision making.

The Delphi method was created and survives, because it is a quick and inexpensive way of getting needed information for making decisions. Such decisions will be made whether or not objective and reliable data are available from other sources. Often, decisions must be made "despite insufficient data, inadequate models, and lack of time and resources for thorough scientific study" (Stewart 1987: 102). The Delphi method helps fill the knowledge gap.

Simulation and Computer Modeling

Simulation and Models Defined

Simulation is a process by which the structure and change of some system, organism, or set of interrelated variables is represented by another, usually manipulable, system or model designed to be similar to the original in some specified and relevant ways. In a simulation, some aspects of reality, such as an economic system, the dynamics of population change in a society, a decision-making system in a factory, a section of a river, or the airflow around the frame of a miniature model of an automobile or an airplane, are imitated or reproduced, usually in microcosm, within a model.

Such models can be merely verbal, pictured in diagrams, actual analogs, or mathematical (see "Gaming" below). Computer models are merely "mathematical models that have been translated into the special, usually digital, languages that computers can read" (Meadows and Robinson 1985: 5).

Of course, in an informal and unsystematic way people use simulation and models in their daily lives all the time. All of us have more-or-less accurate mental models that contain generalizations and assumptions about the nature of the physical and social worlds within which we live. We couldn't make our way in the world effectively—or even as imperfectly as we often do—without them. Through empathy, introspection, imagination, abstraction, generalization, and speculative leaps, people constantly prospectively construct alternative courses and consequences of their possible action and the reactions of others to them as they search their repertoires of behavior to select how they will behave in particular times and places. Of course, such everyday mental models tend to be unconsciously held, unexpressed, unexamined, nebulous, and imprecise (Meadows and Robinson 1985: 2).

A "formal model" is, like an everyday mental model, a set of generalizations or assumptions about the world, but, unlike an everyday mental model, it is explicitly stated in some form. Formal models, compared with everyday mental models, have several possible advantages, including greater rigor (e.g., they are explicit and precise), comprehensiveness (e.g., more information can be included), logical coherence (e.g., conclusions can be error-free within the system because strict deducibility is possible), and accessibility (e.g., other people can see what is being done and can examine the procedures for reliability and validity) (Meadows and Robinson 1985). But, as we shall see below, these advantages are only potential and are not always actually achieved in specific cases of computer modeling.

The primary purpose of simulation, of course, is to discover what would really happen in the real world if thus or so were done by doing it first on a small scale using a model. What would happen if certain assumptions, conditions and parameters were changed, for example, or if certain policy decisions rather than others were made and implemented. To conduct a simulation is to perform an experiment of sorts in which the behavior of variables can be observed in the model as an estimate of how similar variables might be expected to behave in the larger real life and real space-time system under similar conditions.

Simulating reality and experimenting with various policies using a model, of course, can be much less expensive and time-consuming than carrying out the same experiment full scale. Simulation is a way of making and correcting errors, and learning from them without the large costs and risks of actual application (Brewer and Shubik 1979: 9). For example, actually trying out different economic policy choices on a nation's economy to see what would happen if various policies were followed might have disastrous consequences on the economy and on people's lives. It might never allow us to return to the initial conditions when we began the experiment. Conducting a computer simulation, however, could reveal unintended and unfavorable consequences of a policy without harm to either the economy or the people who drew their livelihood from it. In other areas, for example, testing a new design of an airplane wing or a bridge to make certain that they would not disintegrate or collapse under the stresses of anticipated uses, simulation allows experimentation that if carried out by trial and error full scale in real life might result in injury or death to individuals. Using a model in such cases puts no one's life in danger.

Meadows and Robinson (1985) identify four different schools of computer simulation and modeling: Input-output analysis, econometrics, optimization, and system dynamics. The last, system dynamics, is the newest school of computer modeling, dating from the 1950s. It simulates complex, nonlinear, multiloop feedback, time-delayed systems and focuses both on the whole and on the interrelationships of the parts. Generally, system-dynamics models are applied to relatively long-term time horizons of thirty years or more.

Beyond the Limits: An Exemplar

General description of World3/91. Let's take a closer look at system-dynamics modeling as used in *The Limits to Growth* that was discussed in chapter 1. We can do so by examining some of the results reported in its sequel, *Beyond the Limits* (Meadows et al. 1992), done twenty years later. At the same time, we can learn something more about some of the serious problems facing humankind today.

Recall that the world model used by Meadows et al. (1972) included five basic sectors: persistent pollution, nonrenewable resources, population, agriculture (food production, land fertility, and land development and loss), and the economy (industrial output, services output, and jobs). Their updated model, World3/91, contained the same variables as the earlier model, but with a few revisions based on the actual historical performance of the variables from 1970 to 1990 (Meadows et al. 1992: 239). Also, in World3/91, technology is assumed to be adaptive (Cole 1993: 816).

In *Beyond the Limits*, the authors report more than a dozen different computer runs using World3/91. Each one begins with the same model structure, but includes some changes to see what might happen if different estimates of real world parameters were made, if different technological developments actually occur or if the world chooses different economic and social policies. Each computer run constitutes a different scenario of future possibilities.

In each case the authors use World3/91 to calculate interactions among all its 225 variables, a new value being calculated every six months in simulated time from the year 1900 to the year 2100. Doing so, the model produces 90,000 numbers for each scenario which must be interpreted and explained. The authors then summarize each scenario by plotting a few key variables on time graphs, giving the same format for each scenario in two basic graphs (Meadows et *al.* 1992: 119).

The future if humans continue as they are now changing. Figure 6.7 shows Scenario 1, formerly called "the standard run," which is the most likely outcome *if* economic and population policies and behavior are similar to those in the past, *if* changes in technologies and values continue as present trends indicate, and *if* the initial data for the model's variables are roughly correct.

The top part of the figure, "State of the World," reports global totals for population, food production, industrial production, relative level of pollution, and remaining nonrenewable resources. This scenario is, as the authors point out, not necessarily the most probable under all conditions, but it is one of many possibilities and is the most probable under the assumed conditions.

In this scenario the growth of the economy stops and reverses. After the year 2000, pollution rises so high that it begins to affect the fertility of the land, land erosion increases, and total food production eventually, after 2015, begins to fall. Between 1990 and 2020 population increases by 50 percent and industrial output grows by 85 percent, the rate of the use of nonrenewable resources doubles and, by 2020, the remaining resources constitute only a thirty-year supply. More capital and energy are required to find, extract, and refine what resources remain and, thus, less output is left to invest in basic capital growth while investment cannot keep up with depreciation (Meadows et al. 1992: 134).

The industrial plant, understandbly, begins to decline, taking with it the service and agricultural sectors and, finally, after a lag, population, too, begins to decrease, and the death rate goes up as food production falls and health services decline (Meadows et al. 1992: 134). These forecasted consequences can be seen in the bottom half of figure 6.7, where eventual declines in four measures of "Material Standard of Living" are shown: life expectancy, consumer goods per person, food per person, and services per person. Thus, based on "business-as-usual" assumptions, overshoot and collapse occur, resulting in a dismal and frightening picture of the future level of living for humankind.

World3/91 under other assumptions. As they did in *The Limits to Growth*, the authors run their world model making a variety of different assumptions, with much the same results. Overshoot and collapse occur time after time, but in different ways and timings. Meadows et al. (1992:

FIGURE 6.7
Scenario 1

State of the world

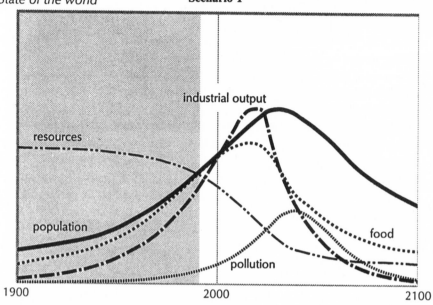

Material standard of living

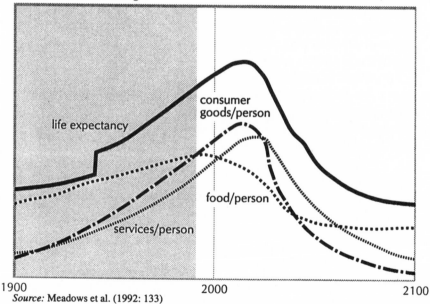

Source: Meadows et al. (1992: 133)

179) conclude that "in a complex, finite world if you remove or raise one limit and go on growing, you encounter another limit. Especially if the growth is exponential, the next limit will show up surprisingly soon." Moreover, the authors claim that postponing reaching a limit through economic and technical fixes increases the eventual likelihood of running into several limits in the future at the same time.

The future if sustainability policies begin in 1995. But let's assume, as do the authors, that by 1995 the world's people put constraints on growth and began applying improved technologies with these results: only two children per couple; industrial output per capita of $350 at 1975 prices (about the average level in material comforts of Europe in 1990); reduced military expenditures and corruption; technologies that increased the efficiency of resource use, decreased pollution emissions per industrial output unit, controlled land erosion, and increased land yields until food per capita reached a desired level (Meadows et al. 1992: 196, 198).

These assumptions are among those built into the computer run that results in Scenario 10 shown in figure 6.8. The global population levels off at about 8 billion people and lives at a desirable material level of living for almost a century. Note that life expectancy levels off at somewhat over 80 years and there are sufficient services and food for everyone. Pollution peaks at a relatively low level and then falls. Nonrenewable resources deplete slowly, with half of them still present in the simulated year 2100 (Meadows et al. 1992: 200).

Although a sustainable world society has been achieved, it is not a static society. It can control pollution, create and spread new knowledge, become more efficient, progress technologically, manage itself better, and become more equitable and diverse (Meadows et al. 1992: 200–01).

There are other possible combinations of variables that would produce sustainable societies, too, but each has its costs because tradeoffs and choices are necessary among the basic variables, for example more people can be supported at a lower level of living. Most important, perhaps, is the timing of the adoption of sustainable policies. If they had been adopted in 1975, then the human future would have been brighter than it is, with more choices. (Meadows et al. 1992: 202).

The future if sustainability policies begin in 2015. What if sustainability policies are not introduced until 2015? What happens then? Making this assumption, Meadows et al. (1992) run World3/91 again

FIGURE 6.8
Scenario 10

State of the world

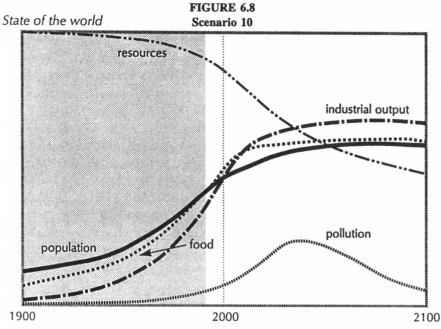

Material standard of living

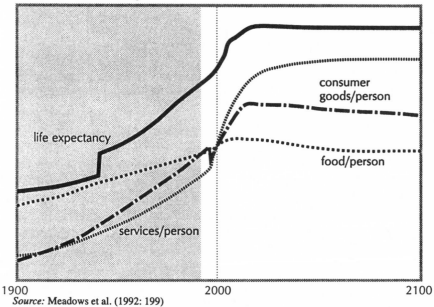

Source: Meadows et al. (1992: 199)

and give their results in Scenario 12 (see figure 6.8). In this future, population reaches 8.7 billion. The twenty-year delay in introducing sustainability policies produces a pollution crisis which, in turn, reduces land yield and food per capita, while increasing death rates. Harsh conditions result in a sharp dip in life expectancy and a decline in total population to 7.4 billion.

With a smaller population and ever-improving technologies, the world eventually recovers. By 2055, pollution begins to go down, food production starts going up, the resource base stops its rapid decline, and eventually industrial output stops its fall and levels off at about $175 per capita, only half of what it was at its peak, providing a far less comfortable level of living than would have been possible had sustainability policies been put into effect in 1995.

Also, by 2055, food per capita and life expectancies begin to rise and services and consumer goods per capita stop declining and level off, but at considerably lower levels than they had reached earlier.

Thus, equilibrium is achieved, finally, even if sustainability policies are not introduced until 2015, but the costs will be high, both in a period of crisis and hard times during which life expectancies are shortened and in an eventually sustainable condition that has a much lower level of living than might have been achieved if sustainability policies had been introduced twenty years earlier.

Beyond the Limits has not created the furor that *Limits* did when it was published. Rather, it has arrived quietly and elicited an "ho-hum" response. One reason for this, Cole (1993: 818) believes, is that "computer simulations and global models are no longer novel." Another and probably more important reason is, as Cole says further, the message had already been "sent, received, and to some extent discredited 20 years ago." The message now seems old hat. Like Cole, I think that this is too bad, because, although we may not believe all the numbers in either *Limits* or *Beyond the Limits*, the "message may still be—perhaps is more—valid."

An Assessment of Simulation and Computer Modeling

Simulation and computer modeling constitute a powerful set of tools for organizing data about the past and present in relatively simple and meaningful ways. They permit the manipulation of data to make fore-

FIGURE 6.9

State of the world   Scenario 12

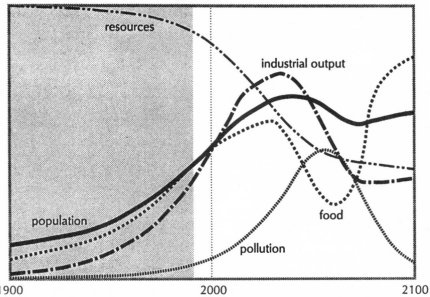

Material standard of living

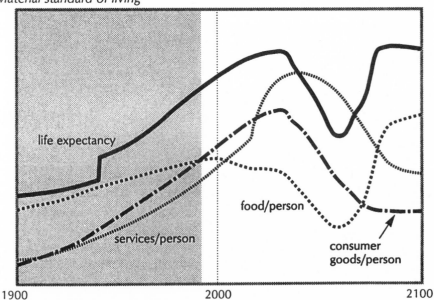

Source: Meadows et al. (1992: 205)

casts about possible alternative futures under different assumptions concerning underlying conditions and policy decisions. Simulation and modeling techniques allow the exploration of the consequences of different policies at a relatively small cost compared to the costs of carrying out full-scale social experiments and they permit consideration of a wide range of possibilities that would be simply impossible to do otherwise. As illustrated by *Beyond the Limits*, with a computer model we can examine the consequences of various economic and population growth policies over a century into the future; we can reflect upon various possible outcomes as they affect the quantity and quality of human life; and we can use such information to help the world community to decide what it ought to do now.

Following Meadows and Robinson (1985), let's review the advantages mentioned earlier.

Indeed, computer models are more rigorous and precise than everyday mental models, because variables must be clearly defined and many assumptions specified explicitly and completely. But an overemphasis on precision can be time wasting. Worse, it can convey a false impression of validity because of its very "precision" when, in fact, important and consequential aspects of real social systems are omitted from the model. In that case, the results of a simulation may precisely describe a totally false image of the future. Many modelers are not trained social scientists and have only a lay person's understanding of psychological and sociological theories.

It is true that computer models can be more comprehensive than everyday mental models in the sense that much more information can be included and processed. They can go well beyond the capabilities of the human mind alone. Yet in practice computer models are rarely complete. Social and psychological factors are seldom included; nor are relevant factors of all kinds for which data are unavailable. Additionally, the oversimplification that mathematization often involves can badly distort reality.

Computer models are more logically coherent than everyday mental models, but their conclusions, nonetheless, are not necessarily error-free. Small measurement errors can have enormous distorting effects when looking at processes that do not average out and do not "forget" initial small events (i.e., that are nonergodic); they can amplify quickly making predictions from such a model little better than guessing (Land 1986:

897). Also, some models may have 80,000 or so parameters and 1,500 or so main equations. For such massive programs mistakes are highly likely. A decimal point can be misplaced, an algebraic sign changed, a ratio inverted, a number transposed—and it is almost impossible to double-check all these entries and operations (Meadows and Robinson 1985: 364). Additionally, anyone who has ever worked with the stacks of papers covered with numbers that a computer print-out produces knows that those stacks have still to be turned into a report. The numbers have to be digested, interpreted, and reduced to tables, figures, diagrams, and prose. Such interpretation and reduction are done by fallible human beings.

In some ways, computer models may be more accessible than everyday mental models, but they, too, are often inaccessible. Not all decisions and actions that lead to a construction of a model in one particular form rather than another are documented, especially when a model is complicated and produced, as they sometimes are, under a tight deadline (Meadows and Robinson 1985: 367). Thus, some assumptions, especially foundational ones, may remain implicit. Moreover, because mathematical and computer skills are required to understand the detailed workings of computer models, such models may be incomprehensible to most people.

Computer models in theory are testable and can be tested by comparing their results with historical data, for example, by running a model in 1990 with data only up to 1970 and then by comparing the computer projections from 1970 to 1990 with the available historical data for the same period. Apart from that, however, Meadows and Robinson (1985: 370) claim that tests tend to be inadequate, because most models' large size makes complete testing too expensive and tedious.

As the case of the users of any method, computer modelers may deliberately falsify results for some perceived personal gain, such as to achieve respect in the eyes of peers, to increase chances of being published or getting additional research funding, or to support some ideological or political view. Thus, every research report invites critical scrutiny.

In sum, computer models are useful tools for many problems, but they may not always be appropriate, at least not as they are currently constructed. Their advantages are clear, when they are used appropriately by skilled, sensitive and honest researchers. But their very advantages can be disadvantages, especially in the hands of harried, clumsy, partially trained or devious practitioners. Yet there are some problems, especially those involving massive data sets that are relevant and necessary to analyze to understand them—as in the case of the problems of world economic and population growth examined here—in which simulation and computer modeling are the most adequate and enlightening methodological techniques to use.

Gaming

What Is Gaming?

Add people as players to a simulation and you have a gaming exercise. Typically, gaming involves a model of some situation or context in which some aspects of reality are simulated and in which human beings play roles in which they make decisions and take actions that affect the outcome of the game being played. Game theory, as Shubik (1982: 3) says, is concerned with "the behavior of independent decision makers whose fortunes are linked in an interplay of collusion, conflict, and compromise." For example, battles between opposing military forces are simulated and commanders' decisions affect outcomes during war games; various aspects of the economic system are simulated in economic games in which people playing roles of businessmen, consumers, and others make plans and choices; and a variety of other phenomena have been the objects of decision makers in gaming studies involving patterns of land use, city planning, international diplomacy (Guetzkow et al. 1963), and other political and social processes. An important criterion of being in a game is the option of leaving, "by saying, 'I am not playing any more'" (Shubik 1975:189).

Not all games, of course, are simulations. For example, many children's games, card games, sports such as football or basketball, and gambling games such as roulette are not simulations of anything. Even chess, although it might be regarded as a simulation of a battlefield, is more accurately described as a model of itself since it is so abstract and formal (Shubik 1975: 6–7). But, generally, gaming involves some simulation or model of a real environment or situation. For example, a sand table can serve as the terrain on which military operations are to be carried out or a computer model can stand in for the relevant firms and industry in a business game. Sometimes, a simulation may not model a real situation, but may be purely hypothethical.

The roles that are played in gaming can be either simulated or not. If a student is asked to play a general in a war game or an ambassador in an international diplomacy game, then he or she is playing a simulated role. If, to the contrary, a general or an ambassador is asked to play himself or herself, then the role is nonsimulated. For training purposes, decision makers in business, the military, education, health services, or some other social or institutional setting may play the roles that they actually occupy in their organizational units, acting out their parts in simulated situations similar to those that they may in fact confront in real life. A newly appointed U.S. ambassador to the Republic of Trinidad and Tobago and his advisers, for example, may spend some days playing themselves (nonsimulated roles) in a simulated negotiation with other Americans who are playing the simulated roles of the Trinidadian prime minister and his advisers in preparation for future real negotiations with the actual prime minister on the same topics.

Games can be extremely simple and inexpensive, such as so-called "free-form" or manual games. The well-known board game of "Monopoly" is an example. Or games can be complex and expensive as is the case of some games in which the simulations are performed by computers in accordance with a complicated computer program.

Different types of models are used in games to represent the realities being simulated. There are (1) *verbal models* which are simply scenarios described in ordinary or technical language and which are useful in simulating complex, ill defined, qualitative phenomena and in achieving nuance, subtlety, and richness of detail. There are (2) *analytic or mathematical models* which are precise, clearly logical, and parsimonious and which are useful when describing phenomena that are easily measured in quantitative terms, for example, the number of cars passing a given point on a highway at different times of the day. There are (3) *diagrammatic or pictorial models* (including maps and flow charts) which sometimes can depict complex processes difficult to describe in words or equations. There are (4) *analog models* which are physical devices that reproduce in three dimensions aspects of actual systems, such as testing a model of a new aircraft in a wind tunnel or analyzing a flood control plan using a small-scale model of a geographical area. And there are (5) *digital models* in which computer-simulation programs portray very complex situations and processes (Brewer and Shubik 1979: 19–26).

Sometimes, several types of models may be used at some point in a single gaming exercise. For example, all five types may be involved in some pilot-training games where a verbal model may describe the setting of a flight, mathematical models may be involved in procedures such as navigation or accuracy of weapons fired during the game, geographical maps may trace the progress of the flight, analog models are used in the construction of the mockup of the flight deck, and computer-assisted models may control the reaction of the whole system to the pilot's actions (Brewer and Shubik 1979).

To conduct a gaming exercise properly, adequate staff and facilities are required so that detailed records are kept of players' moves and decisions and of the progress of the game. This is necessary for a post-gaming examination, analysis, and evaluation of the moves, decisions, and actions of the various players (Brewer and Shubik 1979: 29). It is during the post-gaming recap that some of the learning from gaming takes place.

As early as 1970, the U. S. Department of Defense alone had between 400 and 500 models, simulations, and games of all types. These can be compared with the nonmilitary models, simulations, and games in the United States in 1975 which totaled about 1,200. Brewer and Shubik (1979: 31) estimate the total development cost of both types of models to be about $100 million. Even if it were two or three times more, they say it is small compared to the possible alternative costs of not carrying out such predecision analysis and making the wrong decisions.

Purposes of Gaming

Shubik (1975) gives six goals of gaming and I have added two more. They are:

1. *Teaching.* Gaming can enhance motivation to learn and reinforces other methods of teaching. Gaming can teach facts, theories, skill in interpersonal relations, and norms and values if they have been made part of a game. Moreover, not only the players learn from playing a game. Also, the builders of the game, the controllers or directors of the game, observers, and sponsors can learn.

2. *Training.* If developing skills—for example, training—can be distinguished from teaching, then gaming also can be used for training. During World War II, for example, the U. S. Navy trained students in celestial navigation in large analogue simulations of a flight deck of an aircraft suspended within a large building with a simulation of the night sky overhead. Student navigators could take measurements on "stars," "planets," and the "moon" and compute three-star fixes while simulating

flying on a mission nearly anywhere on Earth, since the relative position of major astronomical bodies used for navigation and their movements could be simulated for any geographical location. Thus, during a simulation, they often "flew" over terrain and navigated using night skies that they had never actually seen before, experiences that helped train them for real flights over such terrain and through such skies that were to come in the future.

To take another example, in Miramar, CA Naval aviators train in a "weapons-envelope trainer" that is a replica of the Tomcat aircraft cockpit. The pilots experience simulated air combat in a 360-degree environment with a precise illusion of land and sky. They "fly" their aircraft and engage enemy aircraft in combat, attacking and being attacked.

3. *Operational Gaming.* Beyond training, gaming can be used to test out communications and other human systems, plan strategy, test tactics, and explore alternative possible responses to a situation. Instead of sending out a naval task force of actual ships and aircraft, for example, for an expensive shakedown cruise or war manuevers that can only be conducted a few times a year and that tie up equipment, weapons, and personnel, the Navy can simulate war exercises, fleet maneuvers, and "dry runs" numerous times at relatively low cost through gaming facilities.

4. *Experimentation or research.* In experimentation or research, the focus of a game is on testing hypotheses, exploring alternative possibilities, or creating new knowledge. Often, researchers use gaming to learn something fundamental about the phenomenon under study, Axelrod's (1984) research on the conditions under which cooperation will evolve (discussed in chapter 4, volume 2) being an example.

5. *Entertainment.* Another goal of gaming is entertainment, as illustrated above by board games, spectator sports, or participant games. Entertainment, though, is seldom the purpose of scholarly, scientific, and futurist uses of gaming (although playing games, even those designed for training or research, can be fun).

6. *Therapy and diagnosis.* Gaming is also a technique that can be used in group therapy sessions, for example, acting out family, occupational, or peer group roles in an effort to locate and rectify dysfunctional behaviors (Shubik 1975).

Additionally, we can add two more goals of gaming:

7. *Forecasting.* Although Shubik (1975: 36) points out that gaming is not generally aimed at forecasting, there are several ways in which prediction in the broadest sense of making assertions about possible, prob-

able, and preferable futures enters into gaming. First, players consider the possible consequences of their own actions as they contemplate alternative moves or actions that the game allows. Second, they also more or less consider the possible future actions and reactions of other players. Third, there is a predictive element in the contingency planning that gaming encourages, since alternative actions under varying conditions result in possible and probable outcomes that always occur in the future. Fourth, gaming can result in brainstorming, with the players' discovery of new alternative actions and future possibilities. Fifth, gaming can teach the need for good (accurate) forecasting activities during the planning and decision-making processes (Shubik 1975: 36). Sixth, gaming can give some basis for the evaluation of plans and procedures by providing some presumptive evidence that some consequences of a plan of action are more likely than others before the fact of actually putting the plan into effect in real-life situations. Seventh, and finally, gaming based on existing situations and actual data bases can sometimes give useful forecasts of the range of probable future outcomes of those situations as the players play the game from the present situation into various outcomes of the game which represent alternative possible futures. In this last case, consequences of technologies and social conditions that may not be operational for another decade or so can be explored in a preliminary fashion.

8. *Advocacy*. Obviously, the results of gaming can be used as presumptive evidence to support various policy choices. Moreover, they ought to be so used, if they are used honestly and cautiously along with other data from different sources.

But, as Brewer and Shubik (1979: 222) point out, games have sometimes been used unscrupulously for blatant advocacy in the policymaking of the defense industry. Since the assumptions, structure, rules, and other foundations of particular games can affect their outcomes, games can be shaped to support various views about which weapons, tactics, or strategies are superior to others and which ought to be developed or elaborated. Resulting decisions can affect hundreds of millions of dollars of expenditures for particular weapons systems rather than for others.

An Example: War Gaming

War games have been played at the U. S. Naval War College since 1866. Brewer and Shubik (1979: 119) say that Admiral Chester Nimitz during a lecture there said, "The war with Japan had been reenacted in the game room here by so many people and in so many different ways that nothing that happened during the war was a surprise—absolutely nothing except the Kamikaze tactics toward the end of the war; we had not visualized those."

In the beginning, war games were conducted at the Naval War College using models of various geographical areas constructed on a table and small-scale models of ships which were moved by the players to simulate sea battles or coastal artillery engagements with ships. Later, model ships were moved on a larger space on the floor that could be altered to represent different geographical locations.

Then, in 1958 the Navy Electronic Warfare Simulator (NEWS), a computer-assisted analog war-gaming system, was completed. It was designed for students at the Naval War College, enabling them "to maneuver forces, receive intelligence, employ weapons, and communicate with friendly forces and umpires" (McHugh 1966: 1–5). For its time, the physical facilities of NEWS were impressive, would have made even a latter-day electronic-game enthusiast think that he or she was in a futurisitc heaven. It had a major space about the size of a motion picture theater with a large screen at one end on which moving images of all active forces in the game were projected and which was photographed from time to time throughout the game for use in postgame debriefings (Brewer and Shubik 1979: 119). In this room, the director and his staff worked, monitoring the game which was visually projected on the screen as various commanders and their forces made their moves. On either side of the screen was an electronic board giving information about the game, from number of weapons fired and hits and misses to damage done to each side in the battle.

Elsewhere in the building were many rooms constructed to be mockups of control rooms and bridges of ships. In these command centers, the players worked, communicated with other players, received information, and made their decisions about what forces to put where, when to attack or retreat, which weapons to use on what enemy ships, planes, or installations, and so on. Elsewhere, a computer system calculated the commands and controlled the movement of the images of ships and aircraft on the giant screen and computed the probablity of hits when weapons were fired.

Like all such games, NEWS involved several different types of roles. There was a war game director or controller who had the major responsibility for the "preparation, play, and postgame critique or evaluation of the simulation of a conflict situation" (McHugh 1966: 1–7). There was a control group consisting of the director's staff whose function was to make calculations and judgments of outcomes as the game proceeded to supplement the computer decision systems built into the gaming technology, serving as umpires or referees as well as evaluators of the players' decisions. (Not a "control group" in the sense of the group from which treatment is withheld in an experiment, but a group that controlled the game.) There were players who made the decisions and took the actions to move the game forward and who used the imaginary forces and weapons under their commands as if they were real forces and weapons. Often, there also were spectators or observers (McHugh 1966: 1–10).

Without being told so, the players may be asked to begin a game in which they start from the facts of some past situation where actual outcomes are now known. Teams may be set up, for example, with all the facts (ships, aircraft, weapons, deployment positions, etc.) of the Japanese and American navies just prior to the World War II Battle of Midway and the players asked to make wartime, battle decisions, some playing Japanese and some American naval commanders unknowingly. The players' decisions and the results of the game then can be compared with the actual outcomes of the real battle.

Or the parameters of some current situation can be the basis for the start of the game, the Persian Gulf, the Mediterranean, the Adriatic, or the Indian Ocean today, with actual presently existing forces, sides, and diplomatic and military tensions. Different diplomatic and military actions can be introduced by the director and his staff, runnning the game over and over again with controlled variations of parameters, so that a manual of possible, probable, and preferable actions could be written as responses to different contingencies, identifying many factors that are likely to enter into particular crisis situations in the future.

Actual field commanders of U.S. forces in different parts of the world can be brought to the Naval War College to play such games, giving them a wide range of decision-making experience and information about possible actions and outcomes that they could not otherwise get short of actually going into battle.

An Evaluation of Gaming

Brewer and Shubik (1979: 54–55) rightly warn the users of gaming methodology that the results of gaming must always be taken with several grains of salt. Assumptions underlying the game must be questioned; sources of data examined for their cogency; methods of analysis investigated; and the resulting decisions subjected to critique. Otherwise, organizational self-delusion can occur.

Always keep in mind when utilizing the results of gaming, that, first, a game is an abstraction of only part of some reality and may fail to be realistic in important and consequential ways. Check game results against common sense and the relevant findings of other types of research (Brewer and Shubik 1979: 57). Second, games depend on the people who construct them. How reasonable, logically consistent, and structurally relevant they are depends on the knowledge and competence of their designers. Third, the results of gaming depend also on the particular players in the game and on the umpires or referees. Differences in skill, judgment, beliefs, and values of both players and umpires can affect the outcomes of games. And, fourth, the results of games when applied to specific decision situations, just as the results of other methods, need to be examined for their appropriateness, effectiveness, utility, and worth (Brewer and Shubik 1979: 57).

Yet with these cautions in mind, gaming offers a relatively inexpensive, expeditious, and effective method for the purposes of teaching, training, operational analysis and evaluation, research, exploring and discovering alternative possibilities for the future, even forecasting the range of probable futures to a limited degree, and providing evidence to evaluate (or support) a policy choice. Gaming, carried out competently and interpreted with caution, can be a powerful instrument in the kit of futurist methodological tools.

Monitoring

What Monitoring Is

Monitoring is a methdological procedure that aims to assess events in process as they occur or as soon after they occur as possible. In our daily lives, we are all familiar with various forms of monitoring. When we stop at a service station to put gasoline into the gas tank of our car, for example, the flow of gasoline through the pump into the hose and into our car is "monitored"—measured and recorded as it flows—and we can watch the counters as they turn telling us the number of gallons and the total price of the gallons that have been pumped at any moment. Other familiar uses of monitoring include keeping track of flows of electricity, the depletion and replenishment of stock supplies in a supermarket, the number and type of bank transactions, changes in leading economic indicators, the level of consumer confidence, the amount of auto traffic on various roads in a city, the movement of trains on a metropolitan subway system, the location and movement of aircraft flying near an airport, the arrival and departure of tourists from a country, and the paths of hurricanes and typhoons. Many of these monitoring systems also involve constant or periodic projecting of the phenomenon being monitoring into the immediate future.

Monitoring often includes several different activities, such as:

• *scanning* (searching for signals or indicators of events or processes in some designated environment, for example, searching for signs of a hurricane);

• *detecting* (noting what has been found of interest to the monitoring process, for example, locating a hurricane and noting its present position and path);

• *projecting* (forecasting the phenomenon's future if it continues as it has been moving and how it might respond to expected changing conditions, for example, predicting the most probable future direction and speed of the hurricane);

• *evaluating* (if not obvious, judging the meaning of what is detected, for example, estimating the force of the hurricane's winds, the amount of rain, the height of the tides and the potential damage each may cause along its probable forecasted path);

• *reacting* (deciding what treatment or response is desirable given the nature of the phenomenon detected, for example, if a hurricane is forecast, deciding what geographical areas ought to be evacuated and what other precautions ought to be taken to protect people from harm and property from damage); and

• *tracking* (constantly keeping track of the phenomenon, once detected, for example, relocating the position of the hurricane as it moves, updating its force, revising projections of its future path and force, reevaluating its meaning and consequences, and rejudging what reactions are appropriate in the light of new data and projections).

In an intensive-care unit, for example, where a patient's vital signs are being constantly monitored electronically, warning signals alert a nurse or doctor to the detection of a dangerously abnormal reading of any vital sign. In turn, the nurse or doctor immediately takes remedial action in an attempt to restore the vital sign to a normal or near-normal level.

Not all monitoring is highly technical. Quite a few agencies, such as Human Rights Watch, Amnesty International, and Freedom House, try to monitor the existence or violation of human rights and democratic processes in the countries of the world, sometimes being weeks or even months behind on some issues. The United Nations Development Programme each year evaluates most of the countries of the world according to a human development scale that includes indices such as life expectancy, adult literacy, and per capita gross national income.

Also, the Global Environmental Monitoring System (GEMS) has been established by the United Nations Environment Program. Other systems include INFOTERRA and the International Register of Potentially Toxic Chemicals (Weiss 1992: 129). Snyder et al. (1976: 222) have proposed a more ambitious and comprehensive Global Monitoring System for the continuous appraisal of the performance of governments, focused on "the impact of governmental actions on espoused official goals and on the attainment and distribution of basic human values." The results of such monitoring of all countries on Earth could be, the authors continue, "distributed periodically to civic and public sectors throughout the world."

Since 1984, to take another example, the Worldwatch Institute of Washington, D.C. has brought out an annual volume entitled *State of the World* that reports on a variety of variables concerning environmental, economic, and social health around the globe. It is an "early-warning system" and was "among the first to note such trends as global warming, fuel-wood shortages in the Third World, world water shortages, soil erosion, and the ecological disasters in eastern Europe" (Brown et al. 1993).

Naisbitt's Megatrends

One example of the use of monitoring in the futures field can be found in the bestselling *Megatrends* by John Naisbitt which appeared in 1982, although it emphasizes the scanning, detecting, and projecting elements in monitoring and leaves most of the evaluating and reacting up to the readers of the book. As it name announces, the book is an effort to identify dominant and important trends, in this case the trends that will define the future of the United States.

The contents of *Megatrends* were largely taken from a quarterly publication then known as the Naisbitt group "Trend Reports" and from the data base underlying them. The staff, says Naisbitt (1984: xxx), "continually monitors 6,000 local newspapers each month" and aims to "pinpoint, trace, and evaluate the important issues and trends." The data base so created includes more than two million articles about local events in American cities and towns during a single year.

Naisbitt's organization focuses especially on five "key indicator" states: California, Florida, Washington, Colorado, and Connecticut. His researchers count the column inches devoted to particular subject matters and note trends in them as they change over time. The data, thus, concern the immediate past and present; the forecasted future is simply extrapolated from the time series, a mostly atheoretical method (Smith 1987: 55).

The megatrends, which according to Richard A. Slaughter (1993d: 829) were not all "mega" and not always "trends" either, were as follows: The United States was viewed as changing:

1. From an industrial society to an information society (which, no doubt, was—and is—a trend but industrial elements may continue into the future even as additional aspects of the information society are added);
2. From dehumanizing or unpleasant technologies to humanizing and pleasant technologies (although the facts may be considerably more complicated);
3. From a national economy to a global economy (which most authors would agree is indeed both "mega" and a trend);
4. From short-term to the use of long-term planning and time frames (which most futurists would say has not happened, though they wish it would);
5. From centralization to decentralization (which is highly doubtful or at least misleading as a general trend);
6. From institutional help to self-reliance (which is also doubtful);
7. From representative democracy to participatory democracy (which Slaughter [1993d] claims is simply wrong);
8. From hierarchical structures to informal networks (which Slaughter [1993d] regards as a false dichotomy since networking takes place within many hierarchies);
9. Geographical migration of Americans from the North and East to the South and West (which, indeed, was a trend);
10. From an either/or society of limited personal choice to a multiple-option society (which has some basis with regard to data on consumer goods but, as Slaughter [1993d] says, may be more apparent than real given the limits of most consumer options and choices).

Megatrends was followed in 1990 by *Megatrends 2000* which was also based on monitoring and which was a better book in that it was less parochial and more global. But both books, despite their commerical successes, are marred by hyperbole and exuberant, perhaps unwarranted, optimism. Each gives the "image of the bubbly, hip-shooting pop futurist" (*Future Survey Annual* 1991: 3). Neither book seriously considers the problems raised, for example, in *The Limits to Growth* and neither undertakes to assess the questions of what must be done to achieve sustainability. As Slaughter (1993d: 829) says about *Megatrends 2000*, it ignores problems and is, thus, thrown "off-centre before it even starts."

Of course, the books are aimed at a commerical market of "advice-giving," as are considerable numbers of other futurist works. A November 1993 twelve-page flyer advertising "John Naisbitt's Trend Letter," for example, sets a general tone, beginning with the following hype:

> *"Huge Fortunes Will Soon Be Made*
> *by far-sighted individuals and companies who*
> *cash in on the hot trends of tomorrow.*
> *You can be one of them,*
> *providing you know, in advance, which trends to follow.*
> *If you'd like to discover what they are, and be*
> *kept up-to-date on a regular basis...*
> *I'll send you 3 valuable reports to start you off, FREE."*

The breathless prose continues in the contents of the flyer: *"223 Hot, New, Amazing Future Business Trends for the 1990s."* Although the breezy, superficial pitch may help sell Naisbitt's "Trend Letter," it is a symptom of the simpleminded, unsophisticated commercial glitz of the megatrend books. Clearly, professional standards of good work and quality control in futures studies are badly needed (Amara 1981b). Yet, as I write, Naisbitt is widely seen among the public as a leading American futurist (*Future Survey Annual* 1991: 3).

An Evaluation of Monitoring

Although it is not shown in its best light by looking at the the megatrend books, monitoring, competently carried out, is a potentially sound method for studying certain phenomena that have some bearing on the future. Monitoring is most appopriate for phenomena that are of constant interest because of their relevance to human well-being and purposes and that invite close surveillance in order to know when to take some kind of action. The actions taken can vary from protection and adaptation to active intervention to alter the phenomena in order to keep them beneficially stable or changing as desired.

Time scales of monitored phenomena may be quite different, some being extremely short-term as in the case of the detection and evaluation of unidentified flying objects and reaction to them by the Air Defense Command, while others are long-term as in the case of some of the global environmental changes now taking place. One relatively long-term project, unfortunately now in some doubt, was the radio search for extraterrestrial intelligence (SETI). Beginning in October 1992, the entire sky was to be searched from California's Mojave Desert over a ten-year period with unprecedented frequency range. The search was aimed at receiving any message that may have been sent from a planet of any of the 400 billion stars that make up the Milky Way Galaxy. A year later the U.S. Congress decided that SETI was too expensive and eliminated its funding, but there are current efforts to revive the project.

The utility of monitoring depends a great deal upon the accuracy of measurements of the phenomena being monitored, the comprehensiveness with which scanning takes place, the certainty of detecting the target phenomena when they occur, and the existence of adequate technical, conceptual and theoretical tools to evaluate the meaning of the phenomena accurately. Also, monitoring presupposes the ability to react in some way, either by preparing for what appears to be coming no matter what people do to prevent it or by taking effective intervening action, when possible, to control future outcomes. Monitoring, of course, is standard practice today in most countries for the measurement and assessment of the state of the economy and society using social indicators, as we saw in chapter 1.

That part of monitoring known as "scanning" itself has come to be a method in its own right, although its use often includes some of the other aspects of monitoring as well. The increased use of scanning has been an effort to cope with the tremendous increase in information that has become available in the last several decades and with the increasing tempo of technological and social change in some areas. As Marien (1991: 83) says, scanning is a normal human activity and is "absolutely necessary for the intelligent guidance of our individual and collective affairs." It is, as he says further, a fundamental task of futures studies. It has been widely used, for example, by state governments as a part of the planning process (Cook 1990; Meeker 1993). Moreover, scanning underlies the abstracts of *Future Survey* which has become an important tool by which futurists and others keep abreast of the published research and leading ideas in the futures and related fields. As I write, *Future Survey* is "the world's best scan of what's being written about probable, possible, and preferable futures" (Cleveland 1995: v).

Content Analysis

The megatrend books also illustrate a futurist's use of content analysis, a well-known method in the social sciences, especially for the study of mass communications. Content analysis is a technique for the systematic and objective study of specified aspects of messages (Holsti 1969). In the example of Naisbitt and his group the messages whose contents were analyzed were printed texts of American newspapers. But many other sources of messages have been used by content analysts, such as private papers and letters, political party platforms, magazines and books, radio and television broadcasts, and motion pictures.

In order to conduct a content analysis, some set of messages must be specified; a sample drawn from them; some scheme established by which to code the messages into meaningful classifications of interest to the investigators; some people (or machines) trained to scan (for example, read, look at or listen to, depending on the medium) the messages, detect and evaluate the instances to be classified as they occur in the messages, and record them; and some people available who are able to analyze the data, interpret them, and write a report.

For example, *The Lasswell Value Dictionary* (Lasswell and Namenwirth 1968) provides a classificatory scheme by which messages are categorized according to the values being expressed in the message, such as power, rectitude, respect, affection, wealth, well-being, enlightenment, and skill. *The Lasswell Value Dictionary* has been used to analyze the contents of several different kinds of messages, including political party platforms (Namenwirth and Lasswell 1970), speeches given at the opening of each session of the British parliament from the year 1689 to 1972, presidential addresses of scientific associations for three decades of the twentieth century (Namenwirth and Weber 1987), and editorials over time in the major newspapers of several different countries.

Computer approaches to content analysis, such as the above, allow the coding and processing of large quantities of messages since the "computer reads the documents in toto and classifies all words according to specified rules," for example, according to values expressed when using *The Lasswell Value Dictionary* (Namenwirth and Lasswell 1970:9). After classification, the computer can count the class frequencies of words for each document and calculate raw numbers, percentages, correlations between the occurrences of various words, and other statistics (Stone et al. 1966).

Content analysis contains pitfalls of reliability and validity unless it is carried out competently. If human coders are used, the same materials ought to be coded several times by different people independently of each other, the results compared, and a count of the errors made and reported to the consumers of the reports. There are also threats to validity in the process of generalizing the results of a content analysis. For example, what can be said about the values of a society from an analysis of the content of the editorials in major newspapers of that society or some other specified set of documents? Perhaps, the editorial writers reflect the values of the larger society and, perhaps, they do so only imperfectly and to some unknown degree.

Yet content analysis has several advantages over some other methods of research, such as, for example, survey research. Usually, the data are already existing and don't have to be collected by some costly means, such as training and sending out armies of interviewers to interview respondents. Some messages—for example, Parliamentary speeches—have a long history and permit the study of processes through time. Moreover, they may importantly reveal the images of the future underlying leaders' decisions.

Content analysis is relatively unobtrusive. Once the messages have been specified and availability obtained, then research can proceed by researcher and computer without having to observe or interview live human subjects directly. Finally, the messages used in content analysis generally remain available for restudy by other researchers, so analyses can be redone and double-checked (Babbie 1986).

Participatory Futures Praxis

Social Activism and Participation

Some futurists are more directly involved in the activist side of futures studies than others. Such futurists are often on the front lines of practical action, spending much of their time organizing, proposing, criticizing, advocating, or implementing specific policies for action now. They go beyond merely creating the knowledge, forecasts, and designs for action to be used by others and get more or less directly involved in praxis themselves. For them, future studies is "in the borderline between research and politics" (Dahle 1992: 83), often more politics than research. They have created a broad set of procedures, some still emergent, that can be conveniently categorized as methods of "participatory futures praxis."

Participatory futures praxis, however, is less technical than some other parts of the futures field and there is no single set of procedures that defines it that is accepted by everyone. Yet there exist various exemplars and there are two major unifying themes: (1) participatory futurists generally aim to democratize the process of futures thinking by incorporating in some meaningful way the participation of a variety of people in the decision-making processes that affect their own future; (2) they serve as catalysts for taking action, for actually doing something to bring about that future.

Of special concern are problems faced by ordinary people whose everyday lives will be affected by whatever decisions and actions may be under consideration. Participatory futurists often attempt to provide a voice for the usually silent "underdogs" of society whose views and concerns are generally excluded from the dominant networks of communication and influence and who suffer the results of decisions taken by elites who represent established interests and power structures.

For example, I have already described in chapter 2 the Honolulu Electronic Town Meeting (ETM), an effort in Hawaii both to inform and listen to ordinary citizens on important economic issues. The ETM is only one of the many participatory projects of James A. Dator and the Manoa School of Futures Studies, which include among their major aims increasing the participation of people in decision making, "to interact with the present and to be an *activist* in the broadest sense of the word" (Jones 1992: 24). In Norway, participatory futures studies became institutionalized in the state-funded Norwegian Alternative Future Project (Dahle 1992).

Another example is what the sociologist William F. Whyte (1989; 1991; and Whyte 1988) calls participatory action research (PAR) which aims to get the subjects of research, including decision makers themselves, incorporated into the research process. PAR is a form of applied social research that contrasts most conspicuously with what Whyte calls the "professional expert" model. In the professional expert model a researcher designs a study to answer the questions posed by the decision makers of some client organization, the aim being to advise the decision makers on what course of action they ought to take. Although Whyte grants that this model can be useful for some problems, it fails often because decision making is a process that develops through time and involves organizational learning, adjustment, and change, not a one-time and forever decision. Also, it can fail because the decision makers have little or no participation in and commitment to the expert's recommendations for action (except that they often pay for them).

In PAR, by contrast, at least some decision makers and workers in the organization being studied team up with professional researchers for the entire research process. They participate in designing the project, specifying and collecting data, analyzing data, interpreting and pondering the findings, and formulating recommendations for action. Thus, some members of the organization being studied are not merely passive informants. Rather, they play active roles in the entire process of research and in considering possible future changes.

As a result, the professional researchers benefit because they have access to the experience and knowledge of the participating members of the organization and, thus, are protected from making serious errors of fact and interpretation in reporting their findings about the organization. Additionally, the practical goals of the project are benefited, because, when the professional researchers leave, some decision makers and workers of the organization, those who have participated in the study, are well informed about the reasons for the recommended actions, since they helped to discover the reasons and to formulate the recommendations. They tend, thus, to be strongly committed to following through with the proposed actions.

PAR can be considered "research" in the conventional sense since new knowledge is often created *and* it is also praxis since it aims to produce practical designs for change that have some reasonable chance of being implemented. PAR has been successfully used in many cases of solving practical problems, resulting in beneficial changes, for example, in the Norwegian shipping industry (Thorsrud 1977), in worker-productivity and job-saving projects at Xerox, and in increasing rank-and-file participation in decision making in worker cooperatives in Spain (Whyte 1989).

Future Workshops

Background and purpose. The late Robert Jungk invented future workshops, another type of participatory futures praxis, to which he devoted much of his life. Jungk had been a victim of Hitler's regime, leaving Germany in 1933 and becoming a political refugee in Switzerland. There, he was overcome by a sense of powerlessness because of his inability to do anything to alter the course of events—or even to persuade journalists to write about the mass murders taking place in Germany (which he believed at the time to be in the thousands and which were, as we now know, in the millions) (Jungk 1987: 5).

Since then, Jungk worked to try to empower people, "to develop," as he said, "hidden, buried or crippled capabilities in countless individuals who have been cheated of their proper development by bad education or social deprivation" (Jungk 1976: 17). "The future," he argued, "belongs to everybody" (Jungk and Müllert 1987: 9). Thus, he worked to prevent the continued colonization of the future by a tiny elite and to open up the shaping of the future to ordinary citizens (Jungk 1987). More than most other efforts at participatory futures praxis, Jungk's future workshops incorporated the least powerful classes of society—the victims, the underclasses, the workers, the smallest consumers, renters and homeless persons, and helpless and dispossessed people. He strove to encourage the democratic participation of such otherwise powerless people.

To do so, he and his followers conducted future workshops, a key element of his "Everyman Project," which he envisioned as a coming worldwide program for enriching democratic participation of men and women and reviving the sense of community in all aspects of collective life. The aims of future workshops, thus, unlike PAR, did not include "research" in the conventional sense. Rather, they were to incorporate the views, ideas, and proposals of people whose lives are affected by some decision. Their views are recorded and given a hearing through communication with decision makers, and sometimes through newspapers, radio, and television.

Jungk ran his first future workshop in 1962. Since then, the future workshop movement has flourished throughout Europe. For example, in Lech, Austria a group worked on the impact of computers on society; in southern France, a group created ways of responding to the energy crisis; in the 15th arrondissement in Paris, people threatened with large-scale business developments in their neighborhood created alternative development plans; residents of Eisenheim, Germany saved their mining village from developers who would have destroyed it; in Vienna a group made recommendations for updating the school system and increasing environmental protection; and ordinary citizens of the Ruhr Valley and Cologne area in Germany constructed images of desirable futures for the area (Jungk and Müllert 1987).

A wide range of projects or issues are suitable foci for the organization of future workshops. Workshops can involve people who are under threat from proposed atomic power stations or nuclear waste sites, new super highways, airports, shopping malls, automation at the workplace, unsafe work conditions, deteriorating housing projects, the relocation of industries, and city development plans. The point is not to be merely reactive, but to be proactive: to encourage people to create their own images of the future and to design actions they can take. There is very little of concern to people that can't be made at least somewhat more manageable—if not solved—by conducting a future workshop on the topic. At the highest level of abstraction, future workshops aim to invent new social institutions, to find new non-violent methods of bringing about beneficial changes, to consider new goals and values, and to build a creative, participatory society (Jungk and Müllert 1987: 11).

How a future workshop works (following Jungk and Müllert 1987). A future workshop is a simple tool; at bare minimum very little cost—except for the time of the participants—is involved. It can take place anywhere where people recognize problems and search for their solutions, anywhere where they can meet and talk with each other about the future.

Preparation for a future workshop includes deciding on a topic and selecting fifteen to twenty-five participants. Select a place to meet, ideally where chairs can be arranged in a semicircle. Obtain large sheets of paper and tape to hang them on the walls of the meeting place and felt pens to write with. During the workshop, keep a record of people's com-

302   Foundations of Futures Studies

ments and suggestions in the form of summary statements writ large for all to see. (Blackboards and chalk can substitute or supplement the sheets of paper.) It's also useful to have a typewriter or computer to list points being made and a photocopier so participants can have their own copies.

In addition to the participants, each workshop has a "facilitator" or group leader, someone who understands the purposes and various stages of the workshop and who serves as a prompter to ask questions and move the workshop along through its various phases, as necessary. When the workshop begins, the facilitator asks everyone to introduce him- or herself, striving to establish an informal and open atmosphere receptive to everyone's ideas. Then, the facilitator briefly explains the background, methods, and objectives of the future workshops in general, describing each of the three phases to come (see below) and giving an example or two of the accomplishments that resulted from past workshops.

After that, the facilitator proceeds by stating the specific problem and objectives of this particular workshop, defining the theme or central topic to be discussed. Also, the facilitator draws up and explains a timetable, such as that given below, as a tentative guide for moving through the three phases of the workshop that is about to begin.

An ideal time for a workshop is three days, although each phase can be compressed or expanded to accomodate a shorter (even as short as one day) or longer period:

Example of a Future Workshop Timetable

Critique Phase —Friday 6 to 9 PM
Fantasy Phase —Saturday 10 AM to 1 PM and 3 to 7 PM
Implementation Phase —Sunday 10 AM to 1 PM and 2 to 4 PM

Each of the three phases of a future workshop is designed to achieve a different set of purposes. In the *critique phase*, the facilitator asks, "What's bothering you? What are your complaints?" The participants, thus, further define and elaborate the problem. Continuing, the facilitator invites the participants to voice all the negative things about the problem they can think of. Participants state their grievances, including giving expression to their emotions—often anger and frustration if they are personally threatened by the problem under discussion. All that can be criticized about the present situation or proposed plans is brought forward. Participants are asked to summarize their points by writing them briefly on the wall sheets.

As complaints dwindle or become repetitious, the facilitator turns to the lists on the wall sheets and asks for clarification of the points and their possible links with others, aiming to classify and organize the results into clusters of similar problem areas. Then, the participants select the areas of greatest interest to them for further consideration by some system of voting or assigning points to different problem areas.

Next, *the fantasy phase* begins. It is useful to begin it by stating the complaints and problems from the Critique Phase in a positive manner. For example, "no space for kids" becomes "plenty of room for kids" and "afraid to walk in the neighborhood" becomes "feel safe and enjoy walking in the neighborhood." Immediately, the tone of the discussion changes, from negative to positive, from undesirable to desirable (p. 62).

Now, the workshop becomes a brainstorming session and participants are asked to be imaginative, innovative, and creative. They are asked to forget reality and practical constraints of any kind, to assume that anything is possible, to break out of their mental ruts, to think the unthinkable, and to open up to their wishes, dreams, desires, and fantasies. "What," they are asked, "are some alternative, fresh solutions to these problems?" With respect to the issues under discussion, "What future would you like if you could have any future you wanted?"

Ideas, however improbable and infeasible, are invited. As before, a record is kept and ideas are summarized and written on the wall sheets in brief form. Cross-fertilization is encouraged and ideas build on ideas. Within thirty minutes in a typical workshop, as many as one hundred suggestions might be recorded (p. 62). Again, the results are mulled over, clarified, combined, and classified. Finally, by voting (usually by each participant assigning a given number of points to different ideas), a few suggestions are selected for further discussion in the implementation phase.

To begin *the implementation phase* the facilitator summarizes the results from the fantasy phase: These, remember, are largely positive, idealistic, innovative, often seemingly impractical hopes for the future and highly desirable solutions and resolutions of the problems defined and elaborated in the critique phase.

Next, the facilitator asks the participants to face the realities of the situation. "Which of the solutions suggested in the fantasy phase," he or she asks, "are or can be made practicable?" "What can actually be done to bring them about?" "What are the obstacles to them and how can they be overcome?" (p. 68).

Consulting with experts or tracking down facts become part of this phase, as participants eliminate unrealistic proposals and create specific tactics and strategies to achieve their desired goals. The result is a plan of action, an implementation proposal, wherein some of the seemingly wild ideas of the fantasy phase are made into real possibilities.

After the workshop, there are several possible followups. Some people write a report. It is sent to each participant with requests for corrections and revisions. After these are incorporated, a final report is sent out. Participants then begin to put the recommendations into effect, including bringing them to the attention of key decision makers. Some of the workshop participants might become an enduring group for follow-up action, continuing to organize for action, making coalitions with likeminded groups, and mobilizing people to support their plan.

Generally, a future workshop results in making participants feel more informed, more effective, and more a part of the community than they had felt before. Moreover, this feeling is often justified: the workshop does tend to spur individual reflection and inquiry and group participation and discussion, and the central problems discussed in the workshop are sometimes at least partially solved. Occasionally, a future workshop turns a participant into a social activist and a group into the beginnings of a social movement (p. 77).

Even in societies with democratic governments, authoritarian elements exist, for example, in the home, the workplace, the bank, the religious hierarchies, the school and the university, the doctor's office and the hospital, the government bureaucracy, the mass media, the housing project, and so on. Future workshops can be a means of democratizing relationships in any sphere of activity, a way of incorporating the reserves of knowledge and imagination of ordinary people into the tasks of designing and building a better future, a way of helping people to move from apathy to action even in the face of the bewildering large-scale and impersonal organizational decision making and social changes going on around them (p. 12).

Future workshops can constitute that special effort that people may need to keep pace with technological changes and their social impacts on them. They are a way that people can constructively face such problems as occupational obsolescence, information overload, increasing complexity, dangerous housing projects, isolation and alienation, and general fear of the future (p. 20). They help to teach people that it is not futile to challenge the status quo.

Future workshops can be regarded as interactive thought experiments, allowing people participating in group discussion to imagine alternative possibilities, to reject the impractical or undesirable ones, and to explore and to act on the practical and desirable ones. They are experiments that move from thought to action. They are thought experiments, moreover, in the small, resting on a handful of people here and there who establish cooperative and supportive relationships with one another in face-to-face groups.

But if such groups were to proliferate? Could they make a difference in the world? Perhaps so. At least, Jungk devoted much of his life to the belief that future workshops can be "the catalysts for social transformation" (Jungk and Müllert 1987: 113). Of course, future workshops by their very existence go some distance toward creating a more desirable society, because they encourage their participants to become caring, thinking, informed, involved, planning, future-oriented, hopeful, and active people.

The drawbacks of future workshops include the possibilities of small groups being collectively uninformed about important relevant facts, inaccurate in forecasting the consequences of alternative actions, and parochial rather than universalistic in their value judgments and goals. Thus, resulting plans of action sometimes may do more damage than good. Future workshops also can result in falsely raising people's hopes, only to have them dashed again if their efforts at social change are squashed. Finally, future workshops can be incompetently or cynically run and produce very little real participation or real change.

Yet future workshops can provide an effective setting for carrying on the kind of critical discourse proposed in this book, if participants are informed by relevant facts, base their planning on presumptively true predictions, and explicitly state and justify the value judgments that they make. Without effective action, however, future workshops may make participants feel good (as a result of a sense of participation and community) even though no actual change occurs.

Social Experiments

Social Experiments Defined

Future workshops are relatively small-scale and inexpensive. In contrast, social experiments tend to be large-scale and very expensive. Social experiments are conducted in the "laboratory of real life" where some people in the society receive some significant treatment or intervention over a period of time. Usually, such experiments, sometimes called "demonstration projects," are designed to find out what would happen if some social policy or plan were to be introduced for an entire region or a whole country by first carrying it out on some part of the population.

The basic idea of an experiment, as described earlier, is quite simple, although carrying one out can become very complicated. It involves selecting a group of people to participate in the experiment and then separating them randomly into two groups, an experimental and a control group. Next, similar measurements of relevant variables are made on the members of each group at time one, before any stimulus is introduced, that is, before any treatment is carried out or any policy implemented. After that, the experimental group is exposed to some treatment or stimulus while the control group is not. The next step requires remeasuring both groups at time two (and perhaps at later times also), after such exposure of the experimental group, to see what effect the stimulus had. This is determined by comparing any change in the experimental group with similar change, if any, in the control group. If the experiment has been designed correctly, then there is a strong presumption that any differences in measurements of change between the experimental and control groups from time one to time two (or later times) is due to the stimulus and not to other factors.

The classic natural scientific experiment usually takes place in a laboratory under highly controlled conditions. But in the case of large-scale experiments aimed at evaluating the possible consequences of social policies, the experimental group consists of human beings who are not in a laboratory in any conventional sense but living in their own communities; the stimulus to which they are exposed is a complex social program; and other conditions that may affect the results are not controlled by the experimenters. Thus, social experiments often turn out to be "quasi-experiments," that is, only approximations to the classic experimental design.

Since the 1960s, there have been several such large-scale social experiments in the United States. They have included studies of the effects of giving people guaranteed incomes, giving parents educational vouchers so they can send their children to the schools of their choice, giving poor families vouchers for housing, providing different health insurance plans to people, giving financial assistance to families (Kershaw 1980), and giving financial aid to criminals when they are released from prisons.

The purpose of such studies is to provide decision makers with information so that they can make better assessments of the possible consequences of alternative social policies and, therefore, reach decisions that will produce more effective and efficient social programs without the costly years of trial and error if a new program begins on a national scale right from its start. Governments, for example, spend many millions—sometimes billions—of dollars on various social programs without knowing in advance whether or not the programs will in fact achieve their goals. Programs involving education, housing, healthcare, school lunches, occupational retraining, and the control of crime, among many others, are sometimes established with relatively little direct evidence about what the consequences will be. With a particular level of funding, will such programs achieve their intended goals? Will they produce unintended consequences as side effects? Are such consequences desirable or undesirable? If they are undesirable, can anything be done to eliminate them? Often, there is a lot of guesswork in trying to answer such questions and sometimes there is acrimonious disagreement about what the possible consequences will be based on anecdotes and analogies, hopes and fears, that are frequently false. In fact, no party to the debate may have much actual evidence to support his or her forecasts.

Seldom are such questions answered by prior experimental studies, even though demonstration projects, if conducted properly, might take much of the guess work out of public policy debates. Pilot programs can be put into operation and monitored to see if they work as planned without costly and undesirable side effects. Even forecasts of the monetary costs of some programs cannot be made with much confidence of accuracy without the results from prior demonstration projects.

The New Jersey Income-Maintenance Experiment: An Exemplar

One such demonstration project was the New Jersey Income-Maintenance Experiment carried out from the late 1960s through the early 1970s. It involved paying a guaranteed income to people, also called, somewhat misleadingly, a "negative income tax." People who receive a "negative income tax" don't pay a tax on their income *to* the government, rather, if their incomes are below a specified level, for example, the poverty line as set by the government, they receive cash payments *from* the government. The aim is to bring people's incomes up above the poverty line.

With funds provided by the U.S. Office of Economic Opportunity, the New Jersey Income-Maintenance Experiment was conducted in four cities in New Jersey (Trenton, Paterson, Passaic, and Jersey City) and in one city in Pennsylvania (Scranton). The major purpose of the experiment was to discover how much work incentive the recipients would lose if they received guaranteed annual incomes. For it was believed that American taxpayers would be unwilling to support a policy based on a guaranteed income if an effect was to reduce the recipients' desire to work.

The subjects of research consisted of two-parent families among the working poor, that included at least one male between the ages of eighteen and fifty-eight who was either working or physically capable of doing so. The selection of families in the five cities was done after screening 30,000 families in a preliminary interview and a pre-enrollment interview of 2,300 families which also was used to collect data for baseline measurements of family composition, income, and other social and economic variables. In the end, more than 1,300 families were involved in the program, although some dropped out before the experiment was completed (Kershaw 1980).

Half of the families were assigned to one of eight different experimental groups, which varied in how much of a reduction in the government payment was made per new dollar earned (30, 50 or 70 percent) and in what percentage of the poverty line was initially set as the guaranteed income (ranging from 50 to 125 percent). The other half of the families were designated the control group and they received no payments whatsoever, although they were studied just as were members of the experimental groups. Families could do what they wished with the payments and they could move anywhere in the United States. Even if a person left the original family unit, he or she still received a proportionate share of the family grant (Kershaw 1980).

Every four weeks, the experimental families sent in forms giving information on their income and every three months they were interviewed about their work experience, finances, health and medical treatments, formal educational experiences, family structure, and political and social participation (Kershaw 1980: 38).

I won't try to summarize the findings of this study here, because they are detailed and lengthy, including separate analyses for whites, blacks, and Spanish-speaking families (Rees 1980). But we can say that, after several years had passed, there was no decline in weekly earnings because of the payments, comparing the experimental with the control group. That is, members of the experimental group did not have any significant drop in their incentive to work. There was, however, a difference of about twelve percent in the number of hours worked, with the experimental group working fewer hours. One possible explanation of this is that the members of the experimental group, since they were receiving a guaranteed income, could afford to take a longer time when looking for work and hold out for a better job.

The "negative income tax," now known as the Earned Income Tax Credit (EITC), was launched under the presidency of Gerald Ford and has been expanded since then. In 1985, it cost the U.S. Treasury $2 billion. By 1995, the cost was $24 billion, and the EITC had become the fastest growing program in the federal budget (Glassman 1995: 27).

An Evaluation of Social Experiments

Social experiments are costly in both time and money. The New Jersey Income-Maintenance Experiment, for example, took more than seven years to complete from design to published findings and cost $34 million, which included not only the costs of research but of income payments as well (Rossi and Freeman 1982: 217).

Other problems include the possibility of the well-known effect of the experiment itself focusing attention on the human subjects of research. They feel special and may react simply to the attention the experimenters give them rather than to the intended stimulus, the so-called Hawthorne effect, although a control group, properly used, cancels this out. Additionally, the subjects of experimental research may react to the aims of the experiment if they know what they are and, hence alter their behavior in ways not directly the result of the stimulus. In the case of the New Jersey study, some critics suggest that the recipients did not take the government payments as seriously as they would have taken earned income and, thus, used the money in atypical ways.

Another criticism is that the repeated questionnaires and interviews sensitized the respondents so that they developed conditioned patterns of responses to subsequent data collections during the experiment. Thus, their answers may have become a function of the fact that they have been questioned and interviewed previously. Moreover, the relatively short duration of the experiment (three years of the actual payments to recipients) may have made responses atypical, for example, as respondents looked toward the end of the experiment and the last of their guaranteed payments (Kershaw 1972: 237).

Difficult problems of attributing effects to specific causes, always a possible vexation in field research, create special problems for large-scale social experiments. Does the treatment (i.e., social policy) create an all-at-once effect that is maintained over time or does the immediate effect disappear after some time? Might there occur a delayed effect of the social policy, perhaps not revealing itself until after the experimental study is concluded? Two or three or even more years of experimentation may not be enough time to answer such questions (Kelly and McGrath 1988).

Also, in social experiments, the uncontrolled larger society can threaten the validity of the results as well. For example, in the New Jersey experiment the comparison between the experimental and control groups was complicated because, after the study was set to begin, the state of New Jersey introduced a new welfare program that included families with male heads of household, then later cut back benefits while the study was underway. At one point in the experiment, 25 percent of the control families and 13 percent of the experimental families were on welfare (Rees 1980: 47). Obviously, the effect of the welfare program could be expected to make the differences between the two groups smaller than it would otherwise have been, because welfare may have led some members of the control group to withdraw from the labor market.

Also, there are other threats to validity from changes in the larger society that face social experiments. What if the United States had gone to war? What if a major economic depression developed? What if some terrible natural disaster took place, such as a major earthquake or a flood? What could we say about the results of the experiment, even if it were not abandoned in midstream?

With respect to the New Jersey experiment, Kershaw (1980: 40) concludes that the social experiment worked: "familes were chosen and assigned to experimental or control groups, money was paid, interviews were conducted, data were assembled, analysis was done and results were sent to Washington, where policymakers used them." The question remains, though, whether or not social experimentation on such a large scale is cost effective. Would it be better simply to begin a large-scale social program including all eligible people from the start, constantly monitor it, and then be prepared to change it as knowledge about how it was working or not working became available? Possibly, but we don't really know the answer.

We could argue "Yes," because in theory the organizational control system of any social program has the potential capacity to receive and evaluate information about how the program is achieving its goals and, if they can be better achieved, to adjust the rules of the program accordingly. In practice, however, designing, legislating, organizing, implementing, and administering a social program are generally very difficult tasks, frought with intellectual, ideological, and political conflict. A program could become a nightmare of confusion and inequity for both bureaucrats and clients if rules, standards, and operating principles changed frequently.

Social experimentation has a place in the futurist's kit of tools. With competent execution and with more than a little luck, it may be a way to obtain information about the probable effects of a proposed new program before it is introduced on a full scale. Even with prior social experimentation, however, no major social program ought to be undertaken without provision for constant monitoring and evaluation, regarding both the achievement of program goals and the creation of unintended and unanticipated consequences that may affect other goals and values.

Finally, some writers have considered the possibility of creating an "experimenting society" in which all social programs and policies would be part of a convention of experimentation and measurement (Campbell 1984a). After all, efforts aimed at designing and implementing social change in natural settings are basically similar to experimentation, though aimed at goals other than knowledge (Campbell 1984a: 22). Such efforts, just as experiments in the laboratory, are willful and deliberate intrusions into ongoing processes.

In totalitarian societies it is highly unlikely that an experimenting society could succeed, because totalitarianism does not allow for the flexibility and openness necessary for the requisite free-flow of information. After all, the purpose of experimentation is to learn, to test ideas, and to change them if they are wrong.

Creating an experimenting society might succeed in democratic societies, although the problem of vicious political partisanship might—probably would—be a serious obstacle to carrying out social experiments properly over the long term. Also, although protections of the rights of every individual would be needed to guarantee human freedom against the threats of heavyhanded social intervention, too many such protections could prevent large-scale social experimentation from taking place. Yet intelligent and effective social policy in the future may require some approximation to the model of the experimenting society, dedicated to addressing social problems with cumulative, gradual, and stepwise social improvements based on testing ideas and programs, paying attention to feedback, and openness to learning.

Ethnographic Futures Research

Description of the Method

Magoroh Maruyama and Arthur M. Harkins (1978), Reed D. Riner (1987, 1991a & b), Robert B. Textor (1979, 1980, 1983, et al. 1984), and other futurist-anthropologists (Razak and Cole 1995) have created what has come to be known as "anticipatory anthropology" or "cultural futures research." Such research is focused not on the cultural developments of the past, although they are not excluded from study as base lines for futures thinking, but on anticipations of the coming sociocultural future (Textor 1990: 141. Quotes and paraphrases that follow are from Textor 1990).

Ethnographic Futures Research (EFR) is one of the major methods used by cultural futures researchers. Textor invented EFR in 1976 and with the help of his colleagues, associates, and numerous students has worked at developing the method ever since, producing a series of revisions of his *A Handbook on Ethnographic Futures Research* and several case studies. Textor has taken the tools of the ethnographer—especially the long, arduous, intensive interaction with cultural informants—and combined them with the purposes of the futures field. While keeping the cultural, holistic, comparative, and macrotemporal perspective of the ethnographer, Textor aims his research at the informants' present images of possible, probable, and preferable future cultures (Textor 1995, forthcoming).

Like other methods of futures research, EFR incorporates the "scenario." Unlike most other methods, EFR makes the scenario its central focus. The scenario as constructed by the use of EFR is complex, tending to deal with a whole range of phenomena between the present and the horizon date of the future being explored, not just a single sector or aspect of cultural life. The changes in the natural and social environment; the impacts of social and cultural changes on people's lives; people's reactions to such impacts, including their individual and collective decisions and policy actions aimed at influencing them, are all taken into account. In EFR "the interviewee is asked to build scenarios about a whole sociocultural system" (p. 144).

At the heart of EFR is lengthy and detailed interviewing of respondents, often repeated interviewing of the same respondents over a period of time. Ethnographic futures researchers only loosely structure their interview sessions. The role of the researcher is "that of an active, sensitive, and sympathetic listener, non-directive stimulator, and careful recorder" who moves the interview along by "showing interest, offering encouragement, and posing questions where needed" (p. 145). By probing where necessary, the researcher aims to help the respondent produce an account that has clarity (intelligibility), comprehensiveness (broad scope and holism), contextualization (the description and role of the sociocultural context within which the change process is projected to occur), and coherence (the grounding of assumed causal processes of change in some sort of explicit reasoning).

The respondent or informant is asked to construct three scenarios, all of which are viewed by him or her as possible, in this case having a "probability judged to be greater than zero" (p. 146). The first is an *Optimistic Scenario*—the possible future that in the judgment of the informant is desirable. The second is a *Pessimistic Scenario*—the possible future judged by the informant to be undesirable.

After the optimistic and pessimistic scenarios are elicited, then the researcher elicits a third scenario, the informant's *Most Probable Scenario*. In constructing it, the informant is asked to forget his or her preferences and value judgments of what is desirable or undesirable and to be as realistic as possible. The ethnographic futures researcher asks the informant: What will most likely really occur? What do you expect the future actually to be, whether you like it or not?

Textor is a longtime student of Thailand, and, therefore, much of his work deals with sociocultural futures of that country. EFR, however, has been applied in other contexts, by Textor as well as others, ranging from studies of images of the alternative futures held by research and development leaders in California's "Silicon Valley" and future sociocultural effects of the microelectronic revolution in Austria (Textor et al. 1983) to the futures of science itself (McKeown 1990) and the future of China (English-Leuk 1990).

The Data

Usually, ethnographic futures research is carried out by several researchers interviewing a relatively small number of informants, for example, three researchers interviewing a total of twenty-four respondents in a pilot inquiry among academics on alternative sociocultural futures for Thailand (Textor et al. 1984) and ten researchers interviewing a total of thirty-two intellectual, policy, and planning elites in a study of future sociocultural effects of the microelectronic revolution in Austria (Textor et al. 1983). In one major study, Dr. Sippanondha's *The Middle Path for the Future of Thailand*, only one informant was interviewed by Textor and three other researchers.

The single interviewee was Dr. Sippanondha Ketudat who was born in Bangkok in 1931 and whose distinguished career began as a Thai Government Scholar as an adolescent. Dr. Sippanondha is a physicist (Ph.D., Harvard); ordained Buddhist monk; onetime director of Thailand's University Development Commission; former deputy director of the Southeast Asian Ministers of Education Secretariat; ex-secretary general of the National Education Commission; sometime minister of education; first president of the National Petrochemical Corporation, Lt.; and Thai patriot. Thus, he has a rich background in science, education, development, and practical experience as well as a commitment to the well-being of the Thai people (Textor 1990: xxvii).

EFR: An Exemplar

Dr. Sippanondha was asked to construct alternative images of the future for Thailand for approximately 2020 A.D. (Buddhist Era 2563). He described the three scenarios—optimistic, pessimistic, and most probable—as EFR is designed to do, covering a wide range of interrelated topics, from population size and technological and economic developments to equitable sharing among all the Thai people, the role of religion,

especially Buddhism and spirit worship, and the effects of modernization, urbanization, tourism, and the world media on the Thai cultural
identity. Most—if not all—of the major components of Thai society and
culture were considered.

Ethnographic futures research emphasizes flexibility and openness,
with considerable room for EFR respondents to shape the structure and
content of the future protocols that they construct. In Dr. Sippanondha's
case, for example, three additions to the standard EFR protocol resulted.
The first was a separate chapter on Dr. Sippanondha's working philosophy of development, especially on the underlying general values and principles that he used in building his scenarios, emphasizing, among others,
the Thai values of harmony, tolerance, freedom, autonomy, and indi-
viduality; the priority on sufficiency and fairness, not luxury, greed, and
inequity; and the relative importance of increasing people's productivity
rather than their consumption.

The second addition resulted from Dr. Sippanondha's discomfort with
the standard EFR's instructions for the most probable scenario. There
was a scenario, he thought, perhaps the most important one for guiding
future policy and decision making, somewhere between his optimistic
and most probable scenarios. Thus, he constructed a fourth scenario,
defined as that future for Thailand that was *reasonably* probable (but
not most probable), *if* Thailand "were to have wise and good leadership,
plus a measure of good luck" (Textor 1990: 151).

The third addition was a list of high priority policy matters that required action now or in the near future (Textor 1990: 151). These included three recommendations (1) attacking challenges rather than merely
reacting to them; (2) developing human resources; and (3) selectively
expanding Thailand's role internationally.

In sum, Ethnographic Futures Research has the advantage over some
other methods in being accessible to a single researcher working alone or
with a small group. It does not take a large research grant or elaborate
organizational machinery to collect, process, and analyze data. It does
take time and it takes training and skill to carry out the probing, loosely
directed interviewing properly. When done with patience and competence, Ethnographic Futures Research results in deeply textured, richly
elaborated, contextually grounded, and creatively imagined scenarios of
probable, possible, and preferable futures, including, in some cases, detailed plans for action.

Conclusion: Scenarios

In this chapter, I have discussed some of the major methods that futurists use to construct images of the future. Among those methods not discussed are the somewhat disappointing efforts of Neustadt and May (1986) to specify a methodology by which to draw lessons from history, technology assessment, relevance trees, issues management, contextual mapping, and the use of chaos theory and fractals. The reader can learn about them in Bright and Schoeman (1973), Coates (1986), Gordon and Greenspan (1988), or Schwarz et al. (1982).

I've saved a discussion of scenarios until the end of this chapter, because, by now, the reader knows that scenarios can be produced by any and all of the specific methods used by futurists. Thus, a scenario can be a way of summarizing the results of futures research, be they based on quantitative methods that produce precise, though delimited, projections and forecasts or on qualitative ones that result in sweeping alternative images of the future of whole societies or civilizations.

Futurists, as Michael (1987) says, tell stories. Their stories objectify alternative possibilities for the future and, thus, permit people to think about them and explore different reactions to them. Moreover, work on some futures topics can only proceed by first making some plausible scenarios about the future of related developments. For example, predicting "the societal impact of climatic changes from increasing concentrations of carbon dioxide" requires first that forecasts about future populations, economy, and technologies be made (Land and Schneider 1987: 23), because scenarios giving the effects of increases in carbon dioxide make sense only if based on scenarios of what would happen without such increases.

Often, futurists create scenarios by using any of the methods described in this chapter. Sometimes, they create scenarios by a variety of other methods. For example, Dator and Rodgers (1991) describe seven futures for the state courts in the United States for the year 2020 by summarizing the various statements made by 300-plus people at a conference on the future and the courts. Sometimes, they do so simply by using their creative imaginations and drawing on their personal experiences, their scholarly knowledge, and their capacities for speculation (Wagar 1991). The "data" on which such scenarios are based are sometimes largely futurists' reflections and probings into their own personal observations, beliefs, and values, often influenced by their understanding of historcial changes or the salient issues and beliefs of their day. At the extreme, scenarios of this sort merge with creative writing, including science fiction and fantasy, and are illustrated by some of the great utopian—or dystopian—writing of all time, such as Thomas More's *Utopia* or George Orwell's *Nineteen Eighty-Four*.

No matter how it is constructed, how full and rich or meager and lean, how factual or fictional, how particularistic or universalistic, the "scenario" gives methodological unity to futures studies. It is used by all futurists in some form or another and is, thus, by far the most widely shared methodological tool of the futures field.

Thus, the end product of all the methods of futures research is basically the same: a scenario, a story about the future, usually including a story about the past and present. Often, it is a story about alternative possibilities for the future, each having different probabilities of occurring under various conditions. Also, it often includes goals and values, evaluating alternative futures as to their desirability or undesirablility. Typically, it includes a description of available possible choices of human action and their anticipated outcomes, and it may include implicit or explicit recommendations regarding what choices and actions ought to be made now in the present to create the most desirable world in the future. Sometimes, a scenario may simply be a single stunning image of the future, either highly desirable or totally abominable, a vision so vivid and compelling that it inspires people to strive to achieve it or a nightmare so dreadful that people will struggle to avoid it or prevent it from happening.

The methods on which it is sometimes based may be codified, but scenario writing in the futures field itself is not. In fact, codification that does full justice to the subtlety, sophistication, and power of the most inspiring scenarios may not be possible. Fortunately, however, there do exist many brilliant and influential classic exemplars of scenarios. A few of them—from Thomas More's vision of Utopia to Karl Marx's image of a future communist society—are described in chapter 1, volume 2.

In describing selected classic utopias, I do so with a focus on preferable futures and the human values by which they can—and ought—to be judged, since utopias are primarily normative in their purposes. So doing, I begin a discussion of the ethical foundations of futures studies, which is the topic of volume 2 of this book.

References

Abaza, M. and G. Stauch. 1988. "Occidental reason, Orientalism, and Islamic fundamentalism." *International Sociology* 3, No. 4 (December): 343–364.

Adam, Barbara. 1988. "Social versus natural time, a traditional distinction re-examined." Pp. 198–226 in M. Young and T. Schuller (eds.), *The Rhythms of Society.* London: Routledge.

Ajzner, Jan. 1988. "Modern society as a moral community (back to Aristotle)." Paper read at the annual meetings of the American Sociological Association, Atlanta, GA (August 24–28).

Alger, Steven F. 1972. "Images of the Future and the Two Cultures." Unpublished Ph.D. dissertation, New Haven, CT: Yale University.

Alonso, William and Paul Starr (eds.). 1987. *The Politics of Numbers.* New York: Russell Sage Foundation.

Amara, Roy. 1981a. "The futures field: searching for definitions and boundaries." *The Futurist* XV (February): 25–29.

———. 1981b. "The futures field: how to tell good work from bad." *The Futurist.* XV (April): 63–71.

———. 1981c. "The futures field: which direction now?" *The Futurist.* XV (June): 42–46.

———. 1984. "New directions for futures research—setting the stage." *Futures* 16 (August): 401–404.

———. 1986. "Letter to the editors." *Futures Research Quarterly* 2, No. 3 (Fall): 5.

Anderson, Elijah. (1976) 1978. *A Place on the Corner.* Chicago: The University of Chicago Press.

Andic, Fuat M. (1985) 1990. "Human resources in the Caribbean." Pp. 153–161 in S.B. Jones-Hendrickson (ed.), *Caribbean Visions.* Frederiksted, VI: Eastern Caribbean Institute.

Andrews, Frank M. (ed.). 1979. *Scientific Productivity: The Effectiveness of Research Groups in Six Countries.* New York: Cambridge University Press, and Paris: UNESCO.

———. (ed.). 1986. *Research on the Quality of Life.* Ann Arbor: The University of Michigan.

Apter, David E. 1963. "Political religion in the new nations." Pp. 57–104 in C. Geertz (ed.), *Old Societies and New States.* New York: The Free Press of Glencoe.

Arblaster, Anthony and Steven Lukes (eds.). 1971. *The Good Society.* New York: Harper & Row, Torchbook.

Arens, W. 1986. *The Original Sin: Incest and Its Meaning.* New York: Oxford University Press.

Argyris, Chris, Robert Putnam, and Diana McLain Smith. 1985. *Action Science.* San Francisco, CA: Jossey-Bass.

319

Arizpe, Lourdes, M. Priscilla Stone, and David C. Major. 1994. "Rethinking the population-environment debate." Pp. 1–9 in L. M. Arizpe, M. P. Stone, and D. C. Major (eds.), *Population & Environment*. Boulder, CO: Westview Press.

Arizpe, Lourdes and Margarita Veláquez. 1994. Pp. 15–40 in L. M. Arizpe, M. P. Stone, and D. C. Major (eds.), *Population & Environment*. Boulder, CO: Westview Press.

Axelrod, Robert. 1984. *The Evolution of Cooperation*. New York: Basic Books.

Babbie, Earl. 1973. *Survey Research Methods*. Belmont, CA: Wadsworth.

————. 1986. *The Practice of Social Research*. Fourth Edition. Belmont, CA: Wadsworth.

Badham, Roger A. 1993. "Constructing a theology of the environment." Pp. 42–58 in H. F. Didsbury, Jr., *The Years Ahead*. Bethesda, MD: World Future Society.

Baier, Annette. 1981. "The rights of past and future persons." Pp. 171–183 in E. Partridge (ed.), *Responsibilities to Future Generations*. Buffalo, NY: Prometheus Books.

Banfield, Edward C. with the assistance of Laura Fasano Banfield. 1958. *The Moral Basis of a Backward Society*. New York: The Free Press.

Barber, Bernard. 1983. *The Logic and Limits of Trust*. New Brunswick, NJ: Rutgers University Press.

————. (ed.). 1987. *Effective Social Science*. New York: Russell Sage Foundation.

Bardis, Panos D. 1986. "Futurology and irenology: will the world be at peace in 2000 A. D.?" *International Journal of World Peace* 3, No. 2 (April-June): 117–124.

Barnes, Barry. 1974. *Scientific Knowledge and Sociological Theory*. London: Routledge & Kegan Paul.

Barney, Gerald O. (study director). 1980. *The Global 2000 Report to the President*. The Council on Environmental Quality and the Department of State. Washington, DC: U.S. Government Printing Office.

Barrett, M. and M. McIntosh. 1985. "Ethnocentrism and socialist-feminist theory." *Feminist Review* No. 20: 23–47.

Barry, Brian. 1977. "Justice between generations." Pp. 268–284 in P.M.S. Hacker and J. Raz (eds.), *Law, Morality, and Society*. Oxford: Clarendon Press.

————. 1978. "Circumstances of justice and future generations." Pp. 204–248 in R. I. Sikora and B. Barry (eds.), *Obligations to Future Generations*. Philadelphia, PA: Temple University Press.

Barry, Brian and Samuel L. Popkin. 1984. "Foreword." In K. Luker, *Abortion and the Politics of Motherhood*. Berkeley and Los Angeles: University of California Press.

Barthes, Roland. (1971) 1976. *Sade, Fourier, Loyola*. Trans. by: Richard Miller. New York: Hill and Wang.

Batson, C. Daniel. 1991. *The Altruism Question*. Hillsdale, NJ: Lawrence Erlbaum Associates.

Bauer, Raymond A. 1966. *Social Indicators*. Cambridge, MA: MIT Press.

Bauman, Zygmunt. 1976. *Socialism: The Active Utopia*. New York: Holmes & Meier.

————. 1987. *Legislators and Interpreters: Modernity, Post-modernity and intellectuals*. Ithaca, NY: Cornell University Press.

Baumeister, Roy. F. 1991. *Meanings of Life*. New York: The Guilford Press.

Bäumer, Bettina Dr. 1976. "Appendix: empirical apperception of time [in India]." Pp. 78–88 in *Cultures and Time*. Paris: The Unesco Press.

Bayes, Thomas. (1764) 1958. "An essay towards solving a problem in the doctrine of chances." *Biometrika* 45: 296–315.

Bayles, Michael D. 1981. *Professional Ethics*. Belmont, CA: Wadsworth.

Beckmann, Petr. 1971. *A History of Pi*. New York: St. Martin's Press.
Beckwith, Burnham P. 1984. *Ideas about the Future*. Palo Alto, CA: B.P. Beckwith.
Behnke, John A. and Sissela Bok (eds.). 1975. *The Dilemmas of Euthanasia* 2nd Edition. New York: Anchor Press/Doubleday.
Bell, Daniel. (1967) 1968. *Toward the Year 2000: Work in Progress*. Boston: Houghton Mifflin. First published as a special issue of *Daedalus* 96 (Summer, 1967).
———. (1973) 1976. *The Coming of Post-Industrial Society: A Venture in Social Forecasting*. New York: Basic Books.
Bell, Wendell. 1954. "A probability model for the measurement of ecological segregation." *Social Forces* 32 (May): 357–364.
———. 1964. *Jamaican Leaders: Political Attitudes in a New Nation*. Berkeley and Los Angeles: University of California Press.
———. (ed.). 1967. *The Democratic Revolution in the West Indies: Studies in Nationalism, Leadership, and the Belief in Progress*. Cambridge, MA: Schenkman.
———. 1971. "Epilogue." Pp. 324–338 in W. Bell and J. A. Mau (eds.), *The Sociology of the Future*. New York: Russell Sage Foundation.
———. 1974. "A conceptual analysis of equality and equity in evolutionary perspective." *American Behavioral Scientist* 18, No. 1 (September/October): 8–35.
———. 1977. "Inequality in independent Jamaica: A preliminary appraisal of elite performance." *Revista/Review Interamericana* 7, No. 2 (Summer): 294–308.
———. 1980. "The futurist as social scientist: From positivism to critical realism." *Futurics* 4, Nos. 3/4: 303–312.
———. 1983. "An introduction to futuristics: Assumptions, theories, methods, and research topics." *Social and Economic Studies* 32, No. 2: 1–64.
———. 1986. "The invasion of Grenada: A note on false prophecy." *The Yale Review* 75, No. 4 (October): 564–586.
———. 1987. "Is the futures field an art form or can it become a science?" *Futures Research Quarterly* 3, No. 1 (Spring): 27–44.
———. 1988. "What is a preferable future? How do we know?" Pp. 293–306 in J. Dator and M. G. Roulstone (eds.), *Who Cares? And How? Futures of Caring Societies*. Honolulu, HI: World Futures Studies Federation, University of Hawaii (Manoa).
———. 1991. "Values and the future in Marx and Marxism." *Futures* 23, No. 2 (March): 146–162.
———. 1993 "Bringing the good back in: Values, objectivity and the future." *International Social Science Journal* 137 (August): 333–347.
Bell, Wendell and Juan J. Baldrich. 1983. "Elites, economic ideologies, and democracy in Jamaica." Pp. 150–187 in M. M. Czudnowski (ed.), *Political Elites and Social Change*, International Yearbook for Studies of Leaders and Leadership, Vol. II. De Kalb, IL: Northern Illinois University Press.
Bell, Wendell and Walter E. Freeman (eds.). 1974. *Ethnicity and Nation-Building*. Beverly Hills, CA: Sage.
Bell, Wendell and James A. Mau. 1971. "Images of the future: Theory and research strategies." Pp. 6–44 in W. Bell and J. A. Mau (eds.), *The Sociology of the Future*. New York: Russell Sage Foundation.
Bell, Wendell and Jeffrey K. Olick. 1989. "An epistemology for the futures field: Problems and possibilities of prediction." *Futures* 21, No. 2 (April): 115–135.
Bell, Wendell and Ivar Oxaal. 1964. *Decisions of Nationhood: Political and Social Development in the British Caribbean*. Denver, CO: Social Science Foundation, University of Denver.

Bellah, Robert N., Richard Madsen, William M. Sullivan, Ann Swidler, and Steven M. Tipton. 1985. *Habits of the Heart: Individualism and Commitment in American Life*. Berkeley: University of California Press.

———. 1991. *The Good Society*. New York: Alfred A. Knopf.

Bennett, Jonathan. 1978. "On maximizing happiness." Pp. 61–73 in R. I. Sikora and B. Barry (eds.), *Obligations to Future Generations*. Philadelphia, PA: Temple University Press.

Bennett, Neil G. 1992. "Demographic methods." Pp. 434–445 in E. F. Borgatta and M. L. Borgatta (eds.), *Encyclopedia of Sociology*. New York: Macmillan.

Berger, Peter L. 1989. "Salvation through sociology." *The New York Times Book Review* (October 22): 34.

Bergmann, Werner. (1983) 1992. "The problem of time in sociology: An overview of the literature on the state of theory and research on the 'sociology of time', 1900–82." *Time and Society* 1, No. 1: 81–134.

Berk, Richard A. and Thomas F. Cooley. 1987. "Errors in forecasting social phenomena." Pp. 247–265 in K. C. Land and S. H. Schneider (eds.), *Forecasting in the Social and Natural Sciences*. Boston: D. Reidel.

Berkman, Lisa F. and Leonard S. Syme. 1979. "Social networks, host resistance, and mortality: A nine-year follow-up study of Alameda County residents." *American Journal of Epidemiology* 109: 186–204.

Berlin, Brent and Paul Kay. 1969. *Basic Color Terms: Their Universality and Evolution*. Berkeley and Los Angeles: University of California Press.

Bernard, Philippe J. (1963) 1966. *Planning in the Soviet Union*. Trans. by I. Nove. Oxford: Pergamon.

Bettelheim, Bruno. 1986. "Their Specialty Was Murder." *The New York Time Book Review* (October 5): 1, 61.

Bhaskar, Roy. 1978. *A Realist Theory of Science*. Sussex, England: Harvester Press.

Bickel, Alexander. 1962. *The Least Dangerous Branch: The Supreme Court at the Bar of Politics*. Indianapolis, IN: Bobbs-Merrill.

Black, Donald. 1976. *The Behavior of Law*. New York: Academic Press.

Blasco, Pedro Gonzales. 1974. "Modern nationalism in old nations as a consequence of earlier state-building: The case of Basque-Spain." Pp. 341–373 in W. Bell and W. E. Freeman (eds.), *Ethnicity and Nation-Building*. Beverly Hills, CA: Sage.

Bloch, Ernst. 1954–1959. *Collected Works*. Frankfurt-am-Main: Suhrkamp.

Bloch, Maurice. 1977. "The past and the present in the present." *Man* 12: 278–292.

Blum, Alan F. 1970. "The corpus of knowledge as a normative order." Pp. 319–336 in J. C. McKinney and E. A. Tiryakian (eds.), *Theoretical Sociology*. New York: Appleton-Century-Crofts.

Boldt, Menno. 1980. "Canadian native Indian leadership: Context and comparison." *Canadian Ethnic Studies* 12, No. 1: 15–33.

———. 1981a. "Enlightenment values, romanticism, and attitudes toward political status: A study of native leaders in Canada." *Canadian Review of Sociology and Anthropology* 18, No. 4: 545–565.

———. 1981b. "Philosophy, politics and extralegal action: Native Indian leaders in Canada." *Ethnic and Racial Studies* 4, No. 2 (April): 205–221.

———. 1982. "Intellectual orientations and nationalism among leaders in an internal colony: A theoretical and comparative perspective." *The British Journal of Sociology* 33, No. 4 (December): 484–510.

———. 1993. *Surviving as* Indians. Toronto: University of Toronto Press.

Boldt, Menno and J. Anthony Long in association with Leroy Little Bear (eds.). 1985. *The Quest for Justice*. Toronto: University of Toronto Press.

Boorman, S. A. and P. R. Leavitt. 1973. "A frequency-dependent natural selection model for the evolution of social cooperation networks." *Proceedings of the National Academy of Science* 70: 187–189.

Botkin, James W., Mahdi Elmadjra and Mircea Malitza. 1979. *No Limits to Learning: Bridging the Human Gap*. Oxford: Pergamon.

Boucher, Wayne I. 1977. "Introduction." Pp 3–13 in W. J. Boucher (ed.), *The Study of the Future: An Agenda for Research*. Washington, DC: U.S. Government Printing Office.

———. 1986. Comments on Early Warning Systems: Current Methods and Future Directions. Conference of the World Future Society, New York (July 14–17).

Boucher, Wayne I. and John V. Helb. 1977. "Appendix: Results from the Survey of Current Forecasting Efforts." Pp. 275–282 in W. I. Boucher (ed.), *The Study of the Future: An Agenda for Research*. Washington, DC: U.S. Government Printing Office.

Boucher, Wayne I. and Katherine H. Willson. 1977. "Monitoring the future." Pp. 210–232 in W. I. Boucher (ed.), *The Study of the Future: An Agenda for Research*. Washington, DC: U.S. Government Printing Office.

Boulding, Elise. 1978. "Futuristics and the imaging capacity of the west." Pp. 7–31 in M. Maruyama and A. M. Harkins (eds.), *Cultures of the Future*. The Hague: Mouton.

Boulding, Kenneth E. 1964. *The Meaning of the Twentieth Century: The Great Transition*. New York: Harper and Row.

———. 1985. *Human Betterment*. Beverly Hills, CA: Sage.

Box, G. E. P. and G. M. Jenkins. 1970. *Time Series Analysis: Forecasting and Control*. San Francisco, CA: Holden Day.

Bracher, Karl Dietrich. (1969) 1971. *The German Dictatorship*. Trans. by J. Steinberg. London: Weidenfeld and Nicolson.

Brand, Myles. 1979. "Causality." Pp. 252–290 in P. D. Asquith and H. E. Kyburg, Jr. (eds.), *Current Research in Philosophy of Science*. East Lansing, MI: Philosophy of Science Association.

Braybrooke, David. 1987. *Meeting Needs*. Princeton, NJ: Princeton University Press.

Brewer, Garry D. 1974. "The policy sciences emerge: To nurture and structure a discipline." RAND Paper Series P-5206 (April). Santa Monica, CA: The RAND Corporation.

Brewer, Garry D. and Peter deLeon. 1983. *The Foundations of Policy Analysis*. Homewood, IL: The Dorsey Press.

Brewer, Garry D. and Martin Shubik. 1979. *The War Game: A Critique of Military Problem Solving*. Cambridge, MA: Harvard University Press.

Brewer, Marilyn B. and Barry E. Collins (eds.). 1981. *Scientific Inquiry and the Social Sciences*. San Francisco, CA: Jossey-Bass.

Bright, James R. and Milton E. F. Schoeman (eds.), *A Guide to Practical Technological Forecasting*. Englewood Cliffs, NJ: Prentice-Hall.

Broad, William J. 1988. "Back into space." *The New York Times Magazine* (July 3): 10 et passim.

Broad, William and Nicholas Wade. 1982. *Betrayers of the Truth*. New York: Simon and Schuster.

Broszat, Martin. (1960) 1966. *German National Socialism, 1919–1945*. Trans. by K. Rosenbaum and I.P. Boehm. Santa Barbara, CA: Clio.

Brown, Donald E. 1991. *Human Universals.* Philadelphia, PA: Temple University Press.

Brown, Harrison. 1954. *The Challenge of Man's Future.* New York: Viking.

Brown, Lester R. et al. 1992. *State of the World 1992.* New York: W. W. Norton.

Brown, Lester R., Hal Hane, Ed Ayres et al. 1993. *Vital Signs 1993.* New York: W. W. Norton.

Brown, Robert. 1963. *Explanation in Social Science.* London: Routledge & Kegan Paul.

Brumbaugh, Robert S. 1966. "Applied metaphysics: Truth and passing time." *Review of Metaphysics* 19 (June): 647–666.

Bruner, Jerome and Carol Fleisher Feldman. 1986. "Under construction." *The New York Review of Books* 33 (March 27): 46–49.

Brutzkus, Boris. 1935. *Economic Planning in Soviet Russia.* Trans. by G. Gardiner. London: George Routledge & Sons.

Buber, Martin. 1957. *Pointing the Way.* Manchester, NH: Ayer.

Bulmer, Martin. 1983. "The British tradition of social administration: Moral concerns at the expense of scientific rigor." Pp. 161–185 in D. Callahan and B. Jennings (eds.), *Ethics, the Social Sciences, and Policy Analysis.* New York: Plenum.

Burchfield, R.W. 1972. "Futurism." P. 1182 in *A Supplement to the Oxford English Dictionary.* London: Oxford University Press.

Burk, James. 1995. "Collective violence and world peace: The social control of armed force." Paper read at the annual meetings of the American Sociological Association, Washington DC, August 19–23.

Caillois, Roger. (1958) 1961. *Man, Play, and Games.* Trans. by M. Barash. New York: The Free Press.

Calder, Nigel. 1969. "Goals, foresight, and politics." Pp. 251–255 in R. Jungk and J. Galtung (eds.), *Mankind 2000.* Oslo: Universitetsforlaget and London: Allen & Unwin.

————. 1983. *Timescale: An Atlas of the Fourth Dimension.* New York: Viking.

Callinicos, Alex. 1982. *Is There a Future for Marxism?* London: The Macmillan Press.

Campbell, Angus, Philip E. Converse, and Willard L. Rodgers. 1976. *The Quality of American Life: Perceptions, Evaluations, and Satisfactions.* New York: Russell Sage Foundation.

Campbell, Donald T. 1965. "Variation and selective retention in socio-cultural evolution." Pp. 19–49 in H. R. Barringer, G. I. Blanksten, and R. W. Mack (eds.), *Social Change in Developing Areas.* Cambridge, MA: Schenkman.

————. 1973. "Ostensive instances and entitavity in language learning." Pp. 1043–1057 in W. Gray and N.D. Rizzo (eds.), *Unity Through Diversity: A Festschrift for Ludwig von Bertalanffy.* New York: Gordon & Breach.

————. 1975. "On the conflicts between biological and social evolution and between psychology and moral tradition." *American Psychologist* 30: 1103–1126.

————. 1977. "Descriptive epistemology: psychological, sociological, and evolutionary." William James Lectures, Harvard University. Excerpts reprinted in M. B. Brewer and B. E. Collins (eds.), *Scientific Inquiry and the Social Sciences.* San Francisco, CA: Jossey-Bass, 1981: 11–17.

————. 1984a. "Can an open society be an experimenting society?" Preliminary draft of a paper read at the International Symposium on the Philosophy of Karl Popper. Madrid, November 6–9.

————. 1984b. "Can we be scientific in applied social science?" Pp. 26–48 in R. F. Conner, D. G. Altman, and C. Jackson (eds.). *Evaluation Studies: Review Annual* Vol. 9. Beverly Hills, CA: Sage.

————. 1986. "Science's social system of validity-enhancing collective belief change and the problems of the social sciences." Pp. 108–155 in D. W. Fiske and R. A. Schweder (eds.), *Metatheory in Social Sciences: Pluralisms and Subjectivities.* Chicago: University of Chicago Press.

Campbell, Joseph. (1949) 1968. *The Hero with a Thousand Faces.* Second Edition. Princeton, NJ: Princeton University Press.

Cantril, Hadley. 1963. "A study of aspirations." *Scientific American* 208 (February): 41–45.

————. 1965. *The Pattern of Human Concerns.* New Brunswick, NJ: Rutgers University Press.

Cantril, Hadley and Lloyd Free. 1962. "Hopes and fears for self and country." Supplement to *The American Behavioral Scientist* 6 (October).

Carneiro, Robert L. 1978. "Political expansion as an expression of the principle of competitive exclusion." Pp. 205–223 in R. Cohen and E. R. Service (eds.), *Origins of the State.* Philadelphia, PA: Institute for the Study of Human Issues.

Caro, Francis G. 1983. "Program evaluation." Pp. 77–93 in H. E. Freeman, R. R. Dynes, P. H. Rossi, and W. F. Whyte (eds.), *Applied Sociology.* San Francisco, CA: Jossey-Bass.

Carroll, John B. (ed.). 1956. *Language, Thought, and Reality: Selected Writings of Benjamin Lee Whorf.* Boston: Technology Press of MIT.

Carter, Robert E. 1984. *Dimensions of Moral Education.* Toronto: University of Toronto Press.

Cassidy, Frederic G. 1961. *Jamaica Talk: Three Hundred Years of the English Language in Jamaica.* London: Macmillan.

Cecil, Andrew R. 1983. *The Foundations of a Free Society.* Austin: The University of Texas at Dallas.

Chandler, David P. 1991. *The Tragedy of Cambodian History.* New Haven, CT: Yale University Press.

Chaplin, George and Glenn D. Paige. 1973. *Hawaii 2000: Continuing Experiment in Anticipatory Democracy.* Honolulu: The University Press of Hawaii.

Charnov, Bruce H. 1987. "The academician as good citizen." Pp. 3–20 in S. L. Payne and B. H. Charnov (eds.), *Ethical Dilemmas for Academic Professionals.* Springfield, IL: Charles C. Thomas.

Chernoff, Herman. 1968. "Decision theory." Pp. 62–66 in D. L. Sills (ed.), *International Encyclopedia of the Social Sciences*, Vol. 4. New York: Macmillan & The Free Press.

Chomsky, Noam. 1972. *Language and Mind.* Enlarged edition. New York: Harcourt, Brace, Jovanovich.

Churchman, C. West. 1971. *The Design of Inquiring Systems.* New York: Basic Books.

————. 1977. "A philosophy for complexity." Pp. 82–90 in H. A. Linstone and W.H. Clive Simmonds (eds.), *Futures Research: New Directions.* Reading, MA: Addison-Wesley.

Clark, Eve V. and Herbert H. Clark. 1978. "Universals, relativity, and language processing." Pp. 225–277 in J. H. Greenberg, C. A. Ferguson, & E. A. Moravcsik (eds.), *Universals of Human Language.* Stanford, CA: Stanford University Press.

Clarke, I.F. 1984. "Journeys through space and time—from the *Santa Maria* to the 'last Columbus.'" *Futures* 16 (August): 425–434.

Clausen, John A. 1986. "Early adult choices and the life course." *Zeitschrift für Sozialisationsforschuung und Erziehungssoziologie* 6 (September): 313–320.

Clayton, Audrey. 1988. "WFS professional members' forum: Winter 1988." *Futures Research Quarterly* 4, No. 4 (Winter): 87–92.

Cleveland, Harlan. 1994. "Foreword." Pp. xi–xiv in R. Kidder, *Shared Values for a Troubled World*. San Francisco, CA: Jossey-Bass Publishers.

———. 1995. "Foreword." P. v in M. Marien (ed.), *World Futures and the United Nations*. Bethesda, MD: World Future Society.

Cliff, Tony. 1974. *State Capitalism in Russia*. London: Pluto.

Clinard, Marshall B. 1990. *Corporate Corruption: The Abuse of Power*. New York: Praeger.

Coates, Joseph F. 1978. "Technology assessment." Pp. 397–421 in J. Fowles (ed.), *Handbook of Futures Research*. Westport, CT: Greenwood Press.

———. 1985. "Scenarios part two: Alternative futures." Pp. 21–46 in J. S. Mendell (ed.), *Nonextrapolative Methods in Business Forecasting*. Westport, CT: Quorum Books.

———. 1987. "Twenty years in the future." Pp. 129–136 in M. Marien and L. Jennings (eds.). *What I Have Learned*. New York: Greenwood Press.

Coates, Joseph F., Vary T. Coates, Jennifer Jarratt, and Lisa Heinz. 1986. *Issues Management*. Mt. Airy, MD: Lomond.

Coates, Joseph F. and Jennifer Jarratt. 1989. *What Futurists Believe*. Mt. Airy, MD: Lomond.

Coddington, Alan. 1975. "Creaking semaphore and beyond: A consideration of Shackle's 'epistemics and economics.'" *The British Journal for the Philosophy of Science* 26 (June): 151–163.

Coe, Michael D. 1994. "The language within us." *The New York Times Book Review* (February 27): 7–8.

Cohen, G. A. 1978. *Karl Marx's Theory of History: A Defence*. Princeton, NJ: Princeton University Press.

Cohen, Joel E. 1982. "How is the past related to the future?" Annual Report, Center for Advanced Study in the Behavioral Sciences. Stanford, CA.

Colby, Anne and Lawrence Kohlberg. 1984. "Invariant sequence and internal consistency in moral judgment stages." Pp. 41–51 in W. M. Kurtines and J. L Gewirtz (eds.), *Morality, Moral Behavior, and Moral Development*. New York: John Wiley & Sons.

Cole, H. S. D., C. Freeman, M. Jahoda, and K. L. R. Pavitt (eds.). 1975. *Models of Doom: A Critique of the Limits to Growth*. New York: Universe Books.

Cole, Sam. 1983. "Models, metaphors and the state of knowledge." Pp. 407–421 in M. Batty and B. Hutchinson (eds.), *Systems Analysis in Urban Policy-Making and Planning*. New York: Plenum Press.

———. 1993. "Learning to love *Limits*." *Futures* 25, No. 7 (September): 814–818.

Coleman, James S. 1977. "Social action systems." Pp. 11–50 in *Problems of Formalization in the Social Sciences*. UNESCO: Division for International Development of Social Sciences and the Polish Academy of Sciences.

———. 1990. *Foundations of Social Theory*. Cambridge, MA: The Belknap Press of Harvard University Press.

Coleman, J.S., E.Q. Campbell, C.J. Hobson, J. McPartland, A. Mood, F.D. Winfeld, and R.L. York. 1966. *Equality of Educational Opportunity*. Washington, DC: U.S. Government Printing Office.

Collard, David. 1978. *Altruism and Economy*. New York: Oxford University Press.
Collins, H.M. 1985. *Changing Order: Representation, and Induction in Scientific Practice*. Beverly Hills, CA: Sage.
Conant, James B. 1951. *On Understanding Science*. New York: Mentor.
Condorcet, Antoine-Nicolas de. (1795) 1955. *Sketch for a Historical Picture of the Progress of the Human Mind*. Trans. by J. Barraclough. London: Weidenfeld and Nicolson.
Cook, Lauren. 1990. "State government foresight in the US." *Futures Research Quarterly* 6, No. 4 (Winter): 27–40.
Cook, Thomas D. and Donald T. Campbell. 1979. *Quasi-Experimentation: Design & Analysis Issues for Field Settings*. Chicago: Rand McNally.
Cornish, Edward S. 1980. Personal communication to W. Bell (July 28).
Cornish, Edward, with the members and staff of the World Future Society. 1977. *The Study of the Future*. Washington, DC: World Future Society.
Cotterrell, Roger. 1984. *The Sociology of Law: An Introduction*. London: Butterworths.
Cowen, Tyler and Derek Parfit. 1992. "Against the social discount rate." Pp. 144–161 in P. Laslett and J. S. Fishkin (eds.), *Justice Between Age Groups and Generations*. New Haven, CT: Yale University Press.
Cox, Harvey. 1985. "Challenges to faith and religion in the age of high technology." The Tenth Sir Winston Scott Memorial Lecture. Bridgetown, Barbados: Central Bank of Barbados.
Creativity: The Human Resource. 1980. Creativity exhibit, New York (August).
Cressey, Donald R. 1953. *Other People's Money*. Glencoe, IL: Free Press.
Crews, Frederick. 1986. "In the big house of theory." *The New York Review of Books* 33 (May 29): 36–42.
Crick, Francis. 1988. *What Mad Pursuit*. New York: Basic.
Cronbach, Lee J., Sueann Robinson Ambron, Sanford M. Dornbusch, Robert D. Hess, Robert C. Hornik, D.C. Phillips, Decker F. Walker, and Stephen S. Weiner. (1980) 1981. *Toward Reform of Program Evaluation*. San Francisco, CA: Jossey-Bass.
Crowley, J. Donald. 1972. "Introduction." Pp. xii–xxi in D. Defoe, *The Life and Strange Surprizing Adventures of Robinson Crusoe, of York, Mariner*. London: Oxford University Press.
Cumper, Gloria. 1972. *Survey of Social Legislation in Jamaica*. Mona, Jamaica: Institute of Social and Economic Research, University of the West Indies.
Cunningham, F. 1973. *Objectivity in Social Science*. Toronto: University of Toronto Press.
Dahle, Kjell. 1991. *On Alternative Ways of Studying the Future: International Institutions, an Annotated Bibliography and a Norwegian Case*. Trans. by Alison Coulthard. Olso: The Alternative Future Project.
———. 1992. "Participatory futures studies: Concepts and realities." *Futures Research Quarterly* 8, No. 4 (Winter): 83–92.
Darnton, Robert. 1984. "Working class Casanova." *The New York Review of Books* 31 (June 28, 1984): 33.
Dator, Jim. 1983. "The 1982 Honolulu electronic town meeting." Pp. 211–220 in W. Page (ed.), *The Future of Politics*. London: Frances Pinter in association with the World Futures Studies Federation.
———. 1984a. "Quantum theory and political design." Pp. 53–65 in R. Homann, E. Masini, and A. Sicinski (eds.), *Changing Lifestyles as Indicators of New and Cultural Values*. Zurich: the Gottlieb Duttweiler Institute.
———. 1984b. Personal communication (July 27).
———. 1993. "President's Report to the General Assembly of the World Futures Studies Federation." Turku, Finland (August 25).

————. 1994a. "Women in futures studies and women's visions of the futures—one man's tentative view." *The Manoa Journal of Fried and Half-Fried Ideas*, Occasional paper 2 (January): 40–57.

————. 1994b. "What is (and what is not) futures studies." *Papers de Prospectiva* (May): 24–47.

Dator, James A., Christopher B. Jones, and Barbara G. Moir. 1986. *A Study of Preferred Futures for Telecommunications in Six Pacific Island Societies*. Honolulu, HI: Pacific International Center for High Technology Research and Social Science Research Institute.

Dator, James A. and Sharon J. Rodgers. 1991. *Alternative Futures for the State Courts of 2020*. Chicago: State Justice Institute and the American Judicature Society.

Davies, Merryl Wyn, Ashis Nandy, and Ziauddin Sardar. 1993. *Barbaric Others*. London: Pluto Press.

Davies, P.C.W. 1977. *Space and Time in the Modern Universe*. Cambridge: Cambridge University Press.

————. 1993. "The holy grail of physics." *The New York Times Book Review* (April 7): 11–12.

Dawes, R. M. 1986. "Forecasting one's own preference." *International Journal of Forecasting* 2: 5–14.

Defoe, Daniel. (1719) 1972. *The Life and Strange Surprizing Adventures of Robinson Crusoe, of York Mariner*. London: Oxford University Press.

deLeon, Peter. 1984. "Futures studies and the policy sciences." *Futures* 16, No. 6 (December): 586–593.

Denton, David E. 1986. "Images, plausibility and truth." *Futures Research Quarterly* 2, No. 2 (Summer): 53–62.

Derathe, Robert. 1968. "Rousseau, Jean Jacques." Pp. 563–571 in D. L. Sills (ed.), *International Encyclopedia of the Social Sciences*, Vol. 13. New York: Macmillan & the Free Press.

Deutsch, K. W. 1963. "Nation-building and national development: Some issues for political research." Pp. 1–16 in K. W. Deutsch and W. J. Foltz (eds.), *Nation-Building*. New York: Atherton Press.

Deutsch, Morton. 1986. "Cooperation, conflict, and justice." Pp. 3–18 in H. W. Bierhoff, R. L. Cohen, and J. Greenberg (eds.), *Justice in Social Relations*. New York: Plenum Press.

De Vos, George and Lola Romanucci-Ross. 1975. "Ethnicity: Vessel of meaning and emblem of contrast." Pp. 363–390 in G. De Vos and L. Romanucci-Ross (eds.), *Ethnic Identity*. Palo Alto, CA: Mayfield.

Dewey, John. 1959. *Moral Principle in Education*. New York: Philosophical Library.

Dewhurst, J.F. and Associates. 1947 and 1955. *America's Needs and Resources*. New York: The Twentieth Century Fund.

Diamond, Jared. 1992. *The Third Chimpanzee: The Evolution and Future of the Human Animal*. New York: HarperCollins.

Dickson, Paul. (1971) 1972. *Think Tanks*. New York: Atheneum (second printing).

————. 1977. *The Future File*. New York: Avon.

Dickson, Lovat. 1969. *H. G. Wells: His Turbulent Life and Times*. New York: Atheneum.

Didsbury, H. F., Jr. (ed.). 1991. *Prep 21 Bulletin* No. 2 (Spring).

Diesing, P. 1971. *Patterns of Discovery in the Social Sciences*. Chicago: Aldine-Atherton.

Dillon, Michele. 1993. *Debating Divorce: Moral Conflict in Ireland*. Lexington: The University Press of Kentucky.

DiMaggio, Paul. 1989. Personal communication (April 21).
Dolbeare, Kenneth M. and Patricia Dolbeare. 1976. *American Ideologies: The Competing Political Beliefs of the 1970s.* Third Edition. Boston: Houghton Mifflin.
Doob, Leonard W. 1988. *Inevitability: Determinism, Fatalism, and Destiny.* New York: Greenwood Press.
Doyal, Len and Ian Gough. 1991. *A Theory of Human Need.* London: Macmillan.
Dror, Yehezkel. 1971. *Ventures in Policy Sciences: Concepts and Applications.* New York: Elsevier.
————. 1975. "Some fundamental philosophical, psychological and intellectual assumptions of futures studies." Pp. 145–165 in *The Future as an Academic Discipline.* Ciba Foundation Symposium 36. Amsterdam: Elsevier.
Drucker, Malka and Gary Block. 1992. *Rescuers: Portraits in Moral Courage in the Holocaust.* New York: Holmes & Meier.
Dublin, Louis I., Alfred J. Lotka, and Mortimer Spiegelman. 1949. *Length of Life: A Study of the Life Table*, rev. ed.: New York: Ronald.
Dunne, J.W. (1927) no date. *An Experiment with Time.* Third Edition. London: Faber and Faber (paper).
Durand, John D. 1960. "Mortality estimates from Roman tombstone inscriptions." *American Journal of Sociology* 65, No. 4 (January): 365–374.
Dworkin, Ronald. 1981. "What is equality? Part I: Equality of welfare." *Philosophy and Public Affairs* 10: 185–246.
————. 1994. "Mr. Liberty." *The New York Review of Books* XLI, No. 14 (August 11): 17–22.
Dye, Thomas R. 1978. *Understanding Public Policy.* Third Edition. Englewood Cliffs, NJ: Prentice-Hall.
Dyson, Freeman. 1995. "The scientist as rebel." *The New York Review of Books* XLII, No. 9 (May 25): 31–33.
Eaton, Ralph. 1931. *General Logic: An Introductory Survey.* New York: C. Scribner's Sons.
Ebenstein, William. 1968. "National socialism." Pp. 45–50 in D. L. Sills (ed.), *International Encyclopedia of the Social Sciences*, Vol. 11. New York: Macmillan & The Free Press.
Edel, Abraham. (1955) 1994. *Ethical Judgment: The Use of Science in Ethics.* New Brunswick, NJ: Transaction.
Edel, Abraham, Elizabeth Flower, and Finbarr W. O'Connor. 1994. *Critique of Applied Ethics.* Philadelphia, PA: Temple University Press.
Edgerton, Robert B. 1992. *Sick Societies: Challenging the Myth of Primitive Harmony.* New York: The Free Press.
Edwards, Ward. 1968. "Decision making: Psychological aspects." Pp. 34–42 in D. L. Sills (ed.), *International Encyclopedia of the Social Sciences*, Vol. 4. New York: Macmillan & The Free Press.
Ehrenberg, Victor. 1951. *The People of Aristophanes: A Sociology of Old Attic Comedy.* Second Edition. Oxford: Basil Blackwell.
Ehrlich, Anne. 1985. "Critical masses." *The Humanist* 45 (July/August): 18–22 et passim.
Ehrlich, Paul R. and Anne H. Ehrlich. 1991. *The Population Explosion.* New York: Touchstone (Simon & Schuster).
Einaudi, Mario. 1968. "Fascism." Pp. 334–341 in D. L. Sills (ed.), *International Encyclopedia of the Social Sciences*, Vol. 5. New York: Macmillan & the Free Press.

Eisenberg, Nancy. 1986. *Altruistic Emotion, Cognition, and Behavior*. Hillsdale, NJ: Lawrence Erlbaum Associates.

Ekman, Paul. 1980. *The Face of Man*. New York: Garland STPM.

———. 1982. *Emotion in the Human Face*. New York: Cambridge University Press.

Elster, Jon. 1978. *Logic and Society: Contradictions and Possible Worlds*. New York: John Wiley.

———. 1985. *Making Sense of Marx*. Cambridge: Cambridge University Press.

Encel, Solomon, Pauline K. Marstrand, and William Page (eds.). 1975. *The Art of Anticipation*. London: Martin Robertson.

English-Lueck, J. A. 1990. "China 2020: Looking forward." *Futures Research Quarterly* 6, No. 3 (Fall): 5–12.

Etzioni, Amitai. 1968. *The Active Society*. New York: Free Press.

———. 1988. *The Moral Dimension*. New York: Free Press.

———. 1992. "How to fix the pharmaceuticals." *The New York Times* (February 23): F13.

Eulau, Heinz. 1958. "H. D. Lasswell's developmental analysis." *The Western Political Quarterly* XI (June): 229–242.

Falk, R. A. 1975. *A Study of Future Worlds*. New York: Free Press.

———. 1977. "Contending approaches to world order." *Journal of International Affairs* 31: 171–198.

———. 1983. *The End of World Order*. New York: Holmes and Meier.

Fanon, Frantz. 1966. *The Wretched of the Earth*. New York: Grove (First Evergreen Edition).

———. 1967. *Black Skin, White Masks*. New York: Grove.

Farrell, Warren. 1993. *The Myth of Male Power*. New York: Simon & Schuster.

Ferguson, Charles A. 1978. "Talking to children: A search for universals." Pp. 203–224 in J. H. Greenberg, C. A. Ferguson, & E. A. Moravscik (eds.), *Universals of Human Language*. Stanford, CA: Stanford University Press.

Ferkiss, Victor. 1974. *The Future of Technological Civilization*. New York: George Braziller.

———. 1977. "Futurology: Promise, performance, prospects." *The Washington Papers* V, No. 50. Beverly Hills, CA: Sage.

Ferrarotti, Franco. 1986. *Five Scenarios for the Year 2000*. New York: Greenwood.

Feyerabend, Paul. 1975. *Against Method: Outline of an Anarchistic Theory of Knowledge*. London: NLB.

Fink, Arlene and Jacqueline Kosecoff. 1985. *How to Conduct Surveys*. Beverly Hills, CA: Sage.

Flechtheim, Ossip K. 1966. *History and Futurology*. Meisenheim-am-Glan, Germany: Verlag Anton Hain.

———. (1969) 1971. "Is futurology the answer to the challenge of the future?" Pp. 264–269 in R. Jungk and J. Galtung (eds.), *Mankind 2000*. Oslo: Universitetsforlaget and London: Allen & Unwin.

Foerster, Heinz von. 1977. "The curious behavior of complex systems: Lessons from biology." Pp. 104–113 in Harold A. Linstone and W.H. Clive Simmonds (eds.), *Futures Research: New Directions*. Reading, MA: Addison-Wesley.

Fogel, R.W. 1964. *Railroads and American Economic Growth*. Baltimore: Johns Hopkins Press.

Folger, Robert. 1986. "Rethinking equity theory: A referent cognition model." Pp. 145–162 in H. W. Bierhoff, R. L. Cohen, and J. Greenberg (eds.), *Justice in Social Relations*. New York: Plenum Press.

Forrester, Jay W. 1971. *World Dynamics*. Cambridge, MA: Wright-Allen Press.

Foss, Dennis C. 1977. *The Value Controversy in Sociology*. San Francisco, CA: Jossey-Bass.

Fowles, Jib. 1978. "Preface." Pp. ix–xi in J. Fowles (ed.), *Handbook of Futures Research*. Westport, CT: Greenwood.

Fraisse, Paul. 1968. "Time: Psychological aspects." Pp. 25–30 in D. L. Sills (ed.), *International Encyclopedia of the Social Sciences*, Vol. 16. New York: Macmillan & the Free Press.

Fraser, J. T. 1982. *The Genesis and Evolution of Time*. Amherst: The University of Massachusetts Press.

Freeman, Christopher. 1975. "Malthus with a computer." Pp. 5–13 in H. S. D. Cole, C. Freeman, M. Jahoda, and K. L. R. Pavitt (eds.), *Models of Doom: A Critique of the Limits to Growth*. New York: Universe Books.

Freeman, Derek. 1983. *Margaret Mead and Samoa: The Making and Unmaking of an Anthropological Myth*. Cambridge, MA: Harvard University Press.

Frey, James H. 1989. *Survey Research by Telephone*. 2nd Printing. Newbury Park, CA: Sage.

Friedman, Lawrence M. 1987. Review of "The Court and the Constitution." *The New York Times Book Review*. September 20: 3.

Friedrichs, Robert W. 1970. *A Sociology of Sociology*. New York: Free Press.

Furbank, P. N. 1993. "Leave it to chance." *The New York Review of Books* XL, No. 19 (November 18): 48–50.

Gabor, Dennis. 1964. *Inventing the Future*. New York: Alfred A. Knopf.

Gabor, Thomas. 1986. *The Prediction of Criminal Behavior*. Toronto: University of Toronto Press.

Galtung, Johan. 1980. *The True Worlds*. New York: Free Press.

Galtung, Johan and Robert Jungk. (1969) 1971. "Postscript: A warning and a hope." P. 368 in R. Jungk and J. Galtung (eds.), *Mankind 2000*. Oslo: Universitets forlaget and London: Allen & Unwin.

Gappert, Gary. 1979. *Post Affluent America*. New York: Franklin Watts.

———. 1982. "Future urban America: Post-affluent or advanced industrial society?" Pp. 9–34 in G. Gappert and R. V. Knight (eds.), *Cities in the 21st Century*. Beverly Hills, CA: Sage.

Garcia, John and Robert A. Koelling. 1966. "Relation of cue to consequence in avoidance learning." *Psychonomic Science* 4: 123–124.

Gardet, Louis. 1976. "Moslem views of time and history: An essay in cultural typology." Pp. 197–227 in *Cultures and Time*. Paris: The Unesco Press.

Gaston, Jerry. (1970) 1973. *Originality and Competition in Science*. Chicago: The University of Chicago Press.

Geertz, Clifford. 1968. "Religion: Anthropological study." Pp. 398–406 in D. L. Sills (ed.), *International Encyclopedia of the Social Sciences*, Vol. 13. New York: Macmillan & The Free Press.

———. 1983. *Local Knowledge*. New York: Basic.

Gell, Alfred. 1992. *The Anthropology of Time*. Providence, RI: Berg.

Gellner, Ernest. 1981. "General introduction: Relativism and universals." Pp. 1–20 in B. Lloyd and J. Gay (eds.), *Universals of Human Thought: Some African Evidence*. Cambridge: Cambridge University Press.

———. 1985. *Relativism and the Social Sciences*. Cambridge: Cambridge University Press.

Geras, Norman. 1989. "The controversy about Marx and justice." Pp. 211–267 in A. Callinicos (ed.), *Marxist Theory*. Oxford: Oxford University Press.

Gereffi, Gary. 1983. *The Pharmaceutical Industry and Dependency in the Third World*. Princeton, NJ: Princeton University Press.

Gert, Bernard. 1988. *Morality: A New Justification of the Moral Rules*. New York: Oxford University Press.

Gewirth, Alan. 1978. *Reason and Morality*. Chicago: University of Chicago Press.

Gibbs, Jack P. 1989. *Control: Sociology's Central Notion*. Urbana and Chicago: University of Illinois Press.

Gibson, James William. 1986. *The Perfect War*. Boston: The Atlantic Monthly Press.

Giddens, Anthony. 1973. *The Class Structure of the Advanced Societies*. New York: Barnes & Noble.

———. (ed.). 1974. *Positivism and Sociology*. London: Heinemann.

Gifford, James C. 1978. "The prehistory of *Homo sapiens*: Touchstone for the future." Pp. 71–100 in M. Maruyama and A. M. Harkins (eds.), *Cultures of the Future*. The Hague: Mouton.

GilFillan, S. Colum. 1920. "Successful Social Prophecy in the Past." Unpublished M. A. Thesis. New York: Columbia University (Faculty of Political Science, Department of Sociology).

———. 1935. *The Sociology of Invention*. Chicago: Follet.

Gilligan, Carol. 1982. *In a Different Voice*. Cambridge, MA: Harvard University Press.

Gilmore, David D. 1990. *Manhood in the Making*. New Haven, CT: Yale University Press.

Glaser, Barney G. 1965. "The constant comparative method of qualitative analysis." *Social Problems* 12 (Spring): 436–445.

Glaser, Barney G. and Anselm L. Strauss. 1967. *The Discovery of Grounded Theory*. Chicago: Aldine.

Glassman, James K. 1995 "A program gone bonkers." *The Washington Post Weekly Edition* (October 16–22): 27.

Glazer, Myron Peretz and Penina Migdal Glazer. 1989. *Whistleblowers*. New York: Basic Books.

Glock, Charles Y. (1951) 1955. "Some applications of the panel method to the study of change." Pp. 242–250 in P. F. Lazarsfeld and M. Rosenberg (eds.), *The Language of Social Research*. Glencoe, IL: Free Press.

Goldschmidt, Walter. 1990. *The Human Career*. Cambridge, MA: Basil Blackwell.

Goldthorpe, John H. 1971. "Theories of industrial society: Reflections on recrudescence of historicism and the future of futurology." *Archives Européene de Sociologie* 12, No. 2: 263–288.

Goody, Jack. 1968. "Time: Social organization." Pp. 30–42 in D. L. Sills (ed.), *International Encyclopedia of the Social Sciences*, Vol. 16. New York: Macmillan & the Free Press.

Goodenough, Ward H. 1970. *Description and Comparison in Cultural Anthropology*. Chicago: Aldine.

Gordon, T. J. 1968. "New approaches to Delphi." In J. R. Bright (ed.), *Technological Forecasting for Industry and Government*. Englewood Cliffs, NJ: Prentice Hall.

———. 1990. Personal Communication.

———. 1992. "The methods of futures research." *The Annals of the American Academy of Political and Social Science* 522 (July): 25–35.

Gordon, Theodore J. and David Greenspan. 1988. "Chaos and fractals: New tools for technological and social forecasting." *Technological Forecasting and Social Change* 34: 1–25.

Gordon, Theodore J., Herbert Gerjuoy, and Mark Anderson (eds.). 1979. *Life Extending Technologies*. New York: Pergamon.

Gordon, Ted, Herbert Gerjuoy and Robert Jungk. 1987. "The business of forecasting: A discussion of ethical and practical considerations." *Futures Research Quarterly* 3 (Summer): 21–36.

Gordon, T. J. and J. Hayward. 1968. "Initial experiments with the cross-impact matrix method of forecasting." *Futures* 1, No. 2: 100–116.

Gordon, Theodore and Olaf Helmer. 1964. *Report on a Long-Range Forecasting Study*. Santa Monica, CA: RAND paper P-2982.

———. (1964) 1966. "Report on a long-range forecasting study." Pp. 44–96 in O. Helmer (ed), *Social Technology*. New York: Basic Books.

Gorney, Roderic. (1968) 1979. *The Human Agenda*. Los Angeles: Guild for Tutors Press.

Gottheil, Fred M. 1966. *Marx's Economic Predictions*. Evanston, IL: Northwestern University Press.

Gouldner, Alvin W. 1970. *The Coming Crisis of Western Sociology*. New York: Basic Books.

———. 1985. *Against Fragmentation: The Origins of Marxism and the Sociology of Intellectuals*. New York: Oxford University Press.

Goveia, Elsa V. 1956. *A Study on the Historiography of the British West Indies to the End of the Nineteenth Century*. Mexico: Instituto Panamericano de Geografia e Historia.

Granger, C. W. J. 1980. *Forecasting in Business and Economics*. New York: Academic.

Granger, Gilles-Gaston. 1968. "Condorcet." Pp. 213–215 in D. L. Sills (ed.), *International Encyclopedia of the Social Sciences*, Vol. 3. New York: Macmillan & the Free Press.

Greeley, Andrew M. 1989. *Religious Change in America*. Cambridge, MA: Harvard University Press.

Green, Ronald M. (1977) 1981. "Intergenerational distributive justice and environmental responsibility." Pp. 91–101 in E. Partridge (ed.), *Responsibilities to Future Generations*. Buffalo, NY: Prometheus Books.

Griffin, James. 1986. *Well-Being: Its Meaning, Measurement, and Moral Importance*. Oxford: Clarendon Press.

Grisez, Germain and Russell Shaw. 1988. *Beyond the New Morality*. Third Edition. Notre Dame, IN: University of Notre Dame Press.

Grunberger, Richard. 1971. *A Social History of the Third Reich*. London: Weidenfeld and Nicolson.

Guetzkow, H., C. F. Alger, R. Brody, R. D. Noël, and R. C. Snyder. 1963. *Simulation in International Relations: Developments for Research and Teaching*. Englewood Cliffs, NJ: Prentice-Hall.

Gurevich, A.J. 1976. "Time as a problem of cultural history." Pp. 229–245 in *Cultures and Time*. Paris: The Unesco Press.

Haan, Norma. 1983. "An interactional morality of everyday life." Pp. 218–250 in N. Haan, R. N. Bellah, P. Rabinow, and W. M. Sullivan (eds.), *Social Science as Moral Inquiry*. New York: Columbia University Press.

Habermas, Jürgen. 1970a. "On systematically distorted communication." *Inquiry* 13 (Autumn): 205–218.

———. 1970b. "Towards a theory of communicative competence." *Inquiry* 13 (Winter): 360–375.

————. 1973. *Theory and Practice*. Trans. by J. Viertel. Boston: Beacon Press.
————. (1981) 1984. *The Theory of Communicative Action*. Part I. Trans. by T. McCarth. Boston: Beacon.
Hacking, Ian. 1981. "Introduction." Pp. 1–5 in I. Hacking (ed.), *Scientific Revolutions*. Oxford: Oxford University Press.
————. 1986. "Science turned upside down." *The New York Review of Books* 33 (February 27): 21–26.
Hadley, Arthur T. (1971) 1986. *The Straw Giant*. New York: Random House.
Hahn, Walter A. 1985. "Futures in politics and the politics of futures." *Futures Research Quarterly* 1, No. 4 (Winter): 35–56.
Halfpenny, Peter. 1982. *Positivism and Sociology*. London: George Allen & Unwin.
Hall, Robert T. 1987. *Emile Durkheim: Ethics and the Sociology of Morals*. New York: Greenwood Press.
Halley, Edmund. 1693. "An estimate of the degrees of mortality of mankind." *Philosophical Transactions of the Royal Society of London* 17: 596–610.
Hallo, William W. 1993. "Digging up the future." Unpublished paper, New Haven, CT: Yale University.
Hampshire, Stuart. 1955. "Introduction." Pp. vii–xii in Condorcet, *Sketch for a Historical Picture of the Progress of the Human Mind*. Trans. by J. Barraclough. London: Weidenfeld and Nicolson.
Hanson, N.R. 1958. *Patterns of Discovery*. Cambridge: Cambridge University Press.
Hardin, Garrett. 1993. *Living within Limits*. Oxford: Oxford University Press.
Hare, R. M. 1952. *The Language of Morals*. Oxford: The Clarendon Press.
Harman, Gilbert. 1977. *The Nature of Morality*. New York: Oxford University Press.
Harman, Willis W. and Peter Schwartz. 1978. "Changes and challenges for futures research." Pp. 791–801 in J. Fowles (ed.), *Handbook of Futures Research*. Westport, CT: Greenwood Press.
Harner, L. 1975. "*Yesterday* and *tomorrow*: Development of early understanding of the terms." *Developmental Psychology* 11: 864–865.
Harris, Marvin. 1977. *Cannibals and Kings*. New York: Random House.
————. 1989. *Our Kind*. New York: Harper & Row.
Haste, Helen and Jane Baddeley. 1991. "Moral theory and culture: The case of gender." Pp. 223–249 in W. M. Kurtines and J. L. Gewirtz (eds.), *Handbook of Moral Behavior and Development*, Vol 1 "Theory." Hillsdale, NJ: Lawrence Erlbaum Associates.
Hawken, Paul, James Ogilvy, Peter Schwartz. 1982. *Seven Tomorrows*. New York: Bantam.
Hayashi, Yujiro. 1978. "Futures research in Japan." Pp. 31–38 in J. Fowles (ed.), *Handbook of Futures Research*. Westport, CT: Greenwood.
Heilbroner, Robert. 1993. *21st Century Capitalism*. New York: W. W. Norton.
————. 1995. *Visions of the Future*. New York: Oxford University Press.
Heineman, Kenneth J. 1993. *Campus Wars*. New York: New York University Press.
Helmer, Olaf. 1970. *Report on the Future of the Future-State-of-the-Union Reports* R-14. Middletown, CT: Institute for the Future.
————. 1983. *Looking Forward: A Guide to Futures Research*. Beverly Hills, CA: Sage.
Helmer, O. and N. Rescher. 1960. "On the Epistemology of the Inexact Sciences" R-353. Santa Monica, CA: The Rand Corporation.
Hemming, James. 1974. *Probe: 2. The Values of Survival*. Sydney, Australia: Angus & Robertson.

Hempel, Carl G. 1965. *Aspects of Scientific Explanation*. New York: Free Press.

Henshel, Richard L. 1976. *On the Future of Social Prediction*. Indianapolis, IN: Bobbs: Merrill.

———. 1978. "Self-altering predictions." Pp. 99–123 in J. Fowles (ed.), *Handbook of Futures Research*. Westport, CT: Greenwood Press.

———. 1981. "Evolution of controversial fields: Lessons from the past for futures." *Futures* 13 (October): 401–412.

———. 1982. "Sociology and social forecasting." Pp. 57–79 in R. H. Turner and J. F. Short, Jr. (eds.), *Annual Review of Sociology*, Vol. 8. Palo Alto, CA: Annual Reviews.

———. 1987. "Credibility and confidence: Feedback loops in social prediction." A paper read at the VII International Congress of Cybernetics and Systems, University of London (September).

———. 1990. *Thinking about Social Problems*. San Diego, CA: Harcourt Brace Jovanovich.

———. 1993. "Do self-fulfilling prophecies improve or degrade predictive accuracy? How sociology and economics can disagree and both be right." *The Journal of Socio-Economics* 22, No. 2: 85–104.

Henshel, Richard L. and William Johnston. 1987. "The emergence of bandwagon effects: A theory." *The Sociological Quarterly* 28, No. 4 (December): 493–511.

Hermkens, Piet and David van Kreveld. 1991. "Social justice, income distribution, and social stratification in the Netherlands: A review." Pp. 119–138 in H. Steensma and Riël Vermunt (eds.), *Social Justice in Human Relations*, Vol. 2, "Societal and Psychological Consequences of Justice and Injustice." New York: Plenum Press.

Hick, John. 1989. *An Interpretation of Religion*. New Haven, CT: Yale University Press.

Hill, K. Q. and J. Fowles. 1975. "The methodological worth of the Delphi forecasting technique." *Technological Forecasting and Social Change* 7: 179–192.

Hirschi, Travis and Michael Gottfredson (eds.). 1994. *The Generality of Deviance*. New Brunswick, NJ: Transaction.

Hirschi, Travis and Hanan C. Selvin. 1967. *Delinquency Research: An Appraisal of Analytic Methods*. New York: Free Press.

Hirschman, Albert O. 1970. *Exit, Voice, and Loyalty: Responses to Decline in Firms, Organizations, and States*. Cambridge, MA: Harvard University Press.

Hobsbawm, Eric. 1994. *The Age of Extremes*. New York: Pantheon.

Hogben, Lancelot. (1955) 1968. *The Wonderful World of Mathematics*. Garden City, NY: Doubleday.

Holden, K., D. A. Peel, and J. L. Thompson. 1990. *Economic Forecasting: An Introduction*. Cambridge: Cambridge University Press.

Holsti, Ole. 1969. *Content Analysis for the Social Sciences and Humanities*. Reading, MA: Addison-Wesley.

Hopkins, Terence K. and Immanuel Wallerstein. 1967. "The comparative study of national societies." *Social Science Information sur les Sciences Sociales*, VI, No. 5 (October): 25–58.

Horowitz, Irving Louis. 1993. *The Decomposition of Sociology*. New York: Oxford University Press.

———. 1994. "One day in the life of contemporary sociology." *Partisan Review*, LXI, No. 3: 501–510.

House, James S. 1986. "Social support and the quality and quantity of life." Pp. 253–269 in F. M. Andrews (ed.), *Research on the Quality of Life*. Ann Arbor: The University of Michigan.

Howells, W. W. 1960. "Estimating population numbers through archaeological and skeletal remains." Pp. 158–185 in R. F. Heizer and S. F. Cook (eds.), *The Application of Quantitative Methods in Archaeology*. Viking Fund Publications in Anthropology, No. 28. New York: Wenner-Gren Foundation for Anthropological Research, Inc.

Hughes, H. Stuart. 1958. *Consciousness and Society*. New York: Vintage Books (Random House).

Huber, Bettina J. 1973. "Images of the Future Among the White South African Elite." Unpublished Ph.D. dissertation, New Haven, CT: Yale University.

———. 1974. "Images of the future." Pp. 151–169 in H. W. van der Merwe, M. J. Ashley, N. C. J. Charton, and B. J. Huber, *White South African Elites*. Cape Town: Juta.

Huizinga, J.H. 1976. *Rousseau: The Self-made Saint*. New York: Grossman (Viking).

Huntington, Samuel P. 1993. "The clash of civilizations?" *Foreign Affairs* 72, No. 3 (Summer): 22–49.

Husserl, E. (1887) 1966. *The Phenomenology of Internal Time Consciousness*. Bloomington, IN: Midland Books.

Iacocca, Lee with William Novak. 1984. *Iacocca: An Autobiography*. New York: Bantam.

Inayatullah, Sohail. 1993. "From 'who am I?' to 'when am I?'" *Futures* 25, No. 3 (April): 235–253.

Inglehart, Ronald. 1990. *Culture Shift in Advanced Industrial Society*. Princeton, NJ: Princeton University Press.

Innes, Judith Eleanor. (1975) 1990. *Knowledge and Public Policy*. Second Expanded Edition. New Brunswick, NJ: Transaction.

Israeli, Nathan. 1932a. "The social psychology of time." *Journal of Abnormal and Social Psychology* 27 (July): 209–213.

———. 1932b. "The psychopathology of time." *Psychological Review* 39 (September): 486–491.

———. 1932c. "Wishes concerning improbable future events: Reactions to the future." *Journal of Applied Psychology* 16 (October): 584–588.

———. 1933a. "Attitudes to the Decline of the West." *Journal of Social Psychology* 4 (February): 92–101.

———. 1933b. "Group estimates of the divorce rate for the years 1935–1975." *Journal of Social Psychology* 4 (February): 102–115.

———. 1933c. "Group predictions of future events." *Journal of Social Psychology* 4 (May): 201–222.

———. 1933d. "Measurement of attitudes and reactions to the future." *Journal of Abnormal and Social Psychology* 28 (July): 181–193.

Jackson, Walter A. 1990. *Gunnar Myrdal and America's Conscience*. Chapel Hill: University of North Carolina Press.

Jacob, François. 1988. *The Statue Within: An Autobiography*, as quoted in *The New York Times Book Review* 10 (April).

Jacobs, Robert C. and Donald T. Campbell. 1961. "The perpetuation of an arbitrary tradition through several generations of a laboratory microculture." *Journal of Abnormal and Social Psychology* 62, No. 3: 649–658.

Jaffe, A. J. 1968. "Ogburn, William Fielding." Pp. 277–281 in D. L. Sills (ed.), *International Encyclopedia of the Social Sciences*, Vol. 11. New York: Macmillan & the Free Press.

Jahoda, Marie. 1988. "Time: A social psychological perspective." Pp. 53–94 in M. Young and T. Schuller (eds.), *The Rhythms of Society*. London: Routledge.

Jantsch, Erich. 1967. *Technological Forecasting in Perspective*. Paris: Organisation for Economic Cooperation and Development.

Jennings, Lane (ed.). 1993. *The Futures Research Directory: Organizations and Periodicals 1993–94*. Bethesda, MD: World Future Society.

Johnson, George. 1988. "Taking life three seconds at a time." *The New York Times Book Review* (March 27): 41.

Jonas, Hans. (1972) 1981. "Technology and responsibility: The ethics of an endangered future." Pp. 23–36 in E. Partridge (ed.), *Responsibilities to Future Generations*. Buffalo, NY: Prometheus Books.

Jones, Christopher B. 1992. "The *Manoa School* of futures studies." *Futures Research Quarterly* 8, No. 4 (Winter): 19–25.

Jones, Peter M.S. 1977. "One organization's experience." Pp. 194–209 in H. A. Linstone and W.H.C. Simmonds (eds.), *Futures Research: New Directions*. Reading, MA: Addison-Wesley.

Jones, Thomas E. 1979. "The futurist movement: A brief history." *World Future Society Bulletin* (July-August): 13–25.

———. 1980. *Options for the Future*. New York: Praeger.

Joseph, Earl C. 1985. "Editorial." *Future Trends* 16, No. 6 (September): 1.

———. 1987. "Editorial." *Future Trends* 18, No. 4 (April): 1–4.

Jouvenel, Bertrand de. 1963. "Introduction." Pp. ix–xi in B. Jouvenel (ed.), *Futuribles* I. Geneva: Droz.

———. (1964) 1967. *The Art of Conjecture*. New York: Basic Books.

Judd, Charles M., Eliot R. Smith, and Louise H. Kidder. 1991. *Research Methods in Social Relations*. Sixth Edition. Fort Worth, TX: Holt, Rinehart & Winston.

Jungk, Robert. (1969) 1971. "Preface." Pp. 9–10 in R. Jungk and J. Galtung (eds.), *Mankind 2000*. Oslo: Universitetsforlaget and London: Allen & Unwin.

———. (1973) 1976. *The Everyman Project*. New York: Liveright.

———. 1987. "Introduction to the UK edition." Pp. 5–6 in R. Jungk and N. Müllert, *Future Workshops*. London: Institute for Social Inventions.

Jungk, Robert and Norbert Müllert. 1987. *Future Workshops*. London: Institute for Social Inventions.

Kagame, Alexis. 1976. "The empirical apperception of time and the conception of history in Bantu thought." Pp. 89–116 in *Cultures and Time*. Paris: The Unesco Press.

Kahn, Herman. 1960. *On Thermonuclear War*. Princeton, NJ: Princeton University Press.

———. 1962. *Thinking about the Unthinkable*. London: Weidenfeld and Nicolson.

———. 1975. "On studying the future." Pp. 405–442 in F. I. Greenstein and N. W. Polsby (eds.), *Handbook of Political Science*, Vol. 7, "Strategies of Inquiry." Reading, MA: Addison-Wesley.

Kahn, Herman and Anthony J. Wiener. 1967. *The Year 2000: A Framework for Speculation on the Next Thirty-Three Years*. New York: Macmillan.

Kamenka, Eugene. (1962) 1972. *The Ethical Foundations of Marxism*. London: Routledge & Kegan Paul.

———. 1986. "Why was the Bolshevik terror wrong?" *The New York Times Book Review* (February 2): 20.

Kamm, Henry. 1992. "Sowing the killing fields." *The New York Times Book Review* (January 12): 7.

Kant, Immanuel. 1958. *Groundwork of the Metaphysics of Morals*. Trans. by H. J. Paton. New York: Harper & Row.

Kavka, Gregory. (1978) 1981. "The futurity problem." Pp. 109–122 in E. Partridge (ed.), *Responsibilities to Future Generations*. Buffalo, NY: Prometheus Books.

Kelley, J. and M.D.R. Evans. 1990. "Legitimate inequality: Norms on occupational earnings in eight nations." A paper read at the meetings of the International Social Survey Programme, Graz, Austria (May).

Kelly, Janice R. and Joseph E. McGrath. 1988. *On Time and Method*. Newbury Park, CA: Sage.

Kemp, Martin. 1988. *The Science of Art*. New Haven, CT: Yale University Press.

Kennedy, Paul M. 1993. *Preparing for the Twenty-First Century*. New York: Random House.

Kershaw, David N. 1972. "Issues in income maintenance experimentation." Pp. 221–245 in P.H. Rossi and W. Williams (eds.), *Evaluating Social Programs*. New York and London: Seminar Press.

———. (1972) 1980. "A negative-income tax experiment." Pp. 27–41 in D. Nachmias (ed.), *The Practice of Policy Evaluation*. New York: St. Martin's Press.

Keyfitz, Nathan. 1986. "The social and political context of population forecasting." Pp. 235–258 in W. Alonso and P. Starr (eds.), *The Politics of Numbers*. New York: Russell Sage Foundation.

Kidder, Louise H. and Susan Muller. 1991. "What is 'fair' in Japan?" Pp. 139–154 in H. Steensma and R. Vermunt (eds.), *Social Justice in Human Relations*, Vol. 2 "Societal and Psychological Consequences of Justice and Injustice." New York: Plenum Press.

Kidder, Rushworth M. 1993. "A.H. Halsey: Dialects of a common language." *Insights on Global Ethics* 3, No. 8 (August): 1 et passim.

———. 1994. *Shared Values for a Troubled World*. San Francisco, CA: Jossey-Bass Publishers.

———. 1995. *How Good People Make Tough Choices*. New York: William Morrow.

Kim, Tae-Chang. 1994. "Toward a new theory of value for the global age." Pp. 116–141 in *Why Future Generations Now?* Umeda, Kitaku, Osaka: Future Generations Alliance Foundation.

Kimmel, Allan J. 1988. *Ethics and Values in Applied Social Research*. Newbury Park, CA: Sage.

King, Alexander and Bertrand Schneider. 1991. *The First Global Revolution*. London: Simon & Shuster.

King, Martin Luther, Jr. 1964. *Why We Can't Wait*. New York: Harper and Row.

Kiser, Edgar and Kriss A. Drass. 1987. "Changes in the core of the world-system and the production of utopian literature in Great Britain and the United States, 1883–1975." *American Sociological Review* 52, No. 2 (April): 286–293.

Kizlos, Peter J. 1989. "Faces of mid-life." *Yale Alumni Magazine* LII, No. 8 (Summer): 46–49.

Kluckhohn, Clyde. 1953. "Universal categories of culture." Pp. 507–523 in A. L. Kroeber (ed.), *Anthropology Today*. Chicago: University of Chicago Press.

Kohlberg, Lawrence. (1971) 1981. "From is to ought: How to commit the naturalistic fallacy and get away with it in the study of moral development." Pp. 101–189 in L. Kohlberg (ed.), *The Philosophy of Moral Development*. New York: Harper & Row.

———. 1984. *The Psychology of Moral Development*. San Francisco, CA: Harper & Row.

Kolakowski, Leszek. 1978. *Main Currents of Marxism*, Vol. III "The Breakdown." Trans. by P. S. Falla. Oxford: Clarendon Press.

Kolata, Gina. 1992. "Scientists fluff the answer to a billion-dollar question." *The New York Times* (November 1): E2.

Kondo, Tetsuo. 1990. "Some notes on rational behavior, normative behavior, moral behavior, and cooperation." *The Journal of Conflict Resolution* 34, No. 3 (September): 495–530.

Konner, Melvin. 1990. "Mutilated in the name of tradition." *The New York Times Book Review* (April 15): 5–6.

Kothari, R. 1974. *Footsteps into the Future*. New York: Free Press.

Kott, Jan. 1984. *The Theater of Essence*. Evanston, IL: Northwestern University Press.

Kuhn, Thomas S. 1962. *The Structure of Scientific Revolutions*. Chicago: University of Chicago Press.

———. (1969) 1977a. "Second thoughts on paradigms." Pp. 459–517 in F. Suppe (ed.), *The Structure of Scientific Theories*, Second Edition. Urbana: University of Illinois Press.

———. 1977b. *The Essential Tension*. Chicago: The University of Chicago Press.

———. (1970) 1978. "Reflections on my critics." Pp. 231–278 in I. Lakatos and A. Musgrave (eds.), *Criticism and the Growth of Knowledge*. Cambridge: Cambridge University Press.

Küng, Hans. 1991. *Global Responsibility: In Search of a New World Ethic*. New York: Crossroad.

Kurian, George Thomas and Graham T. T. Molitor (eds.). 1996. *Encyclopedia of the Future*, Vols. 1 and 2. New York: Simon & Schuster Macmillan.

Lakatos, Imre. 1968. *The Problem of Inductive Logic*. Amsterdam: North Holland.

Land, Kenneth C. 1986. "Methods for national population forecasts: A review." *Journal of the American Statistical Association* 81., No. 396 (December): 888–901.

Land, Kenneth C. and Stephen H. Schneider (eds.). 1987a. *Forecasting in the Social and Natural Sciences*. Dordrecht, Holland: D. Reidel.

———. 1987b. "Forecasting in the social and natural sciences: An overview and analysis of isomorphisms." Pp. 7–31 in K.C. Land and S.H. Schneider (eds.), *Forecasting in the Social and Natural Sciences*. Dordrecht, Holland: D. Reidel.

Landes, David S. 1983. *Revolution in Time*. Cambridge, MA: Belknap Press of Harvard University Press.

Lane, Robert E. 1962. *Political Ideology*. New York: The Free Press.

———. 1986. "Market justice, political justice." *American Political Science Review* 80, No. 2 (June): 383–402.

———. 1991. *The Market Experience*. Cambridge: Cambridge University Press.

Langer, Susanne K. 1937. *An Introduction to Symbolic Logic*. Boston: Houghton Mifflin.

Lapham, Lewis H. 1986. "America's armchair generals." *The Wall Street Journal* (October 2): 28.

Larmore, Charles. 1987. "Review of *After Philosophy*." *The New York Times Book Review* (March): 21.

Larre, Claude. 1976. "The empirical apperception of time and the conception of history in Chinese thought." Pp. 35–62 in *Cultures and Time*. Paris: The Unesco Press.

Laslett, Peter. 1992. "Is there a generational contract?" Pp. 24–47 in P. Laslett and J. S. Fishkin (eds.), *Justice Between Age Groups and Generations*. New Haven, CT: Yale University Press.

Laslett, Peter and James S. Fishkin. 1992. "Introduction." Pp. 1–23 in P. Laslett and J. S. Fishkin (eds.), *Justice Between Age Groups and Generations*. New Haven, CT: Yale University Press.

Lasswell, H. D. 1935. *World Politics and Personal Insecurity*. New York: McGraw-Hill.
————. 1937. "Sino-Japanese crisis: The garrison state versus the civilian state." *China Quarterly* XI: 643–649.
————. 1941. "The garrison state." *The American Journal of Sociology* XLVI (January): 455–468.
————. 1948. *The Analysis of Political Behavior: An Empirical Approach*. London: Routledge & Kegan Paul.
————. 1951a. "The policy orientation." Pp. 3–15 in D. Lerner and H.D. Lasswell et al. (eds.), *The Policy Sciences*. Stanford, CA: Stanford University Press.
————. (1946) 1951b. "World organization and society." Pp. 102–117 in D. Lerner and H.D. Lasswell et al. (eds.), *The Policy Sciences*. Stanford, CA: Stanford University Press.
————. 1965. "The world revolution of our time: A framework for basic policy research." Pp. 29–96 in H.D. Lasswell and D. Lerner (eds.), *World Revolutionary Elites*. Cambridge, MA: MIT Press.
————. 1971. *A Pre-View of Policy Sciences*. New York: Elsevier.
————. (1975) 1977. "The scope of the conference: Postconference objectives." Pp. 41–57 in B. Pregel, H.D. Lasswell, and J. McHale (eds.), *World Priorities*. New Brunswick, NJ: Transaction.
Lasswell, Harold D. and J. Z. Namenwirth. 1968. *The Lasswell Value Dictionary* (3 volumes). New Haven, CT: Yale University, mimeo.
Laszlo, Ervin. 1994. *Vision 2020*. Yverdon, Switzerland: Gordon and Breach.
Latour, Bruno and Steve Woolgar. 1979. *Laboratory Life*. Beverly Hills, CA: Sage.
Laudan, Larry. 1981. "A problem-solving approach to scientific progress." Pp. 144–155 in I. Hacking (ed.), *Scientific Revolutions*. Oxford: Oxford University Press.
Lee, Keekok. 1985. *A New Basis for Moral Philosophy*. London: Routledge & Kegan Paul.
————. 1989. *Social Philosophy and Ecological Scarcity*. London: Routledge.
Lemert, Charles C. and Garth Gillan. 1982. *Michel Foucault: Social Theory and Transgression*. New York: Columbia University Press.
Lerner, Daniel and Harold D. Lasswell et al. (eds.). 1951. *The Policy Sciences*. Stanford, CA: Stanford University Press.
Lerner, Melvin J. 1980. *The Belief in a Just World*. New York: Plenum Press.
Leslie, Stuart W. 1993. *The Cold War and American Science*. New York: Columbia University Press.
Lewis, J. David and Andrew J. Weigart. (1981) 1990. "The structures and meanings of social-time." Pp. 77–101 in J. Hassard (ed.), *The Sociology of Time*. London: Macmillan.
Lewis, Paul. 1994. "Rise of the Blue Helmets." *The New York Times Book Review* (November 6): 14–15.
Lewis, W. Arthur. 1968. "Planning, economic: III development planning." Pp. 118–125 in D. L. Sills (ed.), *International Encyclopedia of the Social Sciences*, Vol. 12. New York: Macmillan & the Free Press.
Lieberson, Stanley. 1985. *Making It Count*. Berkeley and Los Angeles: University of California Press.
Lincoln, Yvonna S. and Egon G. Guba. 1985. *Naturalistic Inquiry*. Beverly Hills, CA: Sage.
Linn, Ruth. 1989. *Not Shooting and Not Crying: Psychological Inquiry into Moral Disobedience*. New York: Greenwood Press.

Linstone, Harold A. 1975. "Eight basic pitfalls: A checklist." Pp. 573–586 in H. A. Linstone and M. Turoff (eds.), *The Delphi Method: Techniques and Applications.* Reading, MA: Addison-Wesley.
———. 1977. "Confessions of a forecaster." Pp. 3–12 in H.A. Linstone and W.H.C. Simmonds (eds.), *Futures Research: New Directions.* Reading, MA: Addison-Wesley.
Lipset, Seymour Martin. 1963. *The First New Nation.* New York: Basic Books.
Livermore, W. R. 1898. *The American Kriegspiel.* Second Edition. Boston: W. B. Clarke.
Lloyd, G.E.R. 1976. "Views on time in Greek thought." Pp. 117–148 in *Cultures and Time.* Paris: The Unesco Press.
Lombardi, Louis G. 1988. *Moral Analysis: Foundations, Guides, and Applications.* Albany: State University of New York Press.
Long, John F. and David Byron McMillen. 1987. "A survey of Census Bureau population projection methods." Pp. 141–177 in K. C. Land and S. H. Schneider (eds.), *Forecasting in the Social and Natural Sciences.* Dordrecht, Holland: D. Reidel.
Lonner, Walter J. 1980. "The search for psychological universals." Pp. 143–204 in H. C. Triandis and W. W. Lambert (eds.), *Handbook of Cross-Cultural Psychology,* Vol. 1 "Perspectives." Boston: Allyn and Bacon.
Luker, Kristin. 1984. *Abortion and the Politics of Motherhood.* Berkeley and Los Angeles: University of California Press.
Lukes, Steven. 1985. *Marxism and Morality.* New York: Clarendon Press, Oxford University Press.
Lutz, Wolfgang. 1994a. "World population trends: Global and regional interactions between population and environment." Pp. 41–65 in L. M. Arizpe, M. P. Stone, and D. C. Major (eds.), *Population & Environment.* Boulder, CO: Westview Press.
———. 1994b. "The future of world population." *Population Bulletin* 49, 1 (June). Washington, DC: Population Reference Bureau.
Lynch, Michael. 1985. *Art and Artifact in Laboratory Science.* Boston: Routledge & Kegan Paul.
Lyons, Oren. 1985. "Traditional native philosophies relating to aboriginal rights." Pp. 19–23 in M. Boldt and J.A. Long in association with L. Little Bear (eds.), *The Quest for Justice.* Toronto: University of Toronto Press.
McFalls, Jr., Joseph A. 1991. "Population: A lively introduction." *Population Bulletin* 46, No. 2 (October): 2–41.
McHale, John. 1969. *The Future of the Future.* New York: George Braziller.
———. 1971–72. "A Continuation of the Typological Survey of Futures Research, U.S." Division of Special Mental Health Programs, Center for Studies of Metropolitan Problems, National Institute of Mental Health (mimeographed).
———. 1978. "The emergence of futures research." Pp. 5–15 in J. Fowles (ed.), *Handbook of Futures Research.* Westport, CT: Greenwood.
McHale, John and Magda Cordell McHale. n.d. *Futures Studies: An International Survey.* New York: United Nations Institute for Training and Research.
———. 1977. *The Futures Directory: An International Listing and Description of Organizations and Individuals Active in Future Studies and Long-Range Planning.* Guildford, England: IPC Science and Technology Press and Boulder, CO: Westview.
McNamara, Robert S. with Brian VanDeMark. 1995. *In Retrospect: The Tragedy and Lessons of Vietnam.* New York: Times Books/Random House.

McHugh, Francis J. 1966. *Fundamentals of War Gaming*. Third Edition. Newport, RI: The United States Naval War College.

McKeown, C. Timothy. 1990. "The futures of science: The human context of scientific expectations." *Futures* 22, No. 1 (January/February): 46–56.

MacIntyre, Alisdair. 1977. "Epistemological crises, dramatic narrative and the philosophy of science." *The Monist* 60: 453–472.

———. 1984. *After Virtue*. Notre Dame, IN: University of Notre Dame Press.

Macquarrie, John. 1985. "Clearing the mists from Olympus." *The New York Times Book Review* (September 22): 30.

Madge, Charles. 1968. Planning, social. Introduction." Pp. 125–129 in D. L. Sills (ed.), *International Encyclopedia of the Social Sciences*, Vol. 12. New York: Macmillan & The Free Press.

Maier, Charles S. 1984. "August 1914: The whys of war." *The New York Times Book Review*. (July 29): 1, 22–23.

Maines, David R., Noreen M. Sugrue, and Michael A. Katovich. 1983. "The sociological import of G.H. Mead's theory of the past." *American Sociological Review* 48 (April): 161–173.

Malaska, Pentti. 1995. "The futures field of research." *Futures Research Quarterly* 11, No. 1 (Spring): 79–90.

Malcolm X. (1964) 1968. *The Autobiography of Malcolm X*, with the assistance of Alex Haley. Harmondsworth, England: Penquin.

Malinowski, Bronislaw. (1927) 1961. *Sex and Repression in Savage Society*. Cleveland, OH: World.

Malotki, Ekkehart. 1983. *Hopi Time*. Berlin: Mouton.

Mannermaa, Mika. 1986. "Futures research and social decision making." *Futures* 18, No. 5 (October): 658–670.

Manuel, Frank E. and Fritzie P. Manuel. 1979. *Utopian Thought in the Western World*. Cambridge, MA: Belknap Press of Harvard University Press.

Marano, Louis A. 1973. "A macrohistoric trend toward world government." *Behavior Science Notes* 8: 35–40.

Marcuse, Herbert. (1964) 1970. *One-Dimensional Man*. London: Sphere Books.

Marien, Michael. 1985 "Toward a new futures research: Insights from twelve types of futurists." *Futures Research Quarterly* 1, No. 1 (Spring 1985): 13–35.

———. 1987. "What *is* the nature of our embryonic enterprise? An open letter to Wendell Bell." *Futures Research Quarterly* 3, No. 4 (Winter): 71–79.

———. 1992. "The scope of policy studies: Reclaiming Lasswell's lost vision." Pp. 449–493 in W. N. Dunn and R. M. Kelly (eds.), *Advances in Policy Studies Since 1950*. New Brunswick, NJ: Transaction.

———. 1991. "Scanning: An imperfect activity in an era of fragmentation and uncertainty." *Futures Research Quarterly* 7, No. 3 (Fall): 82–90.

Marien, Michael with Lane Jennings (eds.). 1979 yearly through 1995. *Future Survey Annual: 1979 [through 1995]* Vols. 1 through 15. Bethesda, MD: World Future Society.

Markley, O.W. 1983. "Preparing for the professional futures field: Observations from the UHCLC futures program." *Futures* 16 (February): 47–64.

Marschak, Jacob. 1968. "Decision-making: Economic aspects." Pp. 42–55 in D. L. Sills (ed.), *International Encyclopedia of the Social Sciences*, Vol. 4. New York: Macmillan & The Free Press.

Marshack, Alexander. 1972. *The Roots of Civilization*. New York: McGraw-Hill.

Martino, J. P. 1983. *Technological Forecasting for Decision Making.* Second Edition. New York: North-Holland.

———. 1987. "The Gods of the copybook headings: A caution to forecasters." Pp. 143–152 in M. Marien and L. Jennings (eds.), *What I Have Learned.* New York: Greenwood.

———. 1993. "Technological forecasting: An introduction." *The Futurist* 27, No. 4 (July-August): 13–16.

Maruyama, Magoroh. 1978a. "Introduction." Pp. xvii–xxii in M. Maruyama and A. M. Harkins (eds.), *Cultures of the Future.* The Hague: Mouton.

———. 1978b. "Toward human futuristics." Pp. 33–59 in M. Maruyama and A. M. Harkins (eds.), *Cultures of the Future.* The Hague: Mouton.

Marx, Karl. (1875) 1962. "Critique of the Gotha Program." Pp. 45–65 in *The Communist Blueprint for the Future.* New York: E.P. Dutton.

Marx, Karl and Friedrich Engels. (1848) 1962. "Manifesto of the Communist Party." Pp. 9–44 in *The Communist Blueprint for the Future.* New York: E.P. Dutton.

Masini, Eleonra Barbieri. 1978. "The global diffusion of futures research." Pp. 17–29 in J. Fowles (ed.), *Handbook of Futures Research.* Westport, CT: Greenwood.

———. 1981. "Philosophical and ethical foundations of future studies: A discussion." *World Futures* 17: 1–14.

———. 1982. "Reconceptualizing futures: A need and a hope." *World Future Society Bulletin* (November/December): 1–8.

———. 1988. "Future technology and its social implications." *World Futures Studies Federation Newsletter* 14, No. 1 (March): 17.

———. 1993. *Why Futures Studies?* London: Grey Seal.

Maslow, Abraham H. 1968. *Toward a Psychology of Being.* Rev. ed. Princeton, NJ: Van Nostrand.

Masterman, Margaret. (1970) 1978. "The nature of a paradigm." Pp. 59–89 in I. Lakatos and A. Musgrave (eds.), *Criticism and the Growth of Knowledge.* Cambridge: Cambridge University Press.

Mau, James A. 1967. "Images of Jamaica's future." Pp. 197–223 in W. Bell (ed.), *The Democratic Revolution in the West Indies.* Cambridge, MA: Schenkman.

———. 1968. *Social Change and Images of the Future.* Cambridge, MA: Schenkman.

Mead, George H. 1934. *Mind, Self & Society.* Chicago: The University of Chicago Press.

Mead, Margaret. 1928. *Coming of Age in Samoa.* New York: Morrow.

Meadows, Donella H., Dennis L. Meadows, Jorgen Randers, and William W. Behrens III. 1972. *The Limits to Growth:* New York: Universe.

———. 1975. "A response to Sussex." Pp. 217–240 in H. S. D. Cole, C. Freeman, M. Jahoda, and K. L. R. Pavitt (eds.), *Models of Doom.* New York: Universe Books.

Meadows, Donella H., Dennis L. Meadows, and Jorgen Randers. 1992. *Beyond the Limits.* Post Mills, VT: Chelsea Green Publishing Company.

Meadows, D. H. and J. M. Robinson. 1985. *The Electronic Oracle: Computer Models and Social Decisions.* New York: John Wiley & Sons.

Meadows, D. L. and D. H. Meadows (eds.). 1973. *Toward Global Equilibrium.* Cambridge, MA: Wright-Allen Press.

Meadows, Dennis L., William W. Behrens III, Donella H. Meadows, Roger F. Naill, Jorgen Randers, and Erich K. O. Zahn. 1974. *The Dynamics of Growth in a Finite World.* Cambridge, MA: Wright-Allen Press.

Medawar, Peter. 1984. *Pluto's Republic*. Oxford: Oxford University Press.

Meeker, Heidi. 1993. "Hands-on futurism: How to run a scanning project." *The Futurist* 27, No. 3 (May-June): 22–26.

Meeks, Wayne A. 1989. Personal communication.

———. 1993. *The Origins of Christian Morality*. New Haven, CT: Yale University Press.

Merkl, Peter. (1960) 1966. "Introduction." Pp. 1–8 in M. Broszat, *German National Socialism, 1919–1945*. Trans. by K. Rosenbaum and I.P. Boem. Santa Barbara, CA: Clio.

Merton, Robert K. (1938) 1973. *The Sociology of Science*. Chicago and London: The University of Chicago Press.

Mettler, Peter H. (ed.). 1995. *Science and Technology for Eight Billion People*. London: New European Publications in association with Adamantine Press.

Michael, Donald. 1963. *The New Generation*. New York: Random House, Vintage Books.

———. 1985. "With both feet planted firmly in mid-air: Reflections on thinking about the future." *Futures* 17 (April): 94–103.

———. (1985) 1987. "The futurist tells stories." Pp. 75–86 in M. Marien and L. Jennings (eds.), *What I Have Learned*. New York: Greenwood Press.

Michalos, Alex C. 1979. "Philosophy of social science." Pp. 463–502 in P.D. Asquith and H.E. Kyburg, Jr. (eds.), *Current Research in Philosophy of Science*. East Lansing, MI: Philosophy of Science Association.

Midgley, Mary. (1991) 1993. *Can't We Make Moral Judgements?* New York: St. Martin's Press.

Miles, Ian. 1975. *The Poverty of Prediction*. Westmead, Farnborough, England: Saxon House, D.C. Heath.

———. 1978. "The ideologies of futurists." Pp. 67–97 in J. Fowles (ed.), *Handbook of Futures Research*. Westport, CT: Greenwood Press.

———. 1985. *Social Indicators for Human Development*. London: Frances Pinter.

Miller, James. 1984. *Rousseau: Dreamer of Democracy*. New Haven, CT: Yale University Press.

Millett, Stephen M. and Edward J. Honton. 1991. *A Manager's Guide to Technology Forecasting and Strategy Analysis Methods*. Columbus, OH: Battelle Press.

Mills, Charles W. 1989. Personal communication (April 9).

Mitchell, Robert Cameron and Richard T. Carson. 1989. *Using Surveys to Value Public Goods*. Washington, DC: Resources for the Future.

Mitroff, Ian. 1974. *The Subjective Side of Science*. New York: Elsevier.

Mitroff, Ian I. and Murray Turoff. 1975. "Philosophical and methodological foundations of Delphi." Pp. 17–36 in H.A. Linstone and M. Turoff (eds.), *The Delphi Method: Techniques and Applications*. Reading, MA: Addison-Wesley.

Mitroff, Ian I. and Ralph H. Kilmann. 1978. *Methodological Approaches to Social Science*. San Francisco, CA: Jossey-Bass.

Moghadam, Valentine M. 1994. "Women in societies." *International Social Science Journal* 139 (February): 95–115.

Moll, Peter. 1991. *From Scarcity to Sustainability*. Frankfurt am Main: Peter Lang.

———. 1993. "The discreet charm of the Club of Rome." *Futures* 25, No. 7 (September): 801–805.

Monro, D. H. 1967. *Empiricism and Ethics*. Cambridge: Cambridge University Press.

Montias, John Michael. 1968. "Planning, economic. Eastern Europe." Pp. 110–118 in D. L. Sills (ed.), *International Encyclopedia of the Social Sciences*, Vol. 12. New York: Macmillan & The Free Press.

Moore, Sally Falk. 1986. "Legal systems of the world." Pp. 11–62 in L. Lipson and S. Wheeler (eds.), *Law and the Social Sciences*. New York: Russell Sage Foundation.

Moore, Wilbert E. 1963. *Social Change*. Englewood Cliffs, NJ: Prentice-Hall.

More, Sir Thomas. No date (1516). "Utopia." Pp. 127–232 in *Famous Utopias*. New York: Tudor.

Morowitz, Harold J. and James S. Tefil. 1992. *The Facts of Life: Science and the Abortion Controversy*. New York: Oxford University Press.

Morris, Richard. 1984. *Time's Arrows*. New York: Simon and Schuster.

Morrison, Roy. 1991. *We Build the Road as We Travel*. Philadelphia, PA: New Society Publishers.

Moskos, Charles C., Jr. 1967. *The Sociology of Political Independence*. Cambridge, MA: Schenkman.

Moynihan, Daniel Patrick. 1990. *On the Law of Nations*. Cambridge, MA: Harvard University Press.

Mukherjee, Ramkrishna. 1989. *The Quality of Life*. New Delhi: Sage.

Mumford, Lewis. 1922. *The Story of Utopias*. New York: Boni and Liveright.

Munting, Roger. 1982. *The Economic Development of the USSR*. London & Canberra: Croom Helm.

Murdock, George P. 1945. "The common denominator of cultures." Pp. 123–142 in R. Linton (ed.), *The Science of Man in the World Crisis*. New York: Columbia University Press.

Murphy, Brian. 1992. "Linking present decisions to long-range ethical visions in organizations using stakeholder audits." Pp. 321–329 in M. Mannermaa (ed.), *Linking Present Decisions to Long-Range Visions*, Vol. II. Budapest, Hungary: World Futures Studies Federation.

Musgrave, Alan. 1993. *Common Sense, Science and Scepticism*. Cambridge: Cambridge University Press.

Myrdal, Gunnar with the assistance of Richard Sterner and Arnold Rose. 1944. *An American Dilemma*. New York: Harper & Brothers.

Naisbitt, John. (1982) 1984. *Megatrends: The New Directions Transforming Our Lives*. New York: Warner Books.

Naisbitt, John and Patricia Aburdene. 1990. *Megatrends 2000*. New York: William Morrow and Co.

Namenwirth, J. Zvi and Harold D. Lasswell. 1970. "The changing language of American values." *Sage Professional Papers in Comparative Politics*. Beverly Hills, CA: Sage.

Namenwirth, J. Zvi and Robert Philip Weber. 1987. *Dynamics of Culture*. Boston: Allen & Unwin.

Nanus, Burt. 1984. "Futures research—stage three." *Futures* 16 (August): 405–407.

Naroll, Raoul. 1967. "Imperial cycles and world order." *Peace Research Society Papers* 7: 83–101.

———. 1983. *The Moral Order*. Beverly Hills, CA: Sage.

National Center for Health Statistics. 1993, 1994, and 1995. "Vital statistics of the United States, 1989 (1990 and 1991)" Section 6. Washington, DC: Public Health Service.

Neher, André. 1976. "The view of time and history in Jewish culture." Pp. 148–167 in *Cultures and Time*. Paris: The Unesco Press.

Neisser, Ulric. 1976. *Cognition and Reality*. San Francisco, CA: W. H. Freeman.

Neurath, Otto. (1931/2) 1959. "Sociology and physicalism." Pp. 282–317 in A.J. Ayer (ed.), *Logical Positivism*. Glencoe, IL: The Free Press.

Neustadt, Richard E. and Ernest R. May. 1986. *Thinking in Time: The Uses of History for Decision-Makers*. New York: Free Press.

New Scientist. 1984. No. 1435/1436 (December 20/27): 8.

Newton-Smith, W.H. 1981. *The Rationality of Science*. Boston: Routledge & Kegan Paul.

Nisbet, Robert A. 1966. *The Sociological Tradition*. New York: Basic Books.

Noelle-Neumann, Elisabeth. 1989. "The public as prophet: Findings from continuous survey research and their importance for early diagnosis of economic growth." *International Journal of Public Opinion Research* 1, No. 2 (Summer): 136–150.

———. 1994. "Aussichten für 1995." Allensbach am Bodensee, Germany: Institut für Demoskopie Allensbach (Dezember), originalmanuskript.

Nonet, Philippe and Philip Selznick. 1978. *Law and Society in Transition*. New York: Farrar, Straus and Giroux.

Novak, Maximillian E. 1969. "The economic meaning of *Robinson Crusoe*." Pp. 97–102 in F. H. Ellis (ed.), *Twentieth Century Interpretations of Robinson Crusoe*. Englewood Cliffs, NJ: Prentice-Hall.

Nove, Alec. 1969. *An Economic History of the U.S.S.R.* London: Allen Lane, The Penquin Press.

———. 1977. *The Soviet Economic System*. London: George Allen & Unwin.

Nussbaum, Martha. 1986. "Review of Jane Roland Martin's Reclaiming a Conversation: The Ideal of the Educated Woman." *The New York Review of Books* 33 (January 30): 7–12.

NYT. *The New York Times* (date given in text).

NYTBR. *The New York Times Book Review* (date given in text).

OECD (Organization for Economic Cooperation and Development). 1973. "List of social concerns common to most OECD countries." Paris: OECD Social Indicator Development Programme.

Ogilvy, James. 1992. "Futures studies and the human sciences: The case for normative scenarios." *Futures Research Quarterly* 8, No. 2 (Summer): 5–65.

Oldfield, F. 1975. "Discussion." Pp. 154–165 in *The Future as an Academic Discipline*. Ciba Foundation Symposium 36. Amsterdam: Elsevier.

Olshansky, S. Jay, Bruce A. Carnes, Christine Cassel. 1990. "In search of Methuselah: Estimating the upper limits to human longevity." *Science* 250 (2 November): 634–640.

O'Neill, Molly. 1991. "Am I the diner or am I the dish?" *The New York Times Book Review*. (July 28): 7.

Ono, Ryota and Dan J. Wedemeyer. 1994. "Assessing the validity of the Delphi technique." *Futures* 26, No. 3 (April): 289–304.

Ornauer, H., H. Wiberg, A. Sicinski, and J. Galtung (eds). 1976. *Images of the World in the Year 2000: A Comparative Ten Nation Study*. Atlantic Highlands, NJ: Humanities Press.

Orwell, George. 1949. *Nineteen Eighty-Four*. New York: Harcourt, Brace & World.

Osgood, Charles E., William H. May, and Murray S. Miron. 1975. *Cross-Cultural Universals of Affective Meaning*. Urbana: University of Illinois Press.

Pagels, Heinz R. 1982. *The Cosmic Code*. New York: Simon and Schuster.

Palmore, James A. and Robert W. Gardner. (1983) 1991. *Measuring Mortality, Fertility, and Natural Increase*. Honolulu, HI: The East-West Center.

Papineau, David. 1993. "How to think about science." *The New York Times Book Review* (July 25): 14–15.

Partridge, Ernest. 1991. "On the rights of future generations." Pp. 1–21 in D. Scherer (ed.), *Upstream/Downstream: Issues in Environmental Ethics*. Philadelphia, PA: Temple University Press.

———. 1981a. "Introduction." Pp. 1–16 in E. Partridge (ed.), *Responsibilities to Future Generations*. Buffalo, NY: Prometheus Books.

———. (1980) 1981b. "Why care about the future?" Pp. 203–220 in E. Partridge (ed.), *Responsibilities to Future Generations*. Buffalo, NY: Prometheus Books.

Passant, E.J. 1966. *A Short History of Germany 1815–1945*. Cambridge: University Press.

Passmore, John. (1974) 1981. "Conservation." Pp. 45–59 in E. Partridge (ed.), *Responsibilities to Future Generations*. Buffalo, NY: Prometheus Books.

———. 1985. "An end to science?" *The Times Higher Education Supplement* (April 19).

Pàttaro, Germano. 1976. "The Christian conception of time." Pp. 169–195 in *Cultures and Time*. Paris: The Unesco Press.

Patterson, Edwin W. 1953. *Jurisprudence*. Brooklyn, NY: Foundation Press.

Payne, Stephen L. and Robert A. Desman. 1987. "The academician as a consultant." Pp. 95–115 in S.L. Payne and B.H. Charnov (eds.), *Ethical Dilemmas for Academic Professionals*. Springfield, IL: Charles C. Thomas.

Pearson, K. 1901–1902. "On the change in expectation of life in man during a period of circa 2000 years." *Biometrika* I: 261–264.

Pelikan, Jaroslav. 1992. *The Idea of the University—A Reexamination*. New Haven, CT: Yale University Press.

Perkins, H. Wesley and James L. Spates. 1986. "Mirror images? Three analyses of values in England and the United States." *International Journal of Comparative Sociology* 27, Nos. 1–2: 31–52.

Perry, Ralph Barton. (1954) 1978. "A definition of morality." Pp. 12–22 in P. W. Taylor (ed.), *Problems of Moral Philosophy*. Third Edition. Belmont, CA: Wadsworth.

Perun, Pamela and Denise Del Veuto Bielby. 1979. "Midlife: A discussion of competing models." *Research on Aging* 1: 275–300.

Petersen, William. (1961) 1975. *Population*, Third Edition. New York: Macmillan.

Pettit, Philip. (n.d.). "The philosophies of social science." In R.J. Anderson and W.W. Sharrock (eds.), *Teaching Papers in Sociology*. York, England: Longman.

Phillips, Andrew P. 1964. "The Development of a Modern Labor Force in Antigua." Unpublished Ph.D. dissertation, Los Angeles: University of California, Los Angeles.

———. 1967. "Management and workers face an independent Antigua." Pp. 165–196 in W. Bell (ed.), *The Democratic Revolution in the West Indies*. Cambridge, MA: Schenkman.

Phillips, Derek L. 1973. *Abandoning Method*. San Francisco, CA: Jossey-Bass.

———. 1986. *Toward a Just Social Order*. Princeton, N.J.: Princeton University Press.

Phillips, Dretha M. 1983. *A Bibliography Toward Sociological Futures Research and Instruction*. Public Administration Series: Bibliography. Monticello, IL: Vance Bibliographies.

Piaget, Jean. 1965. *The Moral Judgment of the Child*. Trans. by M. Gabain. New York: John Wiley & Sons.

Pickett, Neil. 1992. *A History of Hudson Institute*. Indianapolis, IN: Hudson Institute.

Pieper, Josef. 1966. *The Four Cardinal Virtues: Prudence, Justice, Fortitude, Temperance* (Trans. by R. Winston, C. Winston et al.). Notre Dame, IN: University of Notre Dame Press.

Plato. no date. *The Republic*. Trans. by B. Jowett. New York: The Modern Library, Random House.

Platt, John Rader. 1966. *The Step to Man*. New York: John Wiley & Sons.

————. 1975. "Discussion." Pp. 154–165 in *The Future as an Academic Discipline*. Ciba Foundation Symposium 36. Amsterdam: Elsevier.

Polak, Frederik L. (1955) 1961. *The Image of the Future: Enlightening the Past, Orientating the Present, Forecasting the Future*, Vols. I and II. New York: Oceana.

Polgar, Steven. 1978. "The possible and the desirable: Population and environmental problems." Pp. 63–70 in M. Maruyama and A.M. Harkins (eds.), *Cultures of the Future*. The Hague: Mouton.

Popper, K.R. (1945) 1952. *The Open Society and Its Enemies*. Second Edition. London: Routledge.

————. 1957. *The Poverty of Historicism*. London: Routledge and Kegan Paul.

————. 1959. *The Logic of Scientific Discovery*. New York: Basic Books.

————. 1965. *Conjectures and Refutations: The Growth of Scientific Knowledge*. Second Edition. New York: Basic Books.

Porter, John. 1965. *The Vertical Mosaic: An Analysis of Social Class and Power in Canada*. Toronto: University of Toronto Press.

Preble, J. F. 1983. "Public sector use of the Delphi technique." *Technological Forecasting and Social Change* 23: 75–88.

President's Research Committee on Social Trends. 1933. *Recent Social Trends in the United States*. Two vols. New York: McGraw-Hill.

Priestley, J.B. 1968. *Man and Time*. New York: Dell, a Laurel edition.

Prigogine, Ilya and Isabelle Stengers. 1984. *Order Out of Chaos*. New York: Bantam.

Public Opinion. 1978. Vol. 1 (July/August): 40.

Pugh, George Edgin. 1977. *The Biological Origin of Human Values*. New York: Basic Books.

Punch, Maurice. 1986. *The Politics and Ethics of Fieldwork*. Beverly Hills, CA: Sage.

Putnam, Hilary (1974) 1981. "The 'corroboration' of theories." Pp. 60–79 in I. Hacking (ed.), *Scientific Revolutions*. Oxford: Oxford University Press.

Quine, W.V. 1951. "Two dogmas of empiricism." *Philosphical Review* 60 (January): 20–43.

Rabinow, Paul. 1983. "Humanism as nihilism: The bracketing of truth and seriousness in American cultural anthropology." Pp. 52–75 in N. Haan, R. N. Bellah, P. Rabinow, and W. M. Sullivan (eds.), *Social Science as Moral Inquiry*. New York: Columbia University Press.

Rappaport, Roy A. 1986. "The construction of time and eternity in ritual." The David Skomp Distinguished Lectures in Anthropology. Bloomington: Indiana University.

Ratner, Steven R. 1995. *The New UN Peacekeeping*. New York: St. Martin's Press.

Rawls, John. 1971. *A Theory of Justice*. Cambridge, MA: Belknap Press of Harvard University Press.

Raz, J. 1977. "Promises and obligations." Pp. 210–228 in P.M.S. Hacker and J. Raz (eds.), *Law, Morality, and Society*. Oxford: Clarendon Press.

Razak, Victoria and Sam Cole (eds.). 1995. "Anthropological Perspectives on the Future of Culture and Society." Special issue of *Futures* 27, No. 4 (May).

Rees, Albert. (1974) 1980. "An overview of the labor-supply results." Pp. 41–63 in D. Nachmias (ed.), *The Practice of Policy Evaluation*. New York: St. Martin's Press.

Reich, Walter. 1993. "Erasing the Holocaust." *The New York Times Book Review* (July 11): 1 et passim.

Reichenbach, Hans. 1951. *The Rise of Scientific Philosophy*. Berkeley and Los Angeles: University of California Press.
Reiman, Jeffrey. 1990. *Justice and Modern Moral Philosophy*. New Haven, CT: Yale University Press.
Reinharz, Shulamit. 1979. *On Becoming a Social Scientist*. San Francisco, CA: Jossey-Bass.
Rest, James R. et al. 1986. *Moral Development*. New York: Praeger.
Riasanovsky, Nicholas V. 1969. *The Teaching of Charles Fourier*. Berkeley and Los Angeles: University of California Press.
Richardson, John M. 1984. "The state of futures research." *Futures* 16 (August): 382–395.
Riner, Reed D. 1987. "Doing futures research—anthropologically." *Futures* 19, No. 3 (June): 311–328.
———. 1991a. "From description to design: Ethnographic futures research methods applied in small town revitalization and economic development." *Futures Research Quarterly* 7, No. 1 (Spring): 17–30.
———. 1991b. "Anthropology about the future: Limits and potentials." *Human Organization* 50, No. 3 (Fall): 297–311.
Robinson, James A. 1968. "Decision making: Political aspects." Pp. 55–62 in D. L. Sills (ed.), *International Encyclopedia of the Social Sciences*, Vol. 4. New York: Macmillan & The Free Press.
Robinson, Robert V. and Wendell Bell. 1978. "Equality, success, and social justice in England and the United States." *American Sociological Review* 43, No. 2 (April): 125–143.
Robinson, William S. 1949. *Class Notes for the Logic of Social Inquiry*. Los Angeles: University of California, unpublished.
Roddenberry, Gene. 1984. "The literary image of the future," a talk given at the Fifth General Assembly and Exposition of the World Future Society, Washington, DC, June 10–14.
Rohner, Ronald P. 1986. *The Warmth Dimension*. Beverly Hills, CA: Sage.
Rohrbaugh, J. 1979. "Improving the quality of group judgment: Social judgment analysis and the Delphi technique." *Organizational Behavior and Human Performance* 24: 79–92.
Rojas, Billy and H. Wentworth Eldredge. 1974. "Appendix. Status Report: Sample Syllabi and Directory of Futures Studies." Pp. 345–399 in A. Toffler (ed.), *Learning for Tomorrow*. New York: Random House.
Rokeach, Milton. 1968. *Beliefs, Attitudes, and Values*. San Francisco, CA: Jossey-Bass.
Rolston, Holmes III. 1981. "The river of life: Past, present, and future." Pp. 123–137 in E. Partridge (ed.), *Responsibilities to Future Generations*. Buffalo, NY: Prometheus Books.
Rorty, Richard. 1979. *Philosophy and the Mirror of Nature*. Princeton, NJ: Princeton University Press.
Rosenau, Pauline Marie. 1992. *Post-Modernism and the Social Sciences*. Princeton, NJ: Princeton University Press.
Rosenbaum, David E. 1994. "Down at the statehouse, legislators have more fun." *The New York Times* (March 20): 3E.
Rossi, Peter. 1987. "Unemployment insurance payments and recidivism among released prisoners." Pp. 107–124 in B. Barber (ed.), *Effective Social Science*. New York: Russell Sage Foundation.

Rossi, Peter H. and Howard E. Freeman with the collaboration of Sonia Rosenbaum. 1982. *Evaluation: A Systematic Approach*. Second Edition. Beverly Hills, CA: Sage.

Rossi, Peter H. and William Foote Whyte. 1983. "The applied side of sociology." Pp. 5–31 in H.E. Freeman, R.R. Dynes, P.H. Rossi, and W. F Whyte (eds.), *Applied Sociology*. San Francisco, CA: Jossey-Bass.

Rossi, Peter H., James D. Wright, and Andy B. Anderson (eds.). 1983. *Handbook of Survey Research*. New York: Academic Press.

Rudel, Thomas K. with Bruce Horowitz. 1993. *Tropical Deforestation*. New York: Columbia University Press.

Rudner, Richard S. 1966. *Philosophy of Social Science*. Englewood Cliffs, NJ: Prentice-Hall.

Ryan, Alan. 1995. "The women in the cowshed." *The New York Review of Books* XLII, 8 (May 11): 24–26.

Sackman, H. 1975. *Delphi Critique*. Lexington, MA: D. C. Heath.

Sagan, Leonard A. 1989. *The Health of Nations*. New York: Basic Books.

Saltman, Juliet. 1975. "Implementing open housing laws through social action." *The Journal of Applied Behavioral Science* 11, No. 1: 39–61.

Sartre, Jean-Paul. (1965) 1978. "The humanism of existentialism." Pp. 646–661 in P. W. Taylor (ed.), *Problems of Moral Philosophy*. Third Ed. Belmont, CA: Wadsworth.

Schaefer, Richard T. and Robert P. Lamm. 1992. *Sociology*. Fourth Edition. New York: McGraw-Hill.

Schattschneider, Doris. 1992. "The universal mind at work." *The New York Times Book Review* (April 19): 15.

Scheele, D. S. 1975. "Reality construction as a product of Delphi interaction." Pp. 37–71 in H.A. Linstone and M. Turoff (eds.), *The Delphi Method: Techniques and Applications*. Reading, MA: Addison-Wesley.

Scheer, Lore. 1980. "Experience with quality of life comparisons." Pp. 145–155 in A. Szalai and F.M. Andrews (eds.), *The Quality of Life*. Beverly Hills, CA: Sage.

Scheffler, Israel. 1967. *Science and Subjectivity*. Indianapolis, IN: Bobbs-Merrill.

Schluchter, Wolfgang. (1980) 1989. *Rationalism, Religion, and Domination*. Trans. by N. Solomon. Berkeley: University of California Press.

Schnaiberg, Allan. 1995. "How I learned to reject recycling, Part 2. paradoxes and contraditions: A contextual framework." *Blazing Tattles* 4, No. 1 (January): 1 et passim.

Schoedinger, Andrew B. (ed.). 1992. *The Problem of Universals*. Atlantic Highlands, NJ: Humanities Press International.

Schuessler, Karl F. 1968. "Prediction." Pp. 418–425 in D. L. Sills (ed.), *International Encyclopedia of the Social Sciences*, Vol. 12. New York: Macmillan & The Free Press.

Schuessler, Karl. 1971. "Continuities in social prediction." Pp. 302–329 in H.L. Costner (ed.), *Sociological Methodology*. San Francisco, CA: Jossey-Bass.

Schwartz, Richard D. 1986. "Law and normative order." Pp. 63–107 in L. Lipson and S. Wheeler (eds.), *Law and the Social Sciences*. New York: Russell Sage Foundation.

Schwarz, Brita, Uno Svedin, and Björn Wittrock. 1982. *Methods in Futures Studies*. Boulder, CO: Westview Press.

Sen, Amartya. 1981. *Poverty and Famines*. New York: Oxford University Press.

Sheldon, Eleanor Bernert and Wilbert E. Moore (eds.). 1968. *Indicators of Social Change: Concepts and Measurements*. New York: Russell Sage Foundation.

Shepher, Joseph. 1983. *Incest: A Biosocial View*. New York: Academic Press.

Sher, George. 1992. "Ancient wrongs and modern rights." Pp. 48–61 in P. Laslett and J. S. Fishkin (eds.), *Justice between Age Groups and Generations*. New Haven, CT: Yale University Press.

Shils, Edward. 1963. "On the comparative study of the new states." Pp. 1–26 in C. Geertz (ed.), *Old Societies and New States*. New York: The Free Press of Glencoe.

———. 1981. *Tradition*. Chicago: The University of Chicago Press.

Shrady, Nicholas. 1992. "Glorious in its very stories." *New York Times Book Review* (March 15): 1, 24–25.

Shubik, Martin. 1975. *The Uses and Methods of Gaming*. New York: Elsevier.

———. 1982. *Game Theory in the Social Sciences*. Cambridge, MA: The MIT Press.

Sikora, R. I. 1978. "Is it wrong to prevent the existence of future generations?" Pp. 112–166 in R.I. Sikora and B. Barry (eds.), *Obligations to Future Generations*. Philadelphia, PA: Temple University Press.

Simmonds, W.H. Clive. 1977. "The nature of futures problems." Pp. 13–26 in H. A. Linstone and W.H.C. Simmonds (eds.), *Futures Research: New Directions*. Reading, MA: Addison-Wesley.

Simon, Julian L. 1981. *The Ultimate Resource*. Princeton, NJ: Princeton University Press.

———. 1995. "Why do we hear prophecies of doom from every side?" *The Futurist* 29, No. 1 (January-February): 19–23.

Simon, Julian. L. and Herman Kahn (eds.). 1984. *The Resourceful Earth: A Response to Global 2000*. Oxford: Basil Blackwell.

Singer, Max and Aaron Wildavsky. 1993. *The Real World Order: Zones of Peace/Zones of Turmoil*. Chatham, NJ: Chatham.

Sinha, Durganand. 1984. "Community as target: A new perspective to research on prosocial behavior." Pp. 445–455 in E. Staub, D. Bar-Tal, J. Karylowski, and J. Reykowski (eds.), *Development and Maintenance of Prosocial Behavior*. New York: Plenum Press.

Sippanondha, Ketudat with Robert B. Textor et al. 1990. *The Middle Path for the Future of Thailand*. Honolulu, HI: East-West Center.

Skagestad, Peter. 1981. "Hypothetical realism." Pp. 77–97 in M.B. Brewer and B.E. Collins (eds.), *Scientific Inquiry and the Social Sciences*. San Francisco, CA: Jossey-Bass.

Slaughter, Richard A. (ed.). 1993a. "Special issue: The knowledge base of futures studies." *Futures* 25, No. 3 (April).

———. 1993b. "The substantive knowledge base of futures studies." *Futures* 25, No. 3 (April): 227–233.

———. 1993c. "Futures concepts." *Futures* 25, No. 3 (April): 289–314.

———. 1993d. "Looking for the real 'megatrends.'" *Futures* 25, No. 8 (October): 827–849.

———. 1996. *The Knowledge Base of Futures Studies*, Vols. 1–3. Hawthorn, Victoria, Australia: DDM Media.

Slottje, Daniel J., Gerald W. Scully, Joseph G. Hirschberg, and Kathy J. Hayes. 1991. *Measuring the Quality of Life Across Countries*. Boulder, CO: Westview Press.

Smart, J. J. C. 1984. *Ethics, Persuasion and Truth*. London: Routledge & Kegan Paul.

Smith, D. Kimball. 1983. "Putting the wings on man and overcoming limited views." *Yale Alumni Magazine and Journal* XLVI (June): 38–42.

Smith, Herbert L. 1984. "The social forecasting industry." Paper prepared for the Social Science Research Council's Conference on Forecasting in the Social and Natural Sciences. Boulder, CO, June 10–13.
———. 1987. "The social forecasting industry." Pp. 35–60 in K.C. Land and S.H. Schneider (eds.), *Forecasting in the Social and Natural Sciences.* Dordrecht, Holland: D. Reidel.
Smith, Steven B. 1987. "Functional analysis and Marxian ideology." *Review of Politics* 49: 287–290.
Snarey, John. 1993. *How Fathers Care for the Next Generation.* Cambridge, MA: Harvard University Press.
Snell, George D. 1988. *Search for a Rational Ethic.* New York: Springer-Verlag.
Snyder, Richard C., Charles F. Hermann, and Harold D. Lasswell. 1976. "A global monitoring system: Appraising the effects of government on human dignity." *International Studies Quarterly* 20, No. 2 (June): 221–260.
Snyder, Richard C., H.W. Bruck, and Burton Sapin. (1954) 1962. "Decison-making as an approach to the study of international politics." Pp. 14–185 in R.C. Snyder, H.W. Bruck, and B. Sapin (eds.), *Foreign Policy Decision-Making.* New York: The Free Press.
Sorokin, Pitirim and Robert Merton. (1937) 1990. "Social-time: A methodological and functional analysis." Pp. 56–66 in J. Hassard (ed.), *The Sociology of Time.* London: Macmillan.
Speer, Albert. (1969) 1970. *Inside the Third Reich.* Trans. by Richard and Clara Winston. New York: Macmillan.
Spence, Jonathan. 1992. "From the stone age to Tiananmen Square." *The New York Times Book Review* (May 24): 2.
Sperber, Dan. 1974. *Rethinking Symbolism.* Cambridge: Cambridge University Press.
Spiro, Melford. 1958. *Children of the Kibbutz.* Cambridge, MA: Harvard University Press.
———. 1982. *Oedipus in the Trobriands.* Chicago: University of Chicago Press.
———. 1992. *Anthropological Other or Burmese Brother?* New Brunswick, NJ: Transaction Publishers.
Sprigge, T. L. S. 1988. *The Rational Foundations of Ethics.* London and New York: Routledge & Kegan Paul.
Staal, Frits. 1988. *Universals: Studies in Indian Logic and Linguistics.* Chicago: University of Chicago Press.
Stafford, Frank P. 1988. "A model of Harmony." *ISR Newsletter* 16, No. 1: 10–11.
Stanford Research Institute. 1962. *International Industrial Development Center Study.* Stanford, CA.
Stark, Rodney and William Sims Bainbridge. 1985. *The Future of Religion.* Berkeley and Los Angeles: University of California Press.
Stehr, Nico. 1978. "The ethos of science revisited: Social and cognitive norms." Pp. 172–196 in J. Gaston (ed.), *Sociology of Science.* San Francisco, CA: Jossey-Bass.
Stephens, Evelyne Huber and John D. Stephens. 1986. *Democratic Socialism in Jamaica.* Princeton, NJ: Princeton University Press.
Stevenson, Charles L. 1963. *Facts and Values.* New Haven, CT: Yale University Press.
Steward, Thomas R. 1987. "The Delphi technique and judgmental forecasting." Pp. 97–113 in K. C. Land and S. H. Schneider (eds.), *Forecasting in the Social and Natural Sciences.* Dordrecht, Holland: D. Reidel.
Stinchcombe, Arthur L. 1982a. "On softheadedness on the future." *Ethics* 93 (October): 114–128.

————. 1982b. "The deep structure of moral categories: Eighteenth-century French stratification, and the revolution." Pp. 66–95 in I. Rossi (ed.), *Structural Sociology*. New York: Columbia University Press.

Stone, Philip J., Dexter C. Dunphy, Marshall S. Smith, Daniel M. Ogilvie and associates. 1966. *The General Inquirer: A Computer Approach to Content Analysis*. Cambridge, MA: The M.I.T. Press.

Sudman, Seymour and Norman M. Bradburn. 1982. *Asking Questions*. San Francisco, CA: Jossey-Bass.

Sullivan, Michael J., III. 1991. *Measuring Global Values*. New York: Greenwood Press.

Sumner, L. W. 1987. *The Moral Foundation of Rights*. Oxford: Clarendon Press.

Sumner, W. G. (1906) 1960. *Folkways*. New York: New American Library.

Suppe, Frederick (ed.), 1977. *The Structure of Scientific Theories*. Second Edition. Urbana: University of Illinois Press.

Sylvan, David and Barry Glassner. 1985. *A Rationalist Methodology for the Social Sciences*. Oxford: Basil Blackwell.

Szalai, Alexander (ed.). 1972. *The Use of Time*. The Hague: Mouton.

————. 1980. "The meaning of comparative research on the quality of life." Pp. 7–21 in A. Szalai and F.M. Andrews (eds.), *The Quality of Life*. Beverly Hills, CA: Sage.

Szalai, Alexander and Frank M. Andrews (eds.). 1980. *The Quality of Life*. Beverly Hills, CA: Sage.

Sztompka, Piotr. 1979. *Sociological Dilemmas*. New York: Academic.

Taeuber, Karl E. and Alma F. Taeuber. 1965. *Negroes in Cities*. Chicago: Aldine.

Talbot, Michael. 1986. *Beyond the Quantum*. New York: Macmillan.

Taylor, Paul W. 1981. "The ethics of respect for nature." *Environmental Ethics* 3 (Fall): 197–218.

Taylor, Shelley E. 1989. *Positive Illusions*. New York: Basic Books.

Textor, Robert B. 1979. "The natural partnership between ethnographic futures research and futures education." *Journal of Cultural and Educational Futures* 1 (April): 13–18.

————. 1980. *A Handbook on Ethnographic Futures Research*. Third Edition: Version A. Stanford University (mimeographed).

————. 1990. "Introduction" and "Methodological appendix." Pp. xxiii–xlvii and 135–152 in K. Sippanondha with R. B. Textor et al., *The Middle Path for the Future of Thailand*. Honolulu, HI: East-West Center.

————. 1995. "The ethnographic futures research method: An application to Thailand." *Futures* 27, 4 (May): 461–471.

————. Forthcoming. *Anticipatory Anthropology and Ethnographic Futures Research*.

Textor, Robert B. et al. 1983. *Austria 2005*. Vienna: Orac Pietsch.

Textor, Robert B. et al. 1984. *Anticipatory Anthropology and the Telemicroelectronic Revolution: A Preliminary Report from Silicon Valley*. Unpublished manuscript.

Textor, Robert B., M.L. Bhansoon, Ladavalya, and Sidthinat Prabudhanitisarn. 1984. *Alternative Sociocultural Futures for Thailand*. Chiang Mai, Thailand: Chiang Mai University.

Thom, W. Taylor, Jr. 1961. "Science and engineering—and the future of man." In H. Boyko (ed.), *Science and the Future of Mankind*. World Academy of Art and Science 1. Den Haag: Vitgeverij Dr. W. Junk.

Thomlinson, Ralph. 1965 (and 1976). *Population Dynamics*. (and Second Edition). New York: Random House.

Thorsrud, Einar. 1977. "Democracy at work: Norwegian experience with non-bureaucratic forms of organization." *Applied Behavioral Science* 13: 410–421.

Thrift, Nigel. 1988. "Vivos voco. Ringing the changes in historical geography of time consciousness." Pp. 53–94 in M. Young and T. Schuller (eds.), *The Rhythms of Society*. London: Routledge.

Tiger, Lionel and Robin Fox. 1971. *The Imperial Animal*. New York: Holt, Rinehart & Winston.

Tinbergen, J. 1968. "Planning, economic." Pp. 102–110 in D. L. Sills (ed.), *International Encyclopedia of the Social Sciences*, Vol. 12. New York: MacMillan & The Free Press.

Toch, Hans H. 1958. "The perception of future events: Case studies in social prediction." *Public Opinion Quarterly* 22 (Spring): 57–66.

Toffler, Alvin. 1970. *Future Shock*. New York: Random House.

———. 1978. "Foreword." Pp. ix–xi in M. Maruyama and A.M. Harkins (eds.), *Cultures of the Future*. The Hague: Mouton.

———. 1981. *The Third Wave*. New York: Bantam.

———. 1984. "Foreword. Science and change." Pp. xi–xxvi in I. Prigogine and I. Stengers, *Order Out of Chaos*. New York: Bantam.

Tong, Rosemarie. 1986. *Ethics in Policy Analysis*. Englewood Cliffs, NJ: Prentice-Hall

Tonn, Bruce E. 1991. "The court of generations: A proposed amendment to the US Constitution." *Futures* 23, No. 5 (June): 482–98.

Tough, Allen. 1991a. "Intellectual leaders in futures studies: A survey." *Futures* 23, No. 4 (May): 436–437.

———. 1991b. *Crucial Questions about the Future*. Lanham, MD: University Press of America.

———. 1993a. "Making a pledge to future generations." *Futures* 25, No. 1 (January/February): 90–92.

———. 1993b. "What future generations need from us." *Futures* 25, No. 10 (December): 1041–1050.

Toulmin, Stephen. 1953. *The Philosophy of Science: An Introduction*. London: Hutchinson.

———. 1972. *Human Understanding*, Vol. 1, Part 1. *The Collective Use and Evolution of Concepts*. Princeton, NJ: Princeton University Press.

———. 1981. "Evolution, adaptation, and human understanding." Pp. 18–36 in M.B. Brewer and B.E. Collins (eds.), *Scientific Inquiry and the Social Sciences*. San Francisco, CA: Jossey-Bass.

Turner, Charles F. and Elizabeth Martin (eds.). 1984. *Surveying Subjective Phenomena*, Vols. 1 and 2. New York: Russell Sage Foundation.

Turner, Victor W. 1968. "Religious specialists: Anthropological study." Pp. 437–444 in D. L. Sills (ed.), *International Encyclopedia of the Social Sciences*, Vol. 9. New York: Macmillan & The Free Press.

Tyler, Tom R. 1990. *Why People Obey the Law*. New Haven, CT: Yale University Press.

U.S. Executive Office of the President, Office of Management and Budget. 1973. *Social Indicators 1973*. Washington, DC: U.S. Government Printing Office.

U.S. National Resources Committee, Science Committee. 1937. *Technological Trends and National Policy, Including the Social Implications of New Inventions*. Washington, DC: Government Printing Office.

Unger, Roberto Mangabeira. 1976. *Law in Modern Society*. New York: Free Press.

United Nations. 1990. *1988 Demographic Yearbook*. New York: Department of International Economic and Social Affairs, Statistical Office.

United Nations. 1991. *The World's Women 1970–1990*. New York: Social Statistics and Indicators, Series K, No. 8.

United Nations. 1994. *1992 Demographic Yearbook*. New York: Department for Economic and Social Information and Policy Analysis.

Van Steenbergen, Bart. 1983. "The sociologist as social architect: A new task for macro-sociology?" *Futures* 15, No. 5 (October): 376–386.

———. 1990. "Potential influence of the holistic paradigm on the social sciences." *Futures* 22, No. 10 (December): 1071–1083.

Varenne, Hervé. 1977. *Americans Together*. New York: Teachers College Press.

Vermunt, Riël and Herman Steensma. 1991. "Introduction." Pp. 1–9 in R. Vermunt and H. Steensma (eds.), *Social Justice in Human Relations*. Vol. 1 "Societal and Psychological Origins of Justice." New York: Plenum Press.

Vickers, Geoffrey. 1977. "The future of culture." Pp. 37–44 in H.A. Linstone and W.H.C. Simmonds (eds.), *Futures Research: New Directions*. Reading, MA: Addison-Wesley.

Vitz, Paul C. and Arnold B. Glimcher. 1984. *Modern Art and Modern Science*. New York: Praeger.

Vught, F. A. van. 1987. "Pitfalls of forecasting: Fundamental problems for the methodology of forecasting from the philosophy of science." *Futures* 19 (April): 184–196.

Wachs, Martin. 1987. "Ethical dilemmas in forecasting for public policy." *Futures Research Quarterly* 3, No. 1 (Spring): 45–57.

Waddington, C.H. 1975. "Discussion." Pp. 154–165 in *The Future as an Academic Discipline*. Ciba Foundation Symposium 36. Amsterdam: Elsevier.

Wagar, W. Warren. 1991. *The Next Three Futures*. New York: Praeger.

Wagschall, Peter H. 1983. "Judgmental forecasting techniques and institutional planning: An example." Pp. 39–49 in J. L. Morrison, W. L. Renfro, and W. I. Boucher (eds.), *Applying Methods and Techniques of Futures Research*. San Francisco, CA: Jossey-Bass.

Wallace, Walter L. 1988. "Toward a disciplinary matrix in sociology." Pp. 23–76 in N. J. Smelser (ed.), *Handbook of Sociology*. Newbury Park, CA: Sage.

Walster, Elaine, G. William Walster, and Ellen Berscheid in collaboration with William Austin, Jane Traupmann, and Mary K. Utne. 1978. *Equity: Theory and Research*. Boston: Allyn and Bacon.

Watt, Ian. 1969. "Robinson Crusoe, individualism and the novel." Pp. 39–54 in F.H. Ellis (ed.), *Twentieth Century Interpretations of Robinson Crusoe*. Englewood Cliffs, NJ: Prentice-Hall.

Weber, Max. (1904, 1905, 1917) 1949. *The Methdology of the Social Sciences*, E.A. Shils and H.A. Finch (trans., eds.). Glencoe, IL: Free Press.

———. (1919) 1958a. "Science as a vocation." Pp. 129–156 in H. H. Gerth and C.W. Mills (trans., eds.), *From Max Weber: Essays in Sociology*. New York: Oxford University Press, a Galaxy Book.

———. (1919) 1958b. "Politics as a vocation." Pp. 77–128 in Gerth and Mills, *ibid*.

Weightman, John. 1993. "The human comedy of the divine Marquis." *The New York Review of Books* (September 23): 6–10.

Weil, Andrew. 1972. *The Natural Mind*. Boston: Houghton Mifflin.

Weimer, Walter B. 1979. *Notes on the Methodology of Scientific Research*. Hillsdale, NJ: Lawrence Erlbaum Associates.

Weiss, Edith Brown. (1988) 1992. *In Fairness to Future Generations.* Tokyo: The United Nations University.

Wells, H. G. (1932) 1987. "Wanted—professors of foresight!" *Futures Research Quarterly* 3, No. 1 (Spring): 89–91.

Wendt, Alexander. 1994. "Collective identity formation and the international state." *American Political Science Review* 88, No. 2 (June): 384–396.

Wescott, Roger W. 1970. "Of guilt and gratitude: Further reflections on human uniqueness." *The Dialogist.* 2, No. 3: 69–85.

————. 1978. "The anthropology of the future as an academic discipline." Pp. 509–528 in M. Maruyama and A.M. Harkins (eds.), *Cultures of the Future.* The Hague: Mouton.

————. 1978. "Traditional Greek conceptions of the future." Pp. 281–291 in M. Maruyama and A.M. Harkins (eds.), *Cultures of the Future.* The Hague: Mouton.

Wesson, Robert (ed.). 1987. *Democracy: World Survey 1987.* New York: Praeger.

WFSF (World Futures Studies Federation) Newsletter. 1985. Vol 11 (March/April).

Whyte, William Foote. 1989. "Advancing scientific knowledge through participatory action research." *Sociological Forum* 4, No. 3 (September): 367–385.

————. 1991. *Social Theory for Action.* Newbury Park, CA: Sage.

Whyte, William F. and Kathleen King Whyte. 1988. *Making Mondragón: The Growth and Dynamics of the Worker Cooperative Complex.* Ithaca, NY: ILR Press.

Whitney, Thomas P. 1962. "Introduction." Pp. vii–xiv in *The Communist Blueprint for the Future.* New York: E.P. Dutton (paper).

Wilcox, L.D., R.M. Brooks, G.M. Beal and G.E. Klonglan. 1972. *Social Indicators and Societal Monitoring: An Annotated Bibliography.* San Francisco, CA: Jossey-Bass.

Wilkening, Eugene A. 1974. "Futurology and quality of life in sociological research." Paper read at the annual meeting of the Rural Sociological Society, Montreal (August).

Williams, Bernard. 1985. *Ethics and the Limits of Philosophy.* Cambridge, MA: Harvard University Press.

Williams, Rhys H. and N. J. Demerath III. 1991. "Religion and political process in an American city." *American Sociological Review* 56, No. 4 (August): 417–431.

Williams, Robin M., Jr. 1970. *American Society.* Third Edition. New York: Alfred A. Knopf.

Wockler, R. 1978. "Perfectible apes in decadent cultures: Rousseau's anthropology revisited." *Daedalus* 107, No. 3 (Summer): 107–134.

Wood, Gordon S. 1984. "History lessons: A review of Barbara W. Tuchman, The March of Folly: From Troy to Vietnam." *The New York Review of Books* 31 (March 29): 8–10.

Woolf, Virginia. 1969. "Robinson Crusoe." Pp. 19–24 in F.H. Ellis (ed.), *Twentieth Century Interpretations of Robinson Crusoe.* Englewood Cliffs, NJ: Prentice-Hall.

Worchel, Stephen. 1984. "The darker side of helping: The social dynamics of helping and cooperation." Pp. 379–395 in E. Staub, D. Bar-Tal, J. Karylowski, and J. Reykowski (eds.), *Development and Maintenance of Prosocial Behavior.* New York: Plenum Press.

World Future Society. 1979. *The Future: A Guide to Information Sources.* Washington, DC: World Future Society.

Wright, James D. 1991. "Review of Culture Shift in Advanced Industrial Society." *Contemporary Sociology* 20, No. 6 (November): 892–894.

Wright, Robert. 1990. "Our animals, our selves." *The New York Times Book Review* (July 29): 27

Wuthnow, Robert. 1993. *Christianity in the Twenty-First Century.* New York: Oxford University Press.

Yalman, Nur. 1968. "Magic." Pp. 521–528 in D. L. Sills (ed.), *International Encyclopedia of the Social Sciences*, Vol. 9. New York: Macmillan & The Free Press.

Yaukey, David. 1985. *Demography.* New York: St. Martin's Press.

Yeager, Peter Cleary. 1991. *The Limits of Law.* Cambridge: Cambridge University Press.

Young, Michael and Tom Schuller. 1988. "Introduction: Towards chronosociology." Pp. 1–16 in M. Young and T. Schuller (eds.), *The Rhythms of Society.* London: Routledge.

Zarnowitz, Victor. 1968. "Prediction and forecasting, economic." Pp. 425–439 in D. L. Sills (ed.), *International Encyclopedia of the Social Sciences*, Vol. 12. New York: Macmillan & The Free Press.

Zavalloni, Marisa. 1980. "Values." Pp. 73–120 in H.C. Triandis and R.W. Brislin (eds.), *Handbook of Cross-Cultural Psychology*, Vol. 5 "Social Psychology." Boston: Allyn and Bacon.

Zerubavel, Eviatar. 1981. *Hidden Rhythms: Schedules and Calendars in Social Life.* Chicago: The University of Chicago Press.

———. 1985. *The Seven Day Circle: The History and Meaning of the Week.* New York: Free Press.

Index

About the Author

Wendell Bell, although born in Chicago, was raised from the age of four in Fresno, California. During World War II, he served as a Naval aviator and did a tour of duty in the Philippine theatre. After the war, he earned his living for a time as a commerical pilot while continuing to fly in the Naval Reserve. He started college under provisions of the G.I. Bill, got married, and contributed to the baby boom. He received his B.A. degree in social science from Fresno State University in California, and his M.A. and Ph.D. degrees in sociology from the University of California, Los Angeles.

He held faculty positions at Stanford University, 1952–54, where he directed the Stanford Survey Research Facility; Northwestern University, 1954–57; and back at UCLA, 1957–63, where he directed the West Indies Study Program. He became a professor at Yale University in 1963 where he remained until his retirement in 1995. At Yale, he served at various times as chairman of the Department of Sociology, director of Undergraduate Studies, director of Graduate Studies, and director of the Comparative Sociology Training Program.

Bell's early work dealt with urban sociology, particularly social area analysis and suburbanization, followed by about two decades of research on political and social change in the former colonies and now politically independent states of the Caribbean, especially in Jamaica. He has been a futurist since about 1960, introducing futures studies courses at Yale beginning in 1967 and co-authoring a book, *The Sociology of the Future*, in 1971. He has served as president of the Caribbean Studies Association, a gubernatorial appointee to the Commission on Connecticut's Future, a fellow of the Center for Advanced Study in the Behavioral Sciences at Stanford, California and of the Institute of Advanced Studies, the Australian National University, Canberra, Australia. He is a member of the World Future Society and the World Futures Studies Federation, and he now works as a consulting futurist.

Other books and monographs authored or co-authored by Bell include *Social Area Analysis*; *Public Leadership*; *Jamaican Leaders: Political Attitudes in a New Nation*; *Decisions of Nationhood: Political and Social Development in the British Caribbean*; *The Democratic Revolution in the West Indies*; and *Ethnicity and Nation-Building*.